A MANUAL of SOIL FUNGI

A MANUAL OF

SOIL FUNGI

By

JOSEPH C. GILMAN

BOTANY DEPARTMENT, IOWA STATE COLLEGE

1945

THE IOWA STATE COLLEGE PRESS

AMES, IOWA

Second Printing 1947

PRINTED AT THE IOWA STATE COLLEGE PRESS
AMES, IOWA, U. S. A.

To

E. V. A.

whose need provided the initial stimulus
that made this volume possible.

PREFACE

That fungi constitute a significant group in that vast and complex conglomerate of living organisms known as the soil microbiota needs but to be mentioned to be accepted. Yet, like many similar biological complexities, the details of the relationships, the activities and the functions of fungi within the complex, are still obscure, and new facts concerning the fungous population and its contributions to agriculture continually are being recorded.

The concept of the soil as a casual or adventitious habitat for fungi has long been abandoned and there is now general agreement among soil microbiologists that the fungi are invariably components of the microbiota of normal soils. The characteristics, metabolic and morphological, of many fungi are such that they come particularly into prominence when crop residues or plant parts such as roots or leaves decompose, and indeed the fungi often appear to be the "storm-troops" in the attacks. As a result of their activities a considerable amount of fungal mycelium is synthesized, and it is coming to be realized that this implies an important place for the fungi in the cycle of nitrogen transformations in soil as well as in the carbon cycle.

Some years ago in our laboratories Dr. E. V. Abbott and the author attempted to survey the fungi present in certain Iowa soils and from that investigation was published a paper, "A Summary of the Soil Fungi," which served as a tool during the investigation. That paper was in such demand, not only by workers in the field of soil microbiology, but by men in industry and in medicine that the supply of reprints was quickly exhausted and the demand has continued to the present. Now, the pressure has reached a point where it seemed desirable not only to reprint this work but to revise it and bring it up to date, by incorporating the great amount of additional material which has been published since 1927, the date of the appearance of that paper.

The present volume is designed to place a tool in the hands of investigators that will enable them to identify the soil fungi which they may encounter in the course of their work. It contains little new data but brings together from many scattered points, the descriptions of the fungi which have been reported as isolated from the soil, together with keys to aid in identifying the fungi in hand. In this sense it is a compilation and makes no pretense of being other than that, and since it is a compilation it has all the drawbacks found in material with which the author has no first hand experience. The field is so large, however, and the necessities so diverse that there are few if any persons with such experience. Further, many of the genera treated are under revision or species reported from the soil have not been included in recent revisions. Conditions of this sort will necessitate changes in treatment in the future but for the sake of completeness of treatment a temporary adjustment or an ultra conservative attitude is forced on the compiler. Two instances will illustrate the situation:

Sabet (1935) reported *Oospora egyptiaca* v. Beyma from the soil. Langeron and Talice (73), Ciferri and Redaelli (24), and others consider that the genus Oospora Sacc. emend. is invalid. They did not dispose of the species which have been assigned to that genus except in the case of *Oospora lactis* which they consider to be synonymous with *Geotrichum candidum*. *Oospora egyptiaca* differs sufficiently from Geotrichum to prevent its inclusion in that genus, and hence is left as Oospora in this volume.

A second instance of the inadequacy of the treatment is the genus Fusarium. The treatment followed is that of Wollenweber and Reinking. Since publication of that excellent book, the genus is being studied by Snyder and Hansen (124), but since their work is as yet far from complete, it seemed better to follow the earlier treatment than to have a genus in which part of the species were arranged from one point of view and part from another. In such cases the best judgment of the writer has not always prevailed against the necessities of the moment. Informed workers will find other situations that have been treated similarly.

In general the arrangement of the fungi has followed the scheme set forth by Lindau in Engler and Prantl, "Die naturlichen Pflanzenfamilien" insofar as orders, families and genera are treated. In certain cases more recent treatments are followed particularly for treatment of species: Zycha for the Mucorales, Coker for the Saprolegniaceae, Thom for Penicillium and Aspergillus, and Wol--lenweber and Reinking for Fusarium. Actinomycetes and Zoopagaceae which are unquestionably important soil fungi have been omitted. The first because of the writer's lack of familiarity and the second because of the additional reason that these forms will not usually appear among the fungi that are found in plate cultures made from soil samples but require a special technique for their study.

The text is not intended as an exhaustive study of the techniques necessary to an investigation of soil fungi but is intended to be helpful in identifying molds already in cultures. Hence a certain degree of knowledge and experience in bacteriological and mycological methods on the part of the reader has been taken for granted.

ACKNOWLEDGMENTS

The author is indebted to the following colleagues and publishers for permission to use their material in this volume:

To Williams and Wilkins for keys and descriptions from Thom, The Penicillia and Thom and Church, the Aspergilli; to the University of North Carolina Press for keys and descriptions from Coker, The Saprolegniaceae; to F. J. Seaver for descriptions of the Pezizales from North American Cup Fungi; to the University of Michigan Press for the treatment of the Blastocladiaceae from Sparrow, Aquatic Phycomycetes. The keys and descriptions of the species in the genus Fusarium were translated from those presented in Wollenweber and Reinking, Die Fusarien by permission of the Alien Property Custodian in the public interest under License No. A-725. Similar keys and descriptions in the Mucorales were translated from Zycha, Mucorineae.

The author wishes to express his appreciation to G. W. Martin, A. G. Norman, and G. C. Kent for reading the manuscript, and their valuable suggestions and criticisms of the presentation. Also to the many others who contributed support, encouragement and stimulation in the course of the preparation of the manuscript and its completion, especial thanks are due in spite of the impossibility of thanking them specifically.

JOSEPH C. GILMAN

AMES, IOWA
AUGUST, 1944

CONTENTS

A MANUAL of SOIL FUNGI

INTRODUCTION

The biology of the soil is always an interesting and important study; a study which will continue to contribute to our knowledge of agriculture and biology. Soil is the depository of all life and the laboratory within which are carried out most of the changes that enable life to continue. In the soil are the roots of our plants, the burrowing mammals, insects, and other animals, as well as bacteria and fungi in countless numbers. To the soil are returned the dead remains of both plants and animals and there these are changed from the form and structures by which they were known when alive to the materials which are again used by plants and animals as sources for further life. In accomplishing these changes, fungi play a significant part, often starting the process on the living plant or animal and continuing to thrive there after its death. In such cases fungi become known as the cause of disease in plants and animals and become the concern of the plant pathologist or the physician. Man also undertakes to bring about certain fungous activities under controlled conditions in order to profit from the products of the mold growth, and the fungi involved become the concern of the microbiologist and the biochemist. Further the spores of these fungi frequently are air-borne, and then, if their numbers are sufficient, they may cause conditions, in sensitive people who breathe them, similar to the symptoms caused by air-borne pollen and their recognition is important to the physician.

Finally as a consequence of the air-borne condition of their spores these fungi reach many moist surfaces and become the cause of spoilage. Stored foods, both for men and animals, wall-board, cork-products, plywood, cloth and leather-goods, and even casein paints have been reported to be subject to mildew under certain conditions of humidity and temperature. Here, also, the recognition of the soil-fungi becomes an important factor in the preservation of these materials. The number of these fungi is large, and their variety is extensive. At present no single volume that would serve as a tool in their identification is available. The purpose of this manual is to fill that gap.

The fungi included in the manual are chiefly those species which have been cultivated artificially on various types of biological media. Excluded are the terrestrial mushrooms, the plant pathogens which are considered to be soil-borne but which have not been isolated directly from the soil, and the forms which have been reported from leaf-mold, decayed wood or other substrates that have not yet become fully incorporated in the complex known as soil.

Such a procedure has its drawbacks; particularly it gives no evidence of the functions that the fungi treated may carry out in the soil nor does it differentiate those species that are constant soil inhabitants from those that are more or less commonly introduced into the soil on crop residues, surface litter and manure. Furthermore the number of geographical citations gives no index of the frequency of distribution nor the relative importance of any given species. Its chief advantage is that it limits the field to be covered and thereby permits a more uniform treatment than would otherwise be possible.

Soil fungi have been recognized chiefly by two methods; direct examination and isolation by cultural methods. The limitations of the first method are obvious, and although it has produced many excellent results, the physical barriers in the way of such a method limit its usefulness, and in most cases such observations as were made by it were confirmed by resort to cultural methods. The combination as used by Cholodny, Kubiena, et al., has contributed to the knowledge of soil microbiology in no uncertain terms.

The second method, that of isolation of cultures, has been more widely used and has produced the greater numbers of species of recognized soil fungi. Nevertheless, the question of whether the fungus in the plate was in reality a soil-inhabitant or merely the result of the accidental lodgment of a spore in the sample from which the isolation was made cannot be answered. That the substrate used in making the isolation is of prime importance in determining the species that will be taken cannot be denied. The remarkable discovery of Coker and his associates that many species of Saprolegniales are seemingly present in a very wide range of soils and localities was the result of the application of the techniques and materials used to isolate this group of fungi from water samples. The media used by previous workers were unsuited to the needs of these fungi and therefore they were not thought of as soil fungi previously. In a similar way by the use of sterilized rabbit dung, species of Ascobolus and Pleurage were isolated from soil by Dr. J. M. Beebe. Although the selective action of the medium used in the cultural work has thus limited the species treated in this manual, the forms presented are probably the predominant group in the soil, as is evinced by their repeated isolation by many investigators in all parts of the world. The fact that the media used today for this type of investigation are rather well standardized, defines sharply the area in which the material in this manual will be useful.

FUNGI

Fungi are organisms with their assimilative (non-reproductive) structure made up of a much-branched system of slender tubes known as mycelium. Such mycelium usually grows in a radial manner from the point of origin, if conditions are equal about that point. Each branch of this structure may be divided by cross-walls (septa) or be continuous, depending on the nature of the species. Ordinarily the growth is confined to a rather narrow zone of plane surface and then the mycelium becomes the circular colony of the petri dish or the fairy ring of the open field.

In reproduction the fungi theoretically produce spores as the result of a cell and nuclear fusion (sexual spores). Under cultural conditions, however, such spores are the exception rather than the rule and the investigator is forced to base identifications on the much commoner asexual spores (conidia) which are the result of division of the cells of the mycelium. Such conidial fungi can be grouped into "form" species, genera and families on the characters of their spores and the hyphae which bear them (conidiophores), without regard to their sexual spore structure. These form genera make up the Fungi Imperfecti.

SEXUAL SPORES

If the members of the fungi be grouped according to the formation of their sexual spores, two groups immediately present themselves: (1), the spore results directly from the union of the sex cells and (2) the spore results after the reduction of the fusion nucleus from the $2n$ to the n condition. The first group comprises the Phycomycetes and is further divided into the Zygomycetes in which like gametangia fuse to form a zygospore, and the Oomycetes in which unlike gametangia (rarely gametes) unite to form an oospore. (In many cases the antherozoids are not free but are directed to the egg through a fertilization tube). The second group likewise is divided into two; one in which the spores are borne in a sac (ascus), ascospores, and the other with spores external to the fusion cell (basidium), basidiospores. The former are known as Ascomycetes, the latter as Basidiomycetes. Because of the reduction divisions the number of ascospores in an ascus and of basidiospores on the basidium is definite, usually eight ascospores in each ascus and four basidiospores on each basidium. Other numbers may occur.

The zygospores and oospores in general are found scattered irregularly on the mycelium; and asci and basidia are more likely to be gathered into special structures formed for their production. Asci occur massed within a closed spherical receptacle, the cleistothecium, or in a layer lining the base of a flask-shaped receptacle, the perithecium, or a cup-like receptacle, the apothecium. Basidia also are often found on special fruiting-bodies, the gills of the mushrooms, or the pores of the bracket fungi, or the glebae of the stink-horns. These differences in the structures that bear the asci and basidia are used to divide the Ascomycetes and Basidiomycetes into sub-groups. Within these groups the arrangement, shape, color and consistency of the perithecium, cleistothecium or apothecium, or of basidiocarps carrying the gills, pores and glebae become the characters for further subdivision into orders, families and genera and these characters are used in the keys.

ASEXUAL SPORES

Asexual spores are also of many types both in regard to their formation and their distribution on the hyphae which bear them. In general they may be divided into those produced internally in sporangia and those produced externally on more or less specialized hyphae (conidiophores). Thick-walled resting cells that are homologous to the cysts of protozoa are known as chlamydospores. They may occur terminally or in the course of the hyphae (intercalary).

The sporangial spores are either motile, (zoospores) or non-motile (aplanospores). Some confusion of terminology has arisen from the fact that in both the Oomycetes and the Zygomycetes the whole sporangium may become detached from its parent hypha and become an organ of dissemination. In the Oomycetes the sporangium may not form zoospores until it reaches its locus of germination and in this manner simulates a conidium of the ascomycetous fungi. The situation is further complicated by the fact that in a considerable number of species these cells germinate either by zoospores or by tube depending upon their environment at the time of germination. Some authors call them conidia in their stages of formation and dissemination; others, mindful of the zoospores which are potentially present, call them sporangia. The former usage has been adopted in this paper. In the Zygomycetes, the sporangia contain fewer and fewer aplanospores until in the more highly specialized forms, Cunninghamella particularly, the sporangium simulates a conidium in that a single multinucleate cell becomes formed externally and is shed as a disseminative body. Here the term pseudoconidium is used although in many publications these cells are called conidia, or sporangia.

In the Ascomycetes also the term conidium often has been applied indiscriminately to structures of quite dissimilar origin morphologically and hence much confusion has resulted. Mason (85) reviewed this situation rather clearly and completely and following Vuillemin (145) has divided asexual spores into conidia vera (phialospores), which are spores borne on "an open growing point" on a hypha, and thallospores which are part of the hyphae bearing them and may be further divided into spores formed from the expanded ends of hyphae (aleuriospores), hyphal segmentations (arthrospores) and buds (blastospores). Although these distinctions are of great importance in arranging the fungi, particularly the Fungi Imperfecti, for the purposes of this volume the general term conidium covering all these types seems adequate.

Within the Fungi Imperfecti, the arrangement of the conidiophores in relation to each other and to the substrate is used to combine the species into form genera, families and orders. Three orders are recognized, the Sphaeropsidales with the conidia and conidiophores confined in definite flask-shaped receptacles, the pycnidia; the Melanconiales in which the conidiophores are associated in disk-like or saucer-shaped determinate groups, the acervuli; and the Moniliales in which the conidiophores are more or less scattered on the substrate. In the last group the conidiophores are sometimes aggregated into definite cushion-like clusters, sporodochia, or united into columns of parallel hyphae, coremia. The sporodochium characterizes the Tuberculariaceae, the coremium the Stilbaceae.

Within the orders, hyphal and spore color becomes of first significance. Hyalin and bright colored forms are separated from those with dark colored hyphae and/or spores and these groups in turn are sectioned on the basis of spore septation. One-celled spore-forms are placed in one section, two-celled in another, many-celled by cross-walls in a third, many-celled by cross and longitudinal walls in a fourth, thread-like spores in a fifth and star-shaped spores as a sixth. The manner of conidiophore branching and the position of the conidia on the conidiophore become the final criteria for generic distinction. Spore and conidiophore size are in general used for distinguishing species.

KEYS

The keys which follow are in general based upon the scheme outlined above but are designed to enable the user to identify his material and hence are highly artificial. They are of the dichotomous type since this mechanism is more readily followed than would be the case where more than two possibilities are presented simultaneously. Already existing keys have been selected with such modifications as were deemed necessary to make them fit the material covered. In the main the scheme is that of Lindau in Engler and Prantl, "Die naturlichen Pflanzenfamilien" for the more general section. Keys to the species, however, have been adopted from the best monographs of the groups at hand, wherever they were available. Such a treatment reflects the diversity of the sources but any difficulty such diversity may cause is more than compensated by the more adequate coverage that is accorded by the monographs.

KEY TO THE CLASSES, ORDERS, AND FAMILIES OF SOIL FUNGI

Page

a. Filaments one-celled, rarely septate; asexual spores usually in sporangia; sex cells, when present, uniting to form resting spores
I. PHYCOMYCETES

b. Asexual spores typically in globose to cylindric sporangia, nonmotile; zygosporous A. **MUCORALES**

c. Sporangia always present, pseudoconidia sometimes present

d. Columellae present; zygospores naked or with a few appendages

e. Wall of the sporangium homogeneous, not cuticularized, usually diffluent

f. Sporangia of one sort
a. *MUCORACEAE* 13

ff. Sporangia of two sorts, primary and secondary
b. *THAMNIDIACEAE* 46

ee. Wall cuticularized and persistent above, thin and diffluent below c. *PILOBOLACEAE* 47

dd. Columellae absent; zygospores enveloped in a dense hyphal covering d. *MORTIERELLACEAE* 48

cc. Sporangia rarely present, pseudoconidia always present

d. Pseudoconidia solitary; or sporangioles on special vesicular, fertile hyphae
e. *CHOANEPHORACEAE* 55

dd. Pseudoconidia in chains, or on special basal cells
f. *CEPHALIDACEAE* 58

Page

aa. Filaments septate; conidia borne on conidiophores, sex cells rarely in evidence

b. Spores in a definite number in a sac: as ascospores

II. ASCOMYCETES

c. Asci enclosed in a definite globose or flask-shaped ascocarp

d. Ascocarps without definite mouths

A. **PLECTASCALES**

e. Asci at various levels in the ascocarp

f. Walls of ascocarps of loose felt-like tissues

a. *GYMNOASCACEAE* 147

ff. Walls of ascocarps membranous

b. *ASPERGILLACEAE* 150

ee. Asci in a bush-like arrangement

c. *PERISPORIACEAE* 151

dd. Ascocarps with definite mouths

e. Perithecia dark colored

B. **SPHAERIALES**

f. Walls of perithecia thin and membranous, asci soon disappearing

g. Perithecia always superficial, with mouth surrounded by long-branched, hooked or spirally curved hairs

a. *CHAETOMIACEAE* 152

gg. Perithecia usually sunken, with only short hairs about the mouth

b. *FIMETARIACEAE* 157

ff. Walls of perithecia leathery to carbonous

g. Perithecia not beaked

c. *SPHAERIACEAE* 160

gg. Perithecia with long beak

d. *CERATOSTOMATACEAE* 161

ee. Perithecia bright colored

C. **HYPOCREALES**

f. Stroma wanting; or when present with perithecia entirely superficial

a. *NECTRIACEAE* 162

Page

cc. Asci lining cup-shaped ascocarps
D. **PEZIZALES**

d. A single family treated
a. *PEZIZACEAE* 164

bb. Spores as conidia on conidiophores of various form, not in asci; or entirely lacking
III. FUNGI IMPERFECTI

c. Conidia present

d. Conidia in globoid, cup-shaped or hysterioid receptacles (pycnidia) A. **SPHAEROPSIDALES**

e. Pycnidia typically membranous to carbonous, dark; brown or black; more or less globose, rarely cylindric a. *SPHAERIOIDACEAE* 169

dd. Conidia not in pycnidia

e. Conidiophores gathered into acervuli
B. **MELANCONIALES**

f. A single family a. *MELANCONIACEAE* 172

ee. Conidiophores not in acervuli
C. **MONILIALES**

f. Conidiophores scattered in more or less loose irregular cottony masses; not in coremia nor sporodochia

g. Conidiophores and conidia clear or bright colored a. *MONILIACEAE* 173

gg. Conidiophores and conidia, dark or sometimes conidia hyalin on dark conidiophores, globose to elongate
b. *DEMATIACEAE* 282

ff. Conidiophores gathered into coremia or sporodochia

g. Conidiophores, fasciculate, parallel into coremia c. *STILBACEAE* 311

Page

gg. Conidiophores forming a more or less globose, sessile body; without parallel arrangement, a sporodochium
d. *TUBERCULARIACEAE* 316

cc. Conidia lacking
IV. MYCELIA STERILIA 363

I. PHYCOMYCETES

Vegetative body either unspecialized and converted as a whole into a reproductive organ at times bearing tapering rhizoids, or mycelial and very extensive. Asexual spores borne in sporangia or as pseudoconidia, motile or nonmotile. Sexual reproduction by conjugation of similar gametangia to form zygospores, by conjugation of planogametes, or by fertilization of an egg cell by motile antherozoids to form oospores. In all cases the sexual spores are the direct result of cell and nuclear fusion.

A. MUCORALES

Wholly terrestrial Phycomycetes, usually with well developed, richly branched and rapidly growing mycelium. The hyphae contain many nuclei and in youth show no septa; septation occurs only with the laying down of special organs or in age, or under unfavorable conditions, as limitations of certain hyphal parts. Most species have a sterile substrate mycelium on which arise the fruiting hyphae above the substrata.

Vegetative reproduction by nonmotile, one- or many-nucleated but one-celled spores which are formed in sporangia. These latter are mostly many-spored, but the number may be reduced to a single spore. Conidia (exogenous spores) may also be present. Further, gemmae, which can be formed on any part of the mycelium, serve to carry the fungus over unfavorable conditions.

Sexual reproduction results from the fusion of two similar many-nucleated cells, which show no differentiation of gametes and are called gametangia. Copulation principally isogamous; in the resulting zygote, many nuclear pairs fuse. The zygote-carrying hyphae (suspensors) often form many appendages, which more or less enclose the zygotes. In the higher Mucorales (Mortierellaceae) the cover becomes so thick that a zygote-fruit is formed.

The classification of the Mucorales is largely that used by Zycha (169). Zycha's keys have been used as a basis for the construction of a key to the soil forms.

a. *MUCORACEAE*

Fungi with columellate sporangia; the columella being formed by the pushing of the crosswall separating the sporangiophore and sporangium into the interior of the latter to form a dome- or vesicular structure. Zygospores are naked or only loosely covered by appendages, never by a felt-like layer to form a fruit.

KEY TO THE GENERA OF THE MUCORACEAE

a. Sporangia pyriform (1) **Absidia**

aa. Sporangia spherical

b. Sporangiophores united into groups on a rhizoidiferous stolon

c. Sporangiophores unbranched, at the nodes opposite the rhizoids. Spores usually longitudinally striate (2) **Rhizopus**

cc. Sporangiophores branched, with terminal sporangium large; side branches in whorls; lateral sporangia smaller than the principal sporangium (3) **Actinomucor**

bb. Sporangiophores emerging singly from the mycelium, not from rhizoidiferous stolons. Spores generally smooth, without longitudinal striae

c. Sporangiophores, with metallic sheen, always unbranched, more than 80 mm. long (4) **Phycomyces**

cc. Sporangiophores shorter

d. Sporangiophores branched, side branches strongly curved (5) **Circinella**

dd. Sporangiophores unbranched or branched, side branches not markedly curved

e. Zygospores on special hyphae, not on sporangiophores (6) **Mucor**

ee. Zygospores on short side branches of the sporangiophores (7) **Zygorhynchus**

(1) **Absidia** van Tieghem

Mycelium formed as in the genus Rhizopus by frequently branched stolons, more or less incurved into arches and producing at the point of contact with the substratum more or less richly branched rhizoids. Sporangiophores straight, rarely single, more often in groups of two to five, occurring at the curve of the stolon (internodal) and not at the point of origin of the rhizoid (nodes). At times there occur erect stolons or branches which bear lateral sporangiferous branches which may be confused with the primary sporangiophores. Sporangia apparently equal, pyriform, erect, furnished with an infundibuliform apophysis.

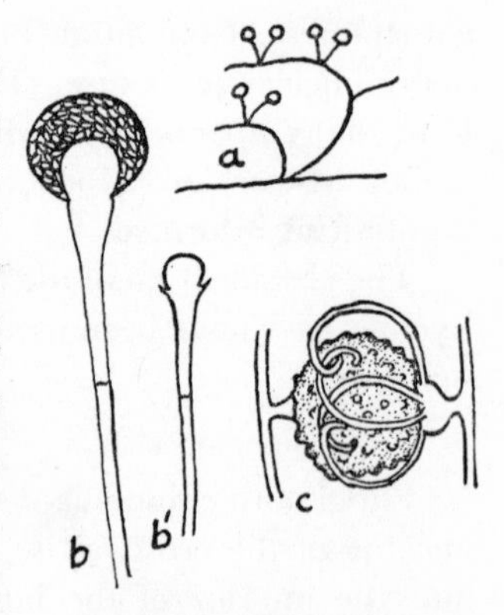

Fig. 1. **Absidia.** *a* habit; *b*, *b′* sporangiophores; *c* zygospore (after Lendner).

Membrane of the sporangium not cuticularized nor incrusted; diffluent, leaving a short basal collarette. Columellae hemispheric, conic or mammiform, more rarely spinescent or terminated by a single long prolongation. They efface themselves in the apophyses. They are cuticularized and their color is more pronounced than that of the sporangiophore. A cross-wall is placed at a definite distance below the sporangium. Spores small, 5–6 μ, round or oval (not angular) with a smooth wall, rarely echinulate, colorless or bluish-black. Zygospores formed on the stolons. They are surrounded by circinate filaments, cutinized, which are borne in a whorl from one or both of the suspensors. Gametangia straight. On germination the zygospores produce either mycelial filaments or sporangiophores. Apparently closely related to the genus Rhizopus, differing from that genus by the fact that the sporangiophores occur on the internodes, by the pyriform sporangia, by the continuance of the columellae into the apophyses and the suspensors having circinate filaments.

KEY TO THE SPECIES OF THE GENUS ABSIDIA

a. Spores elongate cylindric; appendages on only one of the suspensors	
b. Spores 4–5 μ long	1. *A. spinosa*
bb. Spores 6–11 μ long	2. *A. heterospora*
aa. Spores oval or globose; appendages, where known, on both suspensors	
b. Columellae furnished with a single terminal prolongation	
c. Prolongation of columella short, pointed; columella globose, mammiform	3. *A. glauca*
cc. Prolongation longer, rounded at tip, columella turbinate	
d. Spores globose 2.5–3.5 μ in diameter	4. *A. orchidis*
dd. Spores globose, larger 4–7 μ	
e. Sporangia erect	5. *A. coerulea*
ee. Sporangia nodding	6. *A. reflexa*
bb. Columellae smooth or rarely faintly spinescent	
c. Spores generally globose, rarely oval, 3–4 μ in diameter. Columellae generally spinescent	
d. Strong growth at 37° C.	7. *A. lichtheimii*
dd. No growth at 37° C.	8. *A. repens*
cc. Spores irregular; columellae smooth	9. *A. butleri*

1. *Absidia spinosa* Lendner

Syn. *A. cylindrospora* Hagem

Tieghemella spinosa Lendner

Turf very close, the filaments interlacing into a grayish cottony mat, about 2½ cm. above the substratum. Stolons little curved, arched, carrying the sporangia in groups of two or three. Sporangia pear-shaped, bluish, 34μ long, from the apophysis to the end of the sporangium, by 28μ wide. Columellae 20μ wide, swollen, ending in a blunt or rounded spine, reaching ⅓ the length of the columella. Septa present, 25μ below the apophyses separating the sporangia from the sporangiophores. Spores hyalin, oval or short rods, sometimes very slightly constricted in the middle, 2μ in diameter × 4–5μ long. Zygospores spherical or doliform, verrucose, formed by the fusion of the two unequal gametes on a forked hypha. The larger suspensor furnished with circinate appendages.

From soil: Austria (66), Canada (15), Denmark (68), Egypt (115), France (78), Germany (79) (169), Greenland (96), Morocco (169), Norway (55), Switzerland (77), U. S. S. R. (109)

United States: Colorado (76), Hawaii (147), Idaho (106), North Carolina (23)

Absidia fusca Linnemann (79) would seem to be closely related if not the same as the above species.

2. *Absidia heterospora* Ling-Young

Turf thick, dark gray, 5–10 mm. high. Stolons elongate, with thin rhizoids. Sporangiophores at first hyalin, later violet and dark gray, occurring in twos and threes, 0.2–0.5 mm. long, 10μ thick, with a cross-wall below the sporangium. Sporangia at first yellow, then greenish-brown, 35–50μ broad, with diffluent wall. Columellae hemispherical, 25–30μ broad. Spores hyalin, cylindric, constricted in the middle, slightly orange tinted, 2.5–6μ broad and 6–11μ long. Gemmae and zygospores unknown.

In soil: France (78)

3. *Absidia glauca* Hagem

Syn. *Tieghemella glauca* Hagem

Culture on wort gelatin gray-green when young (10 days) then becoming clear yellow-brown. The stolons present the same kind of branching as *Absidia orchidis.* The fertile branches are either isolated, or in groups of two, three, or four. Sporangia pyriform, measuring 40–50μ in diameter × 44–60μ in length. A septum dividing the pedicel from the sporangium is formed a distance equal to half the length of the entire apophysis. Wall incrusted with granules, is diffluent and leaves a very straight collarette. Columella rounded, mammiform, furnished with a very short button; it measures at least 30μ in diameter × 38μ in length. Spores, round, 3–3.5μ, colorless. Heterothallic.

From soil: Austria (66), Canada (15), Czechoslovakia (97), Egypt (115), England (37), France (78), Norway (55), Switzerland (77), U. S. S. R. (108)

United States: Idaho (106), North Carolina (23)

4. *Absidia orchidis* (Vuillemin) Hagem

Primary axes 0.6–10 mm. long, straight or more or less changed into irregular stolons, at times raised and indefinitely elongate; at times curved toward the substratum to which their ends are attached by a tuft of rhizoids; at times erect and ended by a sporangium. These stolons are branched sympodially and bear sterile or fertile branches, the latter occurring singly or in groups of two or three. Sporangiophores simple or bearing at some distance from the tips an oblique branch, shorter than the tip of the principal pedicel. This latter branch also ends in a like sporangium. Septa divide the pedicels at a distance from the infundibuliform apophyses a little greater than the height of the apophyses themselves. Sporangia ovate, 40μ in height × 32μ in diameter (in the case of the large sporangia). Wall incrusted with fine granules, imperfectly diffluent, leaving a straight, rigid, collarette. Columellae conic, rounded, longer than broad, surmounted usually by a knob; attenuated or constricted at the base and remaining upright when, after dehiscence, the columella becomes free. Spores slightly brownish, perfectly spherical, varying from 2.5–3.5μ in diameter. Heterothallic.

From soil: Canada (15), Czechoslovakia (97), Denmark (68), England (36), France (78), Jugoslavia (97) (103), Norway (55)

United States: New Jersey (146) (147), North Carolina (23)

5. *Absidia coerulea* Bainier

Filaments of the thallus bluish-violet, continuous, unequally branched, at times knotted. Sporangiophores single, borne directly on the thallus, attaining 25 mm. in length, ended by an infundibuliform apophysis. Septa 12–24μ from the tip. Sporangia uniform, globular, 36–42μ changing from pale violet to gray, then to brown. Membrane of the sporangium smooth, diffluent, leaving a collarette. Columellae hemispheric or obconic, often ending in a papilla. Spores numerous, small, smooth, pale violet, globose, 4–7μ. Zygospores 60μ, brown, globose, rugose-verrucose. Suspensors straight, enlarging into barrel-shape, furnished with 10–20 circinate appendages, long and thin (7μ in diameter), arranged in a single verticil. Azygospores similar. Chlamydospores smooth, intercalary.

From soil: Holland (100)

6. *Absidia reflexa* van Tieghem

Stolons in steep arches, sporangiophores single, rhizoids scarcely formed. Sporangia nodding, with a septum below the apophysis. Spores globose, 6μ in diameter. Zygospores unknown.

From soil: Austria (131)

7. *Absidia lichtheimii* (Lucet and Costantin) Lendner

Sporangiophores prostrate, branched in corymbs, forming a white felt, woolly. They terminate the corymbiform branching by carrying the sporangia on longer or shorter pedicels. A little below the terminal corymb frequently there occur groups of branches carrying smaller sporangia. Sporangia erect, hyalin, pear-shaped, with an infundibuliform apophysis, becoming attenuate gradually to the sporangiophore. Average diameter 45–60μ, the greatest 70μ, the least 10–20μ. Wall of sporangium colorless, transparent, smooth, diffluent, leaving a basal collarette. Columellae large, hemispheric or globular, 10–20μ, smooth (or furnished with short spines) smoky-gray or brown. The apophysis and pedicel also similarly colored. Spores spheric, subspheric or more rarely oval, colorless, small, usually 2μ in diameter $\times$ 3μ long (sometimes larger, 4–6.5μ). Zygospores not known.

From soil: Czechoslovakia (97), Egypt (115), France (78), Jugoslavia (103), Switzerland (77)

United States: Maine (146) (147), New Jersey (146) (147)

8. *Absidia repens* van Tieghem

Turf white, later brownish-gray, up to 20 mm. high. Stolons stout, extending in flat arches, in older cultures brown. Sporangiophores three to five in a whorl, unbranched, and with a septum below the apophysis. Sporangia gray-brown, about 40μ in diameter. Columellae conical with a papilla-like protrusion. Spores in the primary sporangia, $3 \times 7\mu$, in the smaller secondary sporangia, $3 \times 4\mu$. Zygospores 70μ in diameter, dark brown. Each suspensor carries a whorl of eight to twelve gray-brown appendages.

From soil: Austria (131)

9. *Absidia butleri* Lendner

Syn. *Absidia subpoculata* Paine

Colonies white floccose, aerial hyphae growing to a height of 1.5–2 cm., floccose. Stolons branched with sporangiophores occurring in groups of one to five. Sporangiophores branched, 100–300μ long × 4μ in diameter with a septum 10–12μ below the tip. Sporangia globose 22–24μ in diameter with a smooth diffluent wall, leaving a slight collarette. Columellae oval, slightly constricted at the apophyses; without apophyses 4–7 × 8–9μ, with apophyses 10–20 × 7.5–15μ. Apophyses rounded below into a distinct pouch. Spores oval to spherical to allantoid, 2–2.5 × 3.4μ. Chlamydospores quite numerous, spherical, 4–5μ in diameter.

From soil: United States: Iowa (101), Louisiana (50), Texas (92)

(2) **Rhizopus** Ehrenberg

Mycelium of two kinds, one submerged in the substratum and the other aerial, constituting the arching filaments or stolons. These stolons present from place to place the nodes on which occur the rhizoids, which are implanted in the substratum. At these points the sporangiophores arise. They may be single but usually occur in groups of two, three, or more. The summit of the sporangiophore is enlarged into an apophysis, of the kind that has the columella inserted above the point where the spherical bend attaches into the filament. The sporangia, white at first, become bluish-black at maturity. They are all the same sort, spherical or almost spherical, flattened at the base. Wall not cuticularized, uniformly incrusted and entirely diffluent, without leaving a basal collarette. Columellae broadly subjacent, hemispherical, forming after dehiscence, by collapse, an organ of the shape of the pileus of a mushroom. Spores round or oval, angular, colorless, or colored bluish or brown, with a cuticularized wall, smooth or striate, rarely spinulose. Zygospores naked, formed in the substratum and on the stolons. Suspensors straight, very large and swollen, without appendages.

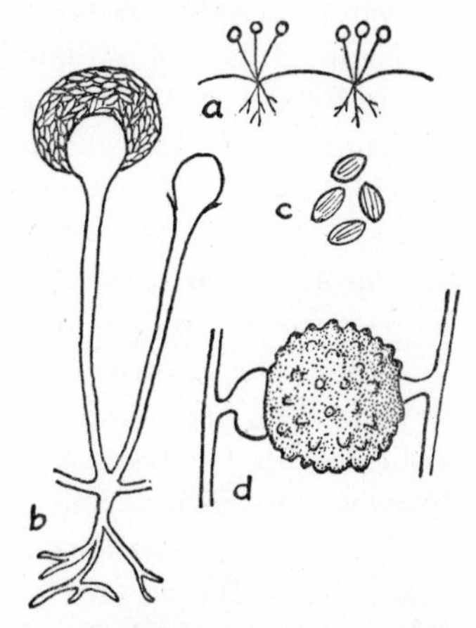

FIG. 2. **Rhizopus.** *a* habit; *b* sporangiophores; *c* sporangiospores; *d* zygospore.

KEY TO THE SPECIES OF THE GENUS RHIZOPUS

a. Rhizoids well developed 1. *R. nigricans*

aa. Rhizoids little developed or lacking

b. Rhizoids rare, pale, short; sporangiophores without swellings

c. Sporangiophores over 150μ high — 2. *R. arrhizus*
cc. Sporangiophores less than 150μ high — 3. *R. cohnii*
bb. Rhizoids rare; sporangiophores and stolons branched and swollen in places — 4. *R. nodosus*

1. *Rhizopus nigricans* Ehrenberg

Stolons creeping, recurving to the substrate in the form of arachnoid hyphae, which are strongly raised and distant from the substrate and implanted at each node by means of rhizoids. The internodes often attain a length of 1–3 cm. and the hyphae are more or less branched. Sporangiophores rarely single, united in groups of three to five or more, 0.5–4 mm. in height × 24–42μ in diameter. Apophyses broad, cuneiform. Sporangia hemispheric 100–350μ. Columellae broad, hemispheric, depressed, 70μ in diameter by 90μ in height (250 × 320μ maximum). Spores unequal, irregular round or oval, angular, striate, 9–12μ long × 7.5–8μ in diameter, of a gray-blue. Zygospores round, or oval, 160–220μ in diameter. Exine brown-black, verrucose. Suspensors swollen, usually unequal. Azygospores present. No chlamydospores.

From soil: Austria (66) (131), Canada (15), China (81), Czechoslovakia (97), Dalmatia (97), Denmark (68), Egypt (115), England (11), France (78), Germany (6) (79), India (21), Italy (97), Japan (132), Norway (55), Switzerland (97), U. S. S. R. (108)

United States: California (146), Colorado (74) (75) (76), Hawaii (147), Idaho (106), Illinois (101), Iowa (1) (3) (147), Louisiana (2) (146), New Jersey (82) (146) (147), New York (67), North Carolina (23), North Dakota (147), Oregon (146) (147), Rhode Island (110), South Carolina (23), Texas (151)

var. *minor*
From soil: England (11)
var. *minutus* (67)
From soil: India (21)
United States: Illinois (122)

2. *Rhizopus arrhizus* Fischer

Differs from *R. nigricans* by its less exuberance. The felt is clearer and it does not extend so far into the substrate. Stolons are little developed and do not form nodes regularly. Rhizoids pale, develop at the nodes and carry sporangia, or are sometimes formed indeterminately. Sporangiophores often prostrate, rarely single, forming umbels or corymbs on their stolons. They measure 0.5–2 mm. in length. All the branches end in sporangia, of greater or less size. Sporangia spherical 120–250μ n diameter. Columellae spherical, flattened on the apophyses, 40–75μ

high × 60–100μ in width, membrane brown, smooth. Spores round or oval, or presenting obtuse angles, grayish brown; walls striated longitudinally, 4.8–7 × 4.8–5.6μ.

From soil: Czechoslovakia (98), Egypt (115), England (36), Germany (79), Hungary (94), India (21) (49)

United States: Colorado (74) (75), North Carolina (23), South Carolina (23)

3. *Rhizopus cohnii* Berlese and de Toni

Turf at first white, later gray-black, up to 30 mm. high. Stolons and sporangiophores as in *R. arrhizus* but shorter; less than 150μ. Sporangia black, 50–100μ in diameter. Spores partly globose, partly more oval, 5–6μ long, slightly angular and striate. Mycelial gemmae up to 40μ long.

From soil: Austria (131)

4. *Rhizopus nodosus* Namyslowski

The mycelium is cottony, white when young, then tinted ochre-yellow. In the midst of the mycelium and on the stolons, branches ending in sporangia occur. These branches 1–2 mm. in height × 12–28μ in diameter have thick, smooth walls, colorless at first, then becoming pale ochre or brown. They are simple or branched, the branches ending in sporangia. The branches may be swollen at any point. When these swellings are terminal they give rise to a group of three to five sporangiophores, each terminating in a sporangium. Sporangiophores 1–2 mm. high, the sporangia are globose 100–200μ in diameter. The spores 6–9μ long × 4–6μ in diameter, striated longitudinally. They may give rise to chlamydospores 16–32μ in diameter. Zygospores 120–140μ occur. They are round, oval or without definite shape. The suspensors are equal or different in size and shape.

From soil: Austria (66), Czechoslovakia (97), Denmark (68), France (78), Jugoslavia (97), Norway (55), Switzerland (77)

United States: New Jersey (146) (147), Oregon (146)

Species of uncertain position:

Rhizopus kasanensis Hanzawa

From soil: Germany (79)

(3) **Actinomucor** Schostakowitsch

Sporangiophores with terminal sporangia or with sterile tips, then serving as hold-fasts. When a holdfast comes in contact with the

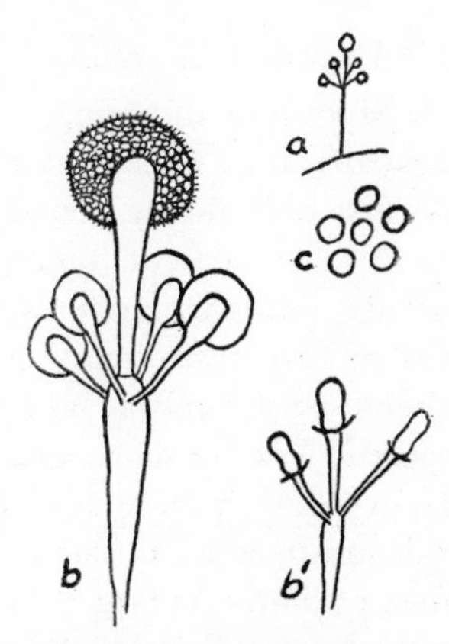

FIG. 3. **Actinomucor.** *a* habit; *b*, *b'* sporangiophores; *c* sporangiospores.

substrate it forms at that point a system of branching rhizoids like those found in the genus Rhizopus. The lateral sporangiophores short, usually stout, whorled, with sporangia smaller but similar to those on primary axis.

A single species treated.

1. *Actinomucor repens* Schostakowitsch

Syn. *Mucor botryoides* Lendner
Mucor glomerula (Bainier) Lendner
Rhizopus elegans Eidam

Sporangiophores erect, very branched. Each erect branch terminated by a very large sporangium, below which occurs a whorl of three to eight secondary filaments, each terminated by a sporangium. These three to eight filaments give rise in their turn to a whorl of three to five sporangiferous filaments. The aerial mycelial filaments usually end in branches carrying nearly sessile sporangioles. Sporangia spherical, hyalin, becoming sienna color when old. Wall roughened by crystals of calcium oxalate, diffluent leaving a collarette. Columellae variable in shape, hemispheric, cylindro-conic, ovoid, sometimes restricted, inserted at the rather suddenly expanded end of the sporangiophore. Spores round and smooth. Aerial chlamydospores round, with thick wall, yellow and spiny. Content oleaginous. Mycelial chlamydospores seemingly submerged but very numerous. Zygospores unknown.

From soil: Canada (15), Czechoslovakia (97), England (37), France (78), Jugoslavia (103), U. S. S. R. (108)

United States: Alaska (147), Colorado (74) (75) (76), Iowa (1) (3), New Jersey (146) (147)

(4) **Phycomyces** Kunze

Mycelium radiating, branching repeatedly into finer and finer hyphae. At first unicellullar, septate in old cultures; cell content slightly yellowish-orange. Sporangiophores erect, greenish to brownish-violet, with metallic sheen. Sporangia, globose, many-spored. Wall of sporangium not cuticularized but incrusted with needles of calcium oxalate; diffluent. Columellae free, pyriform, enlarged toward the tip, sometimes cylindric. Spores ellipsoid, smooth, yellow. Zygospores on the my-

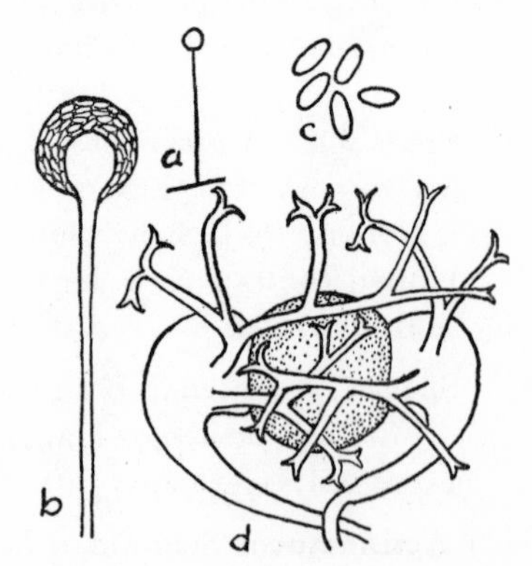

Fig. 4 **Phycomyces.** *a* habit; *b* sporangiophore; *c* sporangiospores; *d* zygospore.

celium, the gametes incurved in the form of pincers. Suspensors furnished with dichotomously divided appendages, brownish-black, encircling the zygospore.

A single species treated.

1. *Phycomyces nitens* Agardh

Sporangiophores forming an olive-green turf, from 7–30 cm. high, about 50–150μ in diameter, without septa. Wall smooth, shining smoky-gray or olive-gray, colorless near the sporangium. Sporangia globose, 0.25–1 mm. in diameter, all at first yellow-orange becoming black at maturity, wall diffluent, without a basal collarette. Columellae free, usually pyriform, campanulate or cylindric; wall smooth, hyalin with slightly yellowish content, measuring in the large sporangia 330μ in length, 130μ in basal diameter, 180μ in tip diameter. Spores elliptic, plano-convex, 8–15μ in diameter × 16–30μ in length, content dark yellow. Zygospores on the mycelium on the surface of the substrate; 300μ in diameter, black, smooth to slightly verrucose. Appendages of suspensors numerous, dichotomously branched, brownish-black. Gemmae unknown. Chlamydospores intercalary.

From soil: Czechoslovakia (97)

(5) **Circinella** van Tieghem and Le Monnier

Mycelium strongly branched, at first nonseptate, becoming divided. The lateral branches become more and more delicate. Sporangiophores erect on the mycelium, branching in sympodia; the tip grows indefinitely, and never terminates in a sporangium. The lateral branches, united into whorls or single, are curved and carry at their tips sporangia of like dimensions. Sporangia many-spored, spherical, with the wall incrusted with calcium oxalate crystals, non-diffluent, but breaking into pieces, leaving an irregular collarette at the base. Columellae large, slightly concrescent at the base, cylindro-conic, sometimes panduriform. Spores spherical or oval, smooth, more or less slate-blue. Zygospores borne on erect hyphae distinct from the sporangiophores. Suspensors without appendages.

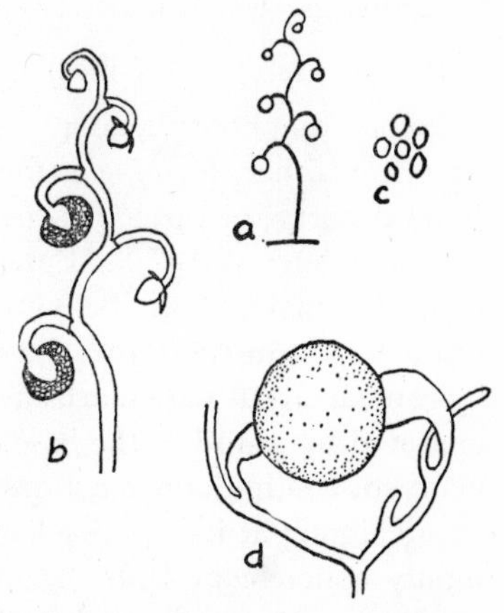

FIG. 5. **Circinella.** *a* habit; *b* sporangiophore; *c* sporangiospores; *d* zygospore (after Zycha).

KEY TO THE SPECIES OF THE GENUS CIRCINELLA

a. Spores 3–5μ
  b. Sporangia brown; turf brown
    c. Turf 2–3 mm. high — 1. *C. simplex*
    cc. Turf up to 20 mm. high — 2. *C. spinosa*
  bb. Sporangia black; turf gray — 3. *C. tenella*
aa. Spores 6–10μ — 4. *C. sydowi*

1. *Circinella simplex* van Tieghem

Sporangiophores erect, forming a short, close turf, brownish, 2–3 mm. high, carrying no sporangia on the lower 0.5 mm. of the basal portion. The sporangiophore carries alternating lateral sporangia, with curved erect pedicels; these latter are unbranched, nonseptate, with a cuticularized wall, brownish-black, incrusted. Sporangia small, spherical, brownish-black. Wall persistent, breaking into pieces, incrusted with oxalate and leaving a basal collarette. Columellae concrescent at the base, spherical, subspherical or campanulate, with a smooth wall. Spores globose, 3μ in diameter, smooth, hyalin, or slowly becoming blue-gray. Zygospores unknown.

From soil: Brazil (109)
United States: Illinois (127)

2. *Circinella spinosa* van Tieghem and Le Monnier

Sporangiophores slender, erect, close and climbing, mutually sustaining one another, forming a turf 2 cm. high. The tip of the sporangiophore is sterile, or rarely bearing a sporangium (the terminal sporangium has a diameter of 147μ). Lateral sporangia in two series along the primary sporangiophore. Below on curved pedicels which are prolonged into a hypha in the form of a spine, above the pedicels without spines. Sporangial wall cuticularized and incrusted, with colorless content, separated by a wall at the base of the filament. Sporangia small globose, incurved, 60μ in diameter, brown, echinulate. Sporangial wall not diffluent, dehiscing at its equator leaving a large basal collarette. Columellae slightly concrescent with the wall of the sporangium, cylindro-conic or globose, wall smooth, slightly brown. Spores globose, 4μ in diameter, gray-brown. Zygospores unknown.

From soil: Czechoslovakia (97), France (78), U. S. S. R. (108)

3. *Circinella tenella* (Ling-Young) Zycha

Syn. *Mucor tenellus* Ling-Young

Turf up to 2 mm. high, bright gray to gray black. Sporangiophores sporadically high and scarcely branched, for the most part low and with strong sympodial branching with curved side-branches. Sporangia dark gray, 40–70μ, with more or less slowly diffluent wall. Columellae globose, blue-gray. Spores globose, 3–4μ, slightly gray. Mycelial gemmae in the substrate.

In soil: France (78)

4. *Circinella sydowi* Lendner

Syn. *Mucor laxorhizus* Ling-Young

Turf white, then gray, 5–7 mm. high. Sporangiophores erect, ending in sterile tips or sporangia. Side branches short, alternating or opposite, sometimes in turn ending in sterile tips. Sporangia 100–150μ in diameter, at first white, later gray and black with dull, smooth, fragile walls. Columellae broadly conical, cylindrical or constricted in the middle. Spores gray, 6–7μ.

In soil: France (78), Germany (79) (169)

(6) **Mucor** Micheli

Mycelium widespread in and on the substratum, but without rhizoids or especial membered stolons; richly branched, with branches always thinner until at last hair-fine; straight or knotted, at first, one-celled; in age with irregular cross-walls, with colorless, infrequently orange-red content; smooth, colorless membrane. Sporangiophores springing singly from the mycelium but usually forming a thick turf, erect, either unbranched with terminal sporangia or branched with like sporangia on all the branch ends; branching in part monopodial, clustered or irregularly panicled or umbelliferous; in part cymose and more or less sympodial, curved, with sporangia also at the tip of the sympodium, never forked. Sporangia erect at all times on sympodial sporangiophores, a few weakly bent, usually all alike, only of different size; many spored, spherical, opening on the sporangiophore, only a few in sympodial forms abscissing while still closed; of various colors. Sporangial wall not cuticularized, incrusted more or less strongly with needles of calcium oxalate, dissolving quickly in water, leaving a collarette, or breaking and then at times persistent. Columellae always present, of various shapes, colorless or colored. Spores spherical or ellipsoid with thin smooth membrane, colorless or colored. Zygospores on the mycelium,

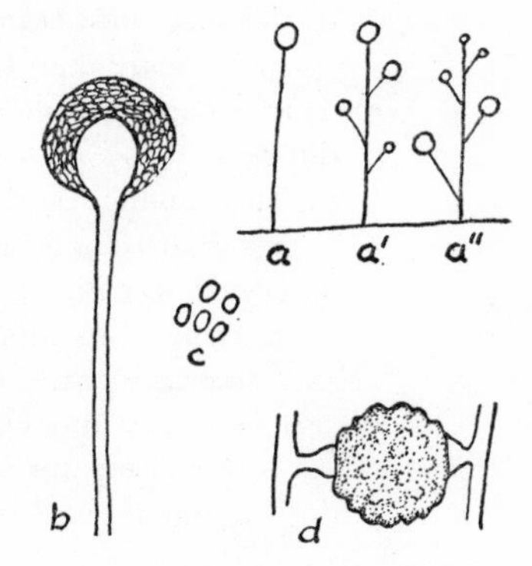

Fig. 6. **Mucor.** *a*, *a'*, *a''* habit; *b* sporangiophore; *c* sporangiospores; *d* zygospore.

not on special branches, naked; suspensors without outgrowths; gametangia straight. Mycelial conidia (stylospores) unknown. Gemmae (chlamydospores) terminal and intercalary, variously formed, colorless, smooth; not in all species.

KEY TO THE SECTIONS OF THE GENUS MUCOR

a. Spores globose — I. Sphaerosporus
aa. Spores not globose
  b. Turf 0.5–3 mm. high, sporangia not over 45μ in diameter — II. Ramannianus
  bb. Turf more than 3 mm. high
    c. Turf delicate, at first white, later gray or brown; sporangiophores richly branched; sporangial walls fragile or slowly diffluent
      d. Gemmae in sporangiophores numerous, spores short oval — III. Racemosus
      dd. Gemmae in sporangiophores scarce or lacking, spores more than twice as long as broad — IV. Fragilis
    cc. Turf remaining white, yellow or gray, spore walls diffluent
      d. Turf usually less than 20 mm. high, sporangia less than 100μ in diameter — V. Hiemalis
      dd. Larger species
        e. Primary sporangiophores more or less sympodially branched — VI. Flavus
        ee. Primary sporangiophores, not sympodially branched, sporangia large, 100–300μ, no gemmae — VII. Mucedo

SECTION I. SPHAEROSPORUS

KEY TO THE SPECIES

a. Thermophilic species — 1. *M. pusillus*
aa. Grow well at room temperatures
  b. Columellae with pointed extensions, spores brown
    c. Sporangiophores less than 2 mm. high — 2. *M. spinescens*
    cc. Sporangiophores more than 1 cm. high — 3. *M. plumbeus*
  bb. Columellae smooth
    c. Spores larger than 8μ in diameter
      d. Giant cells in the substrate — 4. *M. dispersus*
      dd. No giant cells in the substrate
        f. Sporangial wall breaking at maturity — 5. *M. petrinsularis*
        ff. Sporangial wall diffluent at maturity — 6. *M. lamprosporus*
    cc. Spores smaller than 8μ in diameter

d. Turf and sporangia gray to black — 7. *M. jansseni*

dd. Turf and sporangia brown

e. Sporangia 75–120μ, sporangioles absent — 8. *M. globosus*

ee. Sporangia 80–90μ (110μ maximum), sporangioles present — 9. *M. sphaerosporus*

1. *Mucor pusillus* Lindt

Turf thick-felty, up to 2 mm. high, at first white with unbranched sporangiophores, later brown, seldom gray, with strongly branched brown sporangiophores, 6–20μ in diameter and almost always with a septum below the sporangium. The latter, 50–80μ in diameter, bright gray, later brown, with dully shining surface and more or less quickly diffluent wall. Columellae bluish to brown, oval or pear-shaped (reminding of Absidia), large in relation to the sporangium, up to 60μ high, often with a collarette. Spores 2.5–4μ, globose, occasionally oval or paired and biscuit-shaped, often mixed with small crystalline pieces of the sporangial wall. Gemmae and zygotes unknown.

From soil: Greenland (96)

2. *Mucor spinescens* Lendner

Turf very short, 1–2 mm. in height. Sporangiophores branched and short, maximum 1 mm. long by 10μ wide, thinner near the sporangium. They are rarely straight, often slightly incurved. Sporangia globose, variable in size, 60–68μ in diameter. Spores rather large, 7–8μ, rarely smaller, 5–6μ, slightly colored, clear yellow-brown. Columellae at times ovate, at times pear-shaped, or even elongate; often with a varying number of prolongations on their tips. Zygospores unknown. Differs from *M. jansseni* by its spinescent columellae and larger spores; from *M. plumbeus* by its much smaller sporangiophores.

From soil: Denmark (68), France (78), Germany (79), Jugoslavia (103)

United States: Idaho (106)

3. *Mucor plumbeus* Bonorden

Turf close, regular, mouse gray, about 1 cm. deep. Sporangiophores erect, 1 cm. long, branched in groups or in sympodia. All the branches terminated by sporangia. Wall, smooth, colorless. Sporangia 100–300μ in diameter, deep brown or black. Wall diffluent, incrusted, leaving a basal collarette. Columellae free, oval or pear-shaped, furnished at their summit with a variable number of spines (up to twelve or more) irregular, often swollen at the tip, 22–85μ long × 8–65μ wide; they are often colored gray or brown. Spores globose, equal, 5–8μ, exceptionally

9–12μ, gray-blue with a dotted wall. Zygospores globose, yellow-brown; exine furnished with irregular warts in the shape of plates. Chlamydospores formed on the mycelium or the sporangiophores as in *M. racemosus*. Budding cells as in yeasts.

From soil: Austria (66), Czechoslovakia (97), Denmark (68), England (36), Germany (79), Jugoslavia (97), Norway (55), Switzerland (97), U. S. S. R. (108)

United States: California (146), Idaho (106), New Jersey (146) (147), New York (67)

4. *Mucor dispersus* Hagem

Sporangiophores small on slightly thickened turf of various heights. The larger (primary) 2–3 cm. high are widely scattered, very delicate, 5–7μ in diameter, waving here and there, usually soon collapsing, branched in monopodial clusters, with short bent, circinate, often secondarily branched branchlets; the primary as well as secondary branches terminate in sporangia. The smaller (secondary) sporangiophores, 1–2 mm. high, usually circinate, with small sporangia. Sporangia on primary sporangiophores small, about 50μ, with diffluent wall and smaller, usually truncate or broadly globose, seldom oval, 17–19μ high $\times$ 18–21μ broad columellae, with small collarettes. Sporangia of the side branches as well as those of the secondary sporangiophores of various size, 15–45μ, with spiny not diffluent wall and translucent spore mass, frequently very small with only two to four spores but also larger, 30–45μ, with numerous spores. Spores of tolerably different size, 11–13μ, usually round or slightly elongate, some even rounded-angular. Zygospores unknown.

From soil: Canada (15), Czechoslovakia (97), France (78), Germany (79), Norway (57)

United States: North Carolina (23), South Carolina (23)

5. *Mucor petrinsularis* Naumov

Turf about 20 mm. high, at first white, later bright gray or brown. Sporangiophores erect, stout, glassy-hyalin, sympodially branched with a side branch arising immediately below the columella. Sporangia brown, up to 100μ in diameter. At maturity the sporangial wall tears irregularly so that the spores lie as if spread on a dish. Columellae pear-shaped, often brownish, 50μ high. Spores globose, averaging 8–12μ in diameter (seldom oval and then up to 18μ long). Gemmae and zygospores unknown.

From soil: Austria (131)

6. *Mucor lamprosporus* Lendner

Turf about 3 cm. above the surface of the substrate, a dense felt, pale gray. Sporangiophores rather irregularly branched in groups or in sympodia, 3 cm. high, branches alternate, recurved, bearing minute sporangia. Sporangia terminal, globose, 60μ in diameter (90μ maximum). Wall diffluent. Columellae spheric, 20μ in diameter or ovoid, 24μ wide × 28μ long. Lateral sporangia, minute, 30–40μ in diameter, caducous. Spores globose, colorless, hyalin 10μ in diameter (7–12μ, limits).

From soil: Switzerland (77)

United States: North Carolina (23)

7. *Mucor jansseni* Lendner

Turf very short, velvety, becoming yellow to orange with age. Sporangiophores 2–6 mm. high, much branched in corymbs or in sympodia; ending in sporangia. Wall obliquely striate. Sporangia globose, deep bluish black, 50–70μ in diameter. Wall finely granular, not diffluent, but fragile. Columellae sometimes round, with a wide flattened base, subjacent; sometimes elongate and conic, tinted deep blue or gray, 34μ long by 30μ wide; the smaller are proportionately longer, 26μ long × 20μ wide. Spores round, 5–6μ usually, sometimes smaller, 3–4μ in diameter.

From soil: France (78), Switzerland (77)

United States: Idaho (106)

8. *Mucor globosus* Fischer

Turf at first white, later yellow-brown, up to 20 mm. high, usually with a 1–2 mm. high felty part and a higher floccose part. Sporangiophores richly branched, up to 20μ in diameter. Sporangia yellow-brown, 70–120μ, with diffluent wall. Columellae free, steel-blue or brownish, oval or pyriform, about 60μ long. Spores 4–8μ, of a yellowish cast. Mycelium gemmae with oil-drops frequent. Zygospores unknown.

From soil: Czechoslovakia (97), England (11), France (78), Germany (79), Jugoslavia (103), Norway (55), U. S. S. R. (108)

United States: New Jersey (146)

9. *Mucor sphaerosporus* Hagem

Turf short, rather deep brown, 0.5 cm. high. Sporangiophores branched sympodially or in corymbs of three to five branches; about 1 cm. high × 15–18μ wide, with wall colored a pale brownish-red. Sporangia globose, brownish-red, 70–110μ in diameter. Wall diffluent in the case of the large sporangia, fragile and persistent in the case of the small, stippled. Columellae ovoid or round, free at the base or quite campanulate with a flattened base (40–65μ high × 30–55μ broad, Hagem). Spores round, (very exceptionally ovoid), very shining, tardily reddish, and rather equal, 10μ (6–8μ). Sporangioles numerous on the substrate but not caducous. Chlamydospores and oidiospores numerous. Zygospores unknown.

From soil: Czechoslovakia (97), England (11) (36) (37), Germany (97), Jugoslavia (103), Norway (55), U. S. S. R. (108)

United States: Idaho (106), New Jersey (146) (147)

SECTION II. RAMANNIANUS

A single species treated.

10. *Mucor ramannianus* A. Moeller

Turf short, velvety, of a carmine-red-brown. The edge of colony white, becoming gray with age. Sporangiophores unbranched, less than 200μ long by 5–6μ in diameter. Sporangia very small, usually 20μ in diameter (40μ maximum), spherical, flesh-rose to copper-red. Wall, unequally diffluent, smooth. Spores globose, rarely oval, 2–3μ in diameter, colorless (the coloration of the sporangia is probably due to some interstitial substance). Columellae spherical, of variable size, 8–10μ in diameter. Chlamydospores numerous, 10–12μ in diameter, globose or ovoid. Giant-cells are often present. Mycelium and sporangiophores are rather frequently septate.

From soil: Austria (131), Czechoslovakia (97), Denmark (68), England (36), France (78), Germany (79) (104), Jugoslavia (97) (103), Norway (55), Switzerland (97), U. S. S. R. (108)

United States: Iowa (101), Maine (146), Michigan (105), North Carolina (23)

SECTION III. RACEMOSUS

KEY TO THE SPECIES

a. Sporangia less than 100μ in diameter
    b. Young turf white to golden brown
        c. No growth at 37° C. or above
            d. Sporangia very variable, columellae usually globose — 11. *M. racemosus*
            dd. Sporangia of uniform size, columellae elongate — 12. *M. christianiensis*
        cc. Growth at 37° C.

d. Growth at 37° C. limited — 13. *M. praini*
dd. Growth at 37° C. strong
e. Spores 5–7μ long — 14. *M. javanicus*
ee. Spores 4–5μ long — 15. *M. rouxianus*
bb. Young turf gray to gray-brown
c. Sporangia brown — 16. *M. circinelloides*
cc. Sporangia black — 17. *M. griseo-cyanus*
aa. Sporangia more than 100μ in diameter — 18. *M. geophilus*

11. *Mucor racemosus* Fresenius

Sporangiophores erect, close, forming a yellow-brown turf of very variable height, 5–40 mm. high × 8–20μ wide, branched irregularly in groups. All the branches terminated by sporangia which are very unequal. Sporangia small, globose, unequal, 20–70μ in diameter, erect, or at times incurved, pale yellow, then wax-yellow or yellow-brown, hyalin. Walls of sporangium not diffluent, but fragile, persistent, incrusted, leaving a collarette. Columellae free, globose, ovoid or broadly cuneiform, campanulate 17–60μ long × 7–30μ at the base and 9–42μ in greatest diameter. Spores rarely globose, more often elliptic, 5–8μ × 6–10μ, smooth, yellow in mass. Zygospores globose, 70–85μ, brown, with conic warts yellow or red-brown, suspensors straighter than the zygospore and not swollen. Azygospores, chlamydospores always very numerous, the latter formed either on the mycelium or on the sporangiophores and even on the columellae; they are colorless, or yellow, with a smooth membrane, of diverse shapes, 20μ in diameter or 11–20μ in diameter × 20–30μ in length. Budding cells formed in liquid sugars. The mycelium breaks up into oidia.

From soil: Austria (97) (131), Canada (15) (147), Czechoslovakia (97), Denmark (68), England (11) (37), France (78), Germany (6) (97) (104), Greenland (96), Italy (97), Japan (132), Jugoslavia (103), Norway (55), Switzerland (77) (97), U. S. S. R. (108)

United States: Alaska (147), California (147), Colorado (147), Hawaii (147), Iowa (3), Maine (147), New Jersey (23) (30) (146) (147), New York (67), North Carolina (23), Oregon (146) (147), Texas (147) (152), West Virginia (23)

12. *Mucor christianiensis* Hagem

Colonies at first forming thick, superficial, gray mycelial mat, then a sparse formation of sporangiophores. Sporangiophores usually scattered, seldom forming a loose turf, 1.5–2.5 cm. high, exceptionally thin, 6–10μ thick, and quickly collapsing, unbranched when young; monopodially branched later with small, short circinate branches, with numerous chlamydospores on the sporangiophore, at first cylindric, later barrel-shaped or usually globose, a rather regular distance apart, 50–200μ, quickly freed when mature. Sporangia small, usually 40–60μ in diameter, bright yellow when ripe with fragile wall, leaving only a small fragment at the base of the columella. Columellae oval or usually elongate, always longer than wide, 30–45μ long and 25–30μ wide. Spores broad, oval, or some almost spherical, 6–9 × 5–7μ. Zygospores unknown.

From soil: Norway (57)

13. *Mucor praini* Chodat and Nechitch

Sporangiophores attaining 4 cm. in height (on rice), sympodially branched. Sporangia globose, smooth, slightly transparent, from yellow to deep brown. The larger 70–90μ in diameter; the smaller 35μ. Spores subspherical, smooth, hyalin, 6–8μ, up to 10μ, in diameter in the larger sporangia. In the smaller sporangia 3–4μ in diameter. Columellae sometimes globose, sometimes slightly elongate, colorless, smooth, hyalin, with a basal collarette, 50–54μ in diameter. Chlamydospores hyalin, with a thick wall, smooth, variable in form, ellipsoid, ovoid, spherical or irregular. The larger 24μ in diameter. Zygospores unknown.

From soil: Jugoslavia (103)

14. *Mucor javanicus* Wehmer

Turf thick, erect, 10–30 mm. high, soon sinking to the substrate. Sporangia up to 50μ in diameter. Columellae up to 35μ long. Spores 4–5 × 5–7μ, also 3 × 3.5μ. Zygospores 50–60μ, dark brown.

From soil: Austria (66), Czechoslovakia (97)

15. *Mucor rouxianus* (Calmette) Wehmer

Syn. *M. rouxii* (Calmette)

Turf low, at most 4 mm. high, white, light yellow or gray, delicate, loose. Sporangiophores weakly sympodially branched. Sporangia bright yellow or golden-brown, usually 50μ (also 20–100μ) in diameter, only numerous under especially favorable conditions. Columellae up to 40μ high, globose or flattened, often with a colored membrane; spores 4–5μ long. Chlamydospores numerous, of various size, up to 100μ in diameter with walls 7μ thick. Budding observed. Zygotes unknown.

From soil: Greenland (96), Morocco (169)

16. *Mucor circinelloides* van Tieghem

Syn. *M. echinulatus* Paine

Sporangiophores erect, forming a very short turf, close and deep brown, about 1 cm. tall. They are more or less branched in sympodia with branches alternating right and left, short and more or less curved, always terminated by a sporangium. The length of the secondary branches is very variable; they are sometimes so short that the sporangium is seemingly sessile. Sporangia globose, 50–80μ in diameter, gray-brown when walled; erect, or slightly incurved. The larger have a diffluent membrane; in the case of the smaller (the upper) the wall is persistent and the sporangia are caducous. Wall of sporangium incrusted and when diffluent it leaves a basal collarette; but when not incrusted, persistent, firm and smooth. Columellae free, hemispheric or spheric or oval, colorless, smooth. Spores globose or elliptic, 3μ in diameter $\times$ 4–5μ long (Lendner, 4 $\times$ 5–6μ), smooth, colorless when single, but pale gray in mass. Zygospores globose, exine red-brown, covered with very prominent spiny warts, longitudinally striate. Chlamydospores smooth, colorless, deep on the length of the filament. Gemmae as in yeasts and *M. racemosus*.

From soil: Austria (66), Czechoslovakia (97), Denmark (68), Egypt (115), England (11) (36) (37), France (78), Germany (168), Greenland (96), India (21), Japan (132), Jugoslavia (103), Switzerland (77)

United States: Colorado (147), Idaho (106), Illinois (129), Iowa (101), Louisiana (147), Maine (147), New Jersey (146) (147), New York (67), North Dakota (147), Oregon (147)

17. *Mucor griseo-cyanus* Hagem

Turf deep gray-blue, about 1 cm. high. Sporangiophores branched, the longer in groups or in sympodia; the shorter always in sympodia. The lateral branches of the latter are rather circinate. Sporangia globose, 60–80μ in diameter, with a non-diffluent wall, incrusted with very small crystals of calcium oxalate. They are strongly colored gray-blue. Columellae round or ovoid, flattened at the base and concrescent with the membrane of the sporangium, 30–40μ long $\times$ 24–36μ wide, colored a clear fuliginous brown. Spores oval, brown in mass, 5–6μ long $\times$ 4μ wide. Chlamydospores formed on the sporangiophores and the filaments of the mycelium, oval or round, 12–14μ in diameter. Zygospores unknown.

From soil: Czechoslovakia (97), Egypt (115), Germany (169), France (78), Jugoslavia (103), Norway (57), Switzerland (77)

United States: Illinois (129)

18. *Mucor geophilus* Oudemans

Mycelium snow white, very tardily gray, finally pale olive. Sporangiophores simple or branched in cymes, carrying two to three branches. Sporangia globose, at first yellow, then olivaceous, leaving a collarette after the destruction of the membrane, 50–350μ in diameter. Wall with small blunt warts. Columellae globose, voluminous, pale gray. Spores pluriform, globose, round or elliptic, angular, 4.2–6.5μ in diameter, smooth, olive. Chlamydospores on the branches of the mycelium, round, 20μ in diameter, at times in a more or less extended series. Zygospores very like chlamydospores, about 30μ in diameter.

From soil: Holland (100)

United States: Iowa (1) (3)

SECTION IV. FRAGILIS

KEY TO THE SPECIES

a. Turf 1–2 mm. high — 19. *M. ambiguus*
aa. Turf 2–20 mm. high
  b. Spores of various sizes, 6–12μ long — 20. *M. lausannensis*
  bb. Spores smaller, 4–7μ long — 21. *M. fragilis*

19. *Mucor ambiguus* Vuillemin

Sporangiophores erect, forming a blackish turf, 1 mm. in height, branched in sympodia and bearing four to five sporangia, branches short, straight, or slightly incurved. Sporangia globose, 100μ in diameter, gray-black. Wall of sporangia more or less incrusted and more or less diffluent. The successive sporangia are more and more persistent, the wall finally dehiscing by fragmentation. Columellae free, globose, or campanulate. Spores elliptic 4.5μ wide $\times$ 7μ long, with a finely stippled wall. Zygospores unknown.

From soil: United States: Michigan (52)

20. *Mucor lausannensis* Lendner

Sporangiophores erect, little branched, bearing laterally one or two groups of branches. These sporangiophores form a fine compact turf, yellowish, 0.5–1 cm. high (10–14μ in diameter). Sporangia 40–54μ in diameter, often flattened at the base. The wall is not diffluent but fragile as in *M. racemosus*, leaving an irregular basal collarette. Columellae oval or spherical, 30–40μ in diameter $\times$ 50μ long. Spores oval, of very different sizes, the smallest 4 $\times$ 2μ, the largest 12μ long $\times$ 6μ wide. The average size is 8 $\times$ 6μ. They are hyalin, then pale, slowly turning yellowish in mass. Chlamydospores, rather rare, may be formed on either the mycelium or the sporangiophore. They measure on the average 16 $\times$ 14μ, are smooth and granular in content. Zygospores not known.

From soil: Czechoslovakia (97), England (37)

United States: Colorado (74) (76)

21. *Mucor fragilis* Bainier

Turf gray to brown, of various heights: 2–15 mm. Sporangiophores erect about 6–15μ in diameter, usually with marked sympodial branching, helicoid. Sporangia, when moist, yellowish-white to gray, later olive-brown; when dry, beige to gray-brown, later olive-brown; small, 35–86μ, wall more or less slowly fragmenting. Columellae globose to oval, smooth, hyalin, up to 50μ high, with a more or less marked collarette. Spores in mass dark-brown, with regular elliptic-cylindrical shape, twice as long as wide: 2–4 $\times$ 4–8μ. Mycelial gemmae detached. Zygospores numerous in winter and spring, black, spherical, 50μ in diameter.

From soil: Germany (169)

United States: North Carolina (23)

SECTION V. HIEMALIS

KEY TO THE SPECIES

a. Spores long (1:2–3) narrow	
b. Spores cylindric, with rounded ends	22. *M. subtilissimus*
bb. Spores ellipsoid to spindleform	23. *M. luteus*
aa. Spores elongate (1:1.5–2)	
b. Spores rounded cylindric, very uniform up to 4μ long	24. *M. microsporus*
bb. Spores longer than 5μ	
c. Monoecious species	25. *M. genevensis*
cc. Dioecious species	26. *M. hiemalis*
ccc. Zygotes unknown	27. *M. adventitius*
aaa. Spores oval (1:1–1.5)	
b. Gemmae present	
c. Giant cells present in substrate	28. *M. silvaticus*
cc. No giant cells in the substrate	
d. Spores regularly elliptic	29. *M. griseo-lilacinus*
dd. Spores short-oval to globose	30. *M. abundans*
bb. No gemmae present	31. *M. corticolus*
aaaa. Spores various, oval and elongate found in about equal numbers	32. *M. varians*

22. *Mucor subtilissimus* Oudemans

Sporangiophores, colorless, simple, with two to three septa, 210μ long by 4–7μ in diameter. Sporangia globose, colorless, smooth, 40–45μ in diameter. Columellae globose, colorless, 25–35μ in diameter. Spores elliptic, colorless, rounded at the ends, 7 $\times$ 3μ. Zygospores and chlamydospores unknown.

From soil: Austria (66), Cuba (23), Germany (104)

United States: North Carolina (23)

23. *Mucor luteus* Linnemann

Turf 1.5 cm. high, orange-yellow in color. Small side branches in cymes. Sporangiophores very fine, 6–14.5μ in diameter, usually 8–10μ. Mycelium and sporangiophores filled with yellow content. Sporangia globose, 40–70μ in diameter, usually 50–60μ, at first white, then yellow. Sporangial walls diffluent, smooth. Columellae globose, 16–48μ in diameter often somewhat broader than high; colorless or somewhat yellow. Collarette as small recurved membrane. Spores various, ellipsoid to spindle-shaped, of different sizes, 1.5–6μ broad, × 3–16μ long. Chlamydospores not present. Zygospores globose, 40–80μ in diameter usually 60μ. Exine with small warts, black.

From soil: Germany (79)

24. *Mucor microsporus* Namyslowski
Syn. *M. cylindrosporus* Ling-Young

Colonies whitish, with age becoming yellowish, especially when grown on pears; mycelium cottony. Sporangiophores unbranched, up to 2 cm. high, 12–20μ thick, below the columella strongly attenuated. Sporangia brownish, 30–80μ in diameter, mostly 60μ; membrane of young sporangia not diffluent; with age, however, it dissolves. Columellae spherical, somewhat higher than broad, beneath weakly attenuated, always with flat base and short collar, smooth, 20–70μ broad, often filled with brick-red contents. Spores regularly ellipsoidal, hyalin, smooth, 2–3μ long, 1.5μ broad (4μ long maximum), in mass when young ashen-gray, with age bluish.

From soil: Austria (94), Czechoslovakia (97), England (11), France (78), Germany (169), U. S. S. R. (94)

United States: Maine (147), New Jersey (146) (147)

25. *Mucor genevensis* Lendner

Turf close, white, 2 cm. high. Sporangiophores 2 cm. long by 10–15μ wide, little branched in groups, carrying one or two lateral sporangia. Sporangia globose, 66μ in average diameter, but exceptionally reaching 80μ in diameter. Wall diffluent, almost colorless, rather yellow, leaving collarette. Columellae oval or round, free, colorless, 30–36μ in diameter or 24 × 36μ. Spores elongate, planoconvex, 9–10μ long × 3–4μ wide. Chlamydospores frequent, rather than oidiospores which are borne on lateral branches. Zygospores frequent on adjacent branches but not on forks of the same branch, 100μ in diameter. Exine very thick, with conic warts.

From soil: Czechoslovakia (97), France (78), Germany (169)

United States: North Carolina (23)

26. *Mucor hiemalis* Wehmer

Sporangiophores usually unbranched, erect, then prostrate by wilting. Turf about 1 cm. high (0.5–2 cm.) close and fine, cottony, white, rarely grayish-yellow. Sporangia spherical, gray or brownish-yellow, visible to the naked eye, 52μ in diameter. Wall diffluent in young condition, leaving a collarette. Columellae free, spherical or oval, colorless, 28–48μ (spherical) or 25 × 21μ to 36 × 29μ. Spores usually unequal, the majority elongate, ellipsoid or kidney-shaped, 7 × 3.2μ (limits 3–8.4 × 2–5.6μ) smooth, hyalin, with thin membrane. Mycelium comes to resemble that of *M. rouxianus* by the accumulation of oil drops.

From soil: Austria (94) (131), Canada (15), Czechoslovakia (97), Denmark (68), France (78), Germany (79) (104), Greenland (96), Hungary (97), Jugoslavia (97) (103), Norway (55), Switzerland (77), U. S. S. R. (108)

United States: Iowa (101), Louisiana (23), Maine (147), New Jersey (146), New York (67), North Carolina (23)

27. *Mucor adventitius* Oudemans
Syn. *M. humicolus* Raillo

Sporangiophores simple, continuous, hyalin, forming a turf 20 mm. high. Sporangia globose, 80–95μ in diameter, at first hyalin, later light gray, finely echinulate, with a diffluent membrane. Columellae at first globose, later elliptic or campanulate, hyalin with colorless content, 40–48 × 48–64μ and furnished with a basal collarette. Spores elliptic or nearly oblong 8–8.5 × 4.5–5μ smooth, hyalin grayish when in mass. Zygospores and chlamydospores unknown.

From soil: China (81), Germany (169), Holland (100), Japan (132), U. S. S. R. (108)

28. *Mucor silvaticus* Hagem

Turf white or gray formed of thin slightly dense filaments, extending over the surface. Sporangiophores rarely straight, but mostly irregularly incurved, branching near the tip with one or two lateral branches. They reach 1 cm. in height by 10μ in width. Sporangia small, globose not more than 70μ (average 44μ). Wall diffluent, leaving a basal collarette. Spores of a very variable size, oval or subglobose 4 × 2 to 5 × 3μ (8 × 6μ maximum). Columellae globose or oval, 30 × 22 to 20 × 25μ in diameter. At the point of contact with the substrate the chlamydospores are numerous. They are ovoid 16–24μ in diameter. The erect filaments frequently have large swellings which become isolated and form round cells measuring 40–60μ in diameter, rarely longer.

From soil: Canada (15), China (81), Czechoslovakia (97), Germany

(168), Greenland (96), Jugoslavia (103), Norway (55), Switzerland (77)
United States: Louisiana (147), New Jersey (146) (147)

29. *Mucor griseo-lilacinus* Povah

Turf thick, mouse-gray, 10–15 mm. high. Sporangiophores 8–20μ in diameter, at first simple, later branched many times. Sporangia 60–80μ in length, at first yellow, then dark gray to greenish, with diffluent wall and collarette. Columellae fairly globose 27–43μ in diameter, bluish-gray. Spores regularly oval 3–4 $\times$ 4–6μ, bright gray. Gemmae present. Zygospores unknown.

From soil: Illinois (129)

30. *Mucor abundans* Povah

Forming on bread a dense, erect, smoky-gray turf tinged drab 1.5–3.5 cm. tall. Sporangiophores 8–23μ in diameter, at first simple, later with one to three lateral branches which are in turn branched once or twice, with branches always terminating in a sporangium, and with a septum above point of insertion of branch; sporangia globose or sub-globose, smooth or incrusted with very delicate crystals, 56–78μ in diameter, at first yellowish, becoming dark gray with a greenish tinge at maturity. Wall diffluent, leaving a collarette; columellae sub-globose to pyriform, free or slightly adnate 31–40 $\times$ 25–35μ, hyalin or tinged gray; spores variable, globose to short elliptical 3–5μ in diameter or 4–5.5 $\times$ 3–4.5μ. Chlamydospores and yellowish globules in submerged mycelium. Zygospores not found. Related to *M. hiemalis* (sense of Hagem) from which it differs in the shape of the columellae and in the shape and size of the spores.

From soil: Canada (15)
United States: Illinois (129), Michigan (105)

31. *Mucor corticolus* Hagem

Colonies gray or slightly blue-gray. Sporangiophores erect, up to 2 cm. high, sympodially branched with small, long branches, terminating in a sporangium. Lateral branches, long (2–3 times as long as *M. silvaticus*) and usually 600–1500μ long by 10–15μ thick, often more or less curved and terminating with sporangia. Sporangia globose, 50–60μ in diameter, with diffluent wall. Columellae egg-shaped or slightly oval, almost always 3–6μ longer than broad, 27–33μ wide $\times$ 30–36μ long, without or with colorless content and usually with an indistinct collarette. Spores oval or elliptic (larger than *M. silvaticus*) 5–7 $\times$ 3.5–5μ. Zygospores not known.

From soil: Germany (79), Norway (57)
United States: Michigan (105)

32. *Mucor varians* Povah

Turf 1–3.5 cm. tall on bread, ivory-yellow to olive-buff. Sporangiophores 8–20μ in diameter, either little or profusely branched, much coiled, twisted or intertwined, forming a dense tough cottony turf with proliferations of hyphae and columellae often present. Sporangia globose or subglobose, smooth, 60–80μ in diameter, at first yellow or pale orange, then very dark gray tinged with green at maturity; wall diffluent leaving a basal collarette. Columellae free or slightly adnate, very variable in shape, subglobose, hemispherical, flattened hemispherical, oval, cylindrical, elliptical, pyriform, panduriform, cylindroconical, subconical and conical; large columellae hemispherical to conical, small columellae cylindrical to pyriform and panduriform 25–50 × 20–45μ, membrane tinged gray, with or without orange contents. Spores not uniform, oval to subelliptical 4–6 × 3–4μ. Zygospores not found.
From soil: Czechoslovakia (97), Morocco (169)
United States: Illinois (129), Michigan (105)

SECTION VI. FLAVUS

KEY TO THE SPECIES

a. Spores 5–12μ long	
b. Spores oval to globose (1:1–1.3)	33. *M. strictus*
bb. Spores elliptic (1:1.4–1.7)	34. *M. piriformis*
bbb. Spores elongate (1:2)	
c. Turf becoming yellow, not aromatic	35. *M. flavus*
cc. Turf becoming gray, aromatic	36. *M. attenuatus*
aa. Spores 12–20μ long	
b. Turf rose, floccose, sporangiophores quickly wilting	37. *M. rufescens*
bb. Turf white, spores with granulated content	38. *M. oblongisporus*

33. *Mucor strictus* Hagem

Turf grayish-white, about 1 cm. high, ½ below the surface of the substrate and dotted with sporangia; at first white, then pale brown; dark brown with age. Sporangiophores simple or branched sympodially up to 1 cm. high by 16μ wide, slightly incurved at the tip, restricted by the insertion of the sporangia. Wall of sporangiophores striately netted. Sporangia globose, spheric, rather flattened on the side toward the columellae, 70μ high by 88μ broad, up to 170μ in diameter. Wall diffluent but not in all the sporangia. Columellae ovoid, rather flattened at the base, subjacent, 60 × 44μ or 64 × 50μ (140 × 110μ maximum).

Spores subspheric or oval, slightly unequal, 5 × 6μ or 6 × 8μ, rarely 10μ. No chlamydospores. Zygospores unknown.

From soil: France (78), Norway (55)

34. *Mucor piriformis* Fischer

Turf thick, woolly, 20–30 (seldom to 50) mm. high, at first white, later yellowish. Sporangiophores 35–50μ in diameter with a few side-branches. Sporangia at first white, then yellowish-gray to brownish-black, usually 100–300μ in diameter, with quickly dissolving, echinulate wall. Columellae colorless, smooth, oval to pear-shape, occasionally globose, in size according to the sporangium. Spores hyalin, regular, ellipsoid, 4 × 5μ to 8 × 13μ, so that the diameter to length is relatively 1:1.4–1.7. Gemmae have been observed but zygospores are unknown.

From soil: Mexico (23)

United States: Alabama (23)

35. *Mucor flavus* Bainier

Sporangiophores 8 cm. high × 22–24μ wide, little or not branched, at first colorless, then ochre-yellow. Sporangia globose, gray, bluish, then white with a blue tint, 140–160μ in diameter. Wall diffluent, incrusted, leaving a collarette. Spores oval, very variable in size 9.4–12 × 4.2μ, sometimes cylindric or reniform, enclosed in an interstitial mucilaginous substance, very fluid, giving the sporangium a translucent blue appearance. Columellae at first globose, then slightly oval, (Lendner reports them rather pear-shaped, with a flattened base, 110–90μ). In liquid sugar media, budding cells as in yeast occur. Zygospores formed as in *M. racemosus*. Wall formed of numerous plates, brown and deeper at the center of the zygospore, develops echinulations like those of *M. mucedo* rather tardily, 150μ in diameter.

From soil: Austria (66) (97), Czechoslovakia (97), France (78), Germany (104), Jugoslavia (97) (104), Norway (55), Switzerland (77) (97), U. S. S. R. (108)

United States: Louisiana (23), Maine (147), New Jersey (146) (147), North Carolina (23)

36. *Mucor attenuatus* Linnemann

Turf usually bright gray, 1–1.5 cm. high with sweet, fruit-like odor. Primary sporangiophores sturdy, branching in cymes with each successive branch becoming finer; from 100μ to 1 cm. long, up to 40μ in diameter in the primary sporangiophore, the last branches being about 8μ in diameter. Sporangia at first white, later grayish-yellow, 20–240μ in diameter; the primary sporangia 100–200μ, the last smaller sporangia 20–40μ. Columellae (often lacking in the small sporangia) oval, 16.7–120μ in diameter and 25–140μ high, colorless. Sporangial wall finely incrusted, diffluent, leaving a small collarette. Spores broad to long oval, of different sizes, 4.5–5 × 8.5–12μ. Chlamydospores and zygospores unknown.

From soil: Germany (79)

37. *Mucor rufescens* Fischer

Sporangiophores not branched, flaccid, collapsing to form a cottony felt with a reddish color, 2.5 cm. long × 15–25μ in diameter. They are often irregularly divided by septa which separate the collapsed portion from the turgid filaments; wall colorless, contents being furnished with orange-red colored drops. Sporangia large, 120–150μ in diameter, pale yellow, hyalin. Walls of sporangium slowly diffluent, slightly incrusted, colorless, hyalin. Columellae free, globose or elliptical, spherical or sub-spherical, 45–65μ in diameter, with a smooth colorless wall. Contents dense, intensively colored golden-yellow, which is seen through the sporangium wall and gives it its colored appearance. Spores planoconvex, with obtuse tips, and twice as long as broad, 4μ broad up to 10μ long, but may be 8μ in breadth to 21μ in length, colorless, smooth. Zygospores and chlamydospores unknown.

From soil: Czechoslovakia (97), England (36) (37)

38. *Mucor oblongisporus* Naumov

Turf white, 20–25 mm. high. Sporangiophores much or slightly sympodially branched, 20–35μ in diameter. Sporangia 100–120μ in diameter, blue-gray with diffluent wall. Columellae hyalin, globose, to pear-shaped, 70–90μ long. Spores oval to cylindrical, regular, of differing size, imbedded in a slime; 5.5–8 × 10–20μ, containing numerous small oil drops. Gemmae and zygospores unknown.

Differs from *Mucor flavus* by its somewhat larger spores.

From soil: U. S. S. R. (108)

United States: North Carolina (23)

SECTION VII. MUCEDO

KEY TO THE SPECIES

a. Spores 5–15μ long
  b. Spores regularly cylindric to elliptic
    c. Sporangia yellow to gray-blue; spores 8–12μ long — 39. *M. mucedo*
    cc. Sporangia gray to black; spores 6–8μ long — 40. *M. saturninus*
  bb. Spores irregularly oval to globose — 41. *M. albo-ater*
aa. Spores 30μ long — 42. *M. mucilagineus*

39. *Mucor mucedo* (Linne) Brefeld
Syn. *M. brevipes* Riess
M. proliferus Schostakowitsch

Sporangiophores erect, forming a very raised turf up to 15 cm. in height, silvery-gray, shining, not branched, 2–15 cm. high × 30–40μ in diameter, without cross walls. Wall colorless, smooth; content colorless, tardily yellow. (Very rarely branched with very small sporangia). Sporangia large, 100–200μ in diameter; at first yellow, then deep gray or brownish-black. Membrane of sporangium very diffluent, leaving a collarette; it is incrusted with needle-shaped crystals of calcium oxalate. Columellae free, cylindric or campanulate or spherical, 70–140μ long × 50–80μ wide, with colorless wall and red-orange content. Spores elliptic or subcylindric, twice as long as broad, of very various sizes in the same sporangium, 6–12μ long × 3–6μ wide (limits 16.8μ long) with a smooth hyalin wall, content tardily yellow or colorless. Zygospores spherical 90–250μ in diameter. Exine black; thickly, and very strikingly verrucose; hard and fragile. Intine colorless with less striking warts, enclosed in the former. On germination the zygospores give rise to sporangia on unbranched sporangiophores. Chlamydospores not known.

From soil: Austria (94), Czechoslovakia (97), Egypt (115), England (36), France (78), Germany (6), Jugoslavia (103), Norway (55)

United States: New York (67), North Carolina (23)

40. *Mucor saturninus* Hagem

Colonies always more or less dark colored, usually lead-gray or lead-black, but sometimes even blue-black. Sporangiophores of various heights, some low, others high. The lower form, 1–2 mm. high, usually richly branched monopodially or sympodially and forming a lead-black or blue-black turf from the large sporangia. The higher forms, 2–3 cm. high, are more or less scattered, 20–25μ thick and at first erect, later bent, branched monopodially with long branches and of a characteristic bright lead-gray color. Sporangia of the lesser branches at first bright waxy-yellow, then blue-gray, and finally almost black at maturity, of very different sizes, usually 45–180μ in diameter, with a spiny non-diffluent wall. Sporangia of the higher branches with a diffluent wall, leaving only a collarette. Columellae oval, seldom cylindrical, in the sporangia on the higher branches, frequently collapsed on the base, 60–100μ high × 50–90μ broad; the secondary (smaller sporangia) somewhat smaller, 35–70μ high × 25–50μ broad. Spores regularly broad ellipsoid, 6–8 × 4–6μ, as well as a few small globose forms 4–4.5μ in diameter. Zygospores not known.

From soil: Czechoslovakia (97), Germany (78) (169), Norway (57) United States: New Jersey (146)

41. *Mucor albo-ater* Naumov

Turf white, thick, 30–80 mm. high, primary rhizoids rare, secondary numerous in old cultures. Sporangia at first, fawn-yellow, later black, 200–400μ. Sporangia walls rough, more or less easily diffluent. Sporangiophores erect, 20–90μ in diameter, white in the young culture. Columellae typically cylindric to pear-shaped, up to 250 mm. high. Spores irregular, oval to almost globose, 5–15μ long. Zygospores unknown.

In soil: Germany (79), France (78), United States

42. *Mucor mucilagineus* Brefeld

Differs from *M. mucedo* by the brown-black, never yellow sporangia with a somewhat more slowly dissolving wall. Spores 30μ long.

From soil: France (78)

Species of uncertain position:

Mucor heterosporus Fischer

From soil: Czechoslovakia (97)

Mucor hygrophilus Oudemans

From soil: Germany (104)

Mucor wulffii Nielsen

From soil: Greenland (96)

(7) **Zygorhynchus** Vuillemin

Hyphae continuous, branched, unequal, often nodose, immersed, prostrate or forming a cottony aerial turf. Chlamydospores smooth, intercalary or terminal. Sporangiophores solitary or in an irregular sympodial system, bearing typical sporangia or abortive sporangia and zygospores; not apophysate. Sporangia uniform; wall diffluent, with the base concrescent with the columella. Upon its disappearance a collarette remains. Spores numerous, minute, smooth. Zygospores variable, warted. Gametangia very unequal, produced on unequally bifurcated hyphae, one straight and small, the other curved and thicker, at the end a reflexed pear-shape.

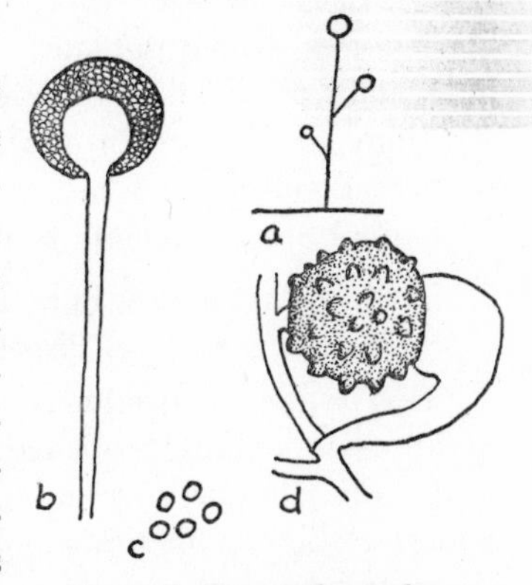

FIG. 7. **Zygorhynchus.** *a* habit; *b* sporangiophore; *c* sporangiospores; *d* zygospore.

KEY TO THE SPECIES OF THE GENUS ZYGORHYNCHUS

a. Spores globose
  b. Spores 2–3μ in diameter — 1. *Z. heterogamus*
  bb. Spores 3.5–7μ in diameter — 2. *Z. exponens*
aa. Spores elongate
  b. Spores 4 × 2μ — 3. *Z. vuilleminii*
  bb. Spores 7 × 4μ
    c. Warts on zygospores 6–7μ long — 4. *Z. macrocarpus*
    cc. Warts on zygospores less than 5μ long
      d. Zygospores averaging 35μ in diameter — 5. *Z. moelleri*
      dd. Zygospores averaging 68μ in diameter — 6. *Z. japonicus*

1. *Zygorhynchus heterogamus* Vuillemin
Syn. *Mucor heterogamus* (Vuillemin) Lendner

Sporangiophores erect, 2 mm. long × 12–15μ wide, sometimes simple, and ending in a sporangium; more often branched, bearing two and at times four branches opposite each other or in whorls, all ending in sporangia. Sporangia equal, globose, 50–60μ in diameter, black. Wall of sporangium diffluent, incrusted, leaving a collarette. At the time of zygospore formation the wall becomes persistent. Columellae spherical, smooth. Spores round, 2–3μ in diameter, smooth. Zygospores formed either on the sporangiophores or on special mycelial filaments, branched sympodially. Gametangia very unequal; on unequally bifurcate filaments, the one straight, slender; the other curved, thicker. Zygospores very variable in size, 45–150μ in diameter. Exine brown, spiny with black points, united in plates. Intine with simple warts. Chlamydospores intercalary or terminal, elliptic or globose (20 × 25μ).

From soil: Canada (15), Czechoslovakia (97), Jugoslavia (103), U. S. S. R. (108).

United States: Rhode Island (110)

2. *Zygorhynchus exponens* Burgeff
Syn. *Z. polygonosporus* Pispek

Turf white or bright gray up to 10 mm. high. Sporangiophores slightly sympodially branched. Sporangia gray, 50–100μ, with diffluent wall, fully disappearing at maturity. Columellae globose to cylindric. Spores globose, sometimes polyhedric, 3.5–7μ in diameter. Zygospores on much branched suspensors, 50–80μ in diameter, brown with reticulate exine.

From soil: Jugoslavia (97), Spain (169)

3. *Zygorhynchus vuilleminii* Namyslowski

Sporangiophores 5–8μ broad, branched. Sporangia globose, not diffluent when young, diffluent at maturity; 30–45μ in diameter (60μ maximum). Sporangia terminal, larger than other species. Columellae broader than high, ovoid, 12–30μ broad (35μ maximum). Spores hyalin, ellipsoid 4μ long $\times$ 2μ broad, often guttulate. Chlamydospores smooth, oval or elongate, of various sizes. Zygospores globose, exine verrucose, brown, 40–50μ in diameter (60μ maximum). Azygospores not rare. Distinguished from *Z. moelleri* by the exine having much smaller warts, commonly aggregated.

From soil: Canada (15) (145), Germany (169), Montenegro (94), U. S. S. R. (108)

United States: Illinois (129), Iowa (1) (3), Louisiana (2), Maine (147), New Jersey (146) (147), North Carolina (23), Oregon (146), Rhode Island (110)

var. *albus* Christenberry

From soil: United States: North Carolina (23)

4. *Zygorhynchus macrocarpus* Ling-Young

Turf at first yellow, later brown-gray, 1–4 mm. high. Sporangiophores 15–20μ thick, often somewhat curved. Sporangia yellow to dark gray, 50–80μ with fragile wall. Columellae flattened globose, up to 45μ broad with collarette. Spores oval, 2.5–3 $\times$ 4.5–5μ, seldom 7μ, yellow. Mycelial gemmae frequent, up to 5μ in size. Zygospores 50–100μ, at first brown, later black with warts, 8–10μ broad and 6–7μ in height.

In soil: France (78)

5. *Zygorhynchus moelleri* Vuillemin

Turf 0.5 cm. high, gray, cottony. Sporangiophores simple or branched and bearing one or two lateral branches, (opposite). Sporangia gray-yellow, slightly wider than long, 48μ long $\times$ 50μ wide. Wall not diffluent. Columellae oval and depressed, wider than long (20–30μ high $\times$ 26–36μ wide), wall smooth. Spores oval, 5μ long $\times$ 3–4μ wide (rarely 4 $\times$ 3μ). Zygospores as in *Z. heterogamus* but smaller, 35μ in diameter, (extremes 20μ and 54μ in diameter).

From soil: Austria (66) (131), Canada (15), Czechoslovakia (97), England (11), France (78), Germany (79) (97), Japan (132), Norway (55), Switzerland (97), U. S. S. R. (108)

United States: Illinois (129), Iowa (101), New York (67)

6. *Zygorhynchus japonicus* Kominami

Turf white as in other species. Sporangiophores up to 10 mm. long and 9–15μ in diameter. Sporangia yellow, 56μ in diameter with diffluent wall. Columellae pear-shaped, seldom broadly globose, up to 45μ in height. Spores long oval, various in size $1.5 \times 3\mu$ to $6 \times 10\mu$. Mycelial gemmae in limited numbers. Zygospores dark brown, averaging 68μ, with very low warts on exine. Suspensors unlike, about $12–35\mu$ thick.

From soil: Japan (71)

b. *THAMNIDIACEAE*

Differs from the Mucoraceae by the predominant formation of short sporangiophores which occur in whorls on simple or slightly branched, long sporangiophores. The latter terminate in a sterile tip or with a large mucor-like primary sporangium, while the great majority of the dichotomously branched short sporangiophores carry only few-spored sporangioles. The sporangioles may be reduced to a single-spored condition. In this case the spores show a true exine (sporangial wall), which is shed on germination. Zygospore formation is like that in the Mucoraceae.

A single genus treated.

(1) **Thamnidium** Link

Sporangiophores erect, terminated by a sporangium resembling that of the genus Mucor. They are formed at definite points on single or verticillate branches, which in turn are dichotomously branched and terminated by small sporangia or sporangioles. The sporangium is terminal, multisporous, with a diffluent membrane, incrusted with calcium oxalate, and possessing a large columella. Sporangioles small, spherical, containing four to ten spores, with an incrusted membrane, persistent, not diffluent, without columellae. They are caducous. The spores are of the same size in both sorts of sporangia; colorless, smooth. Zygospores naked, formed on the mycelium. Suspensors without appendages, gametes straight. Germination not known.

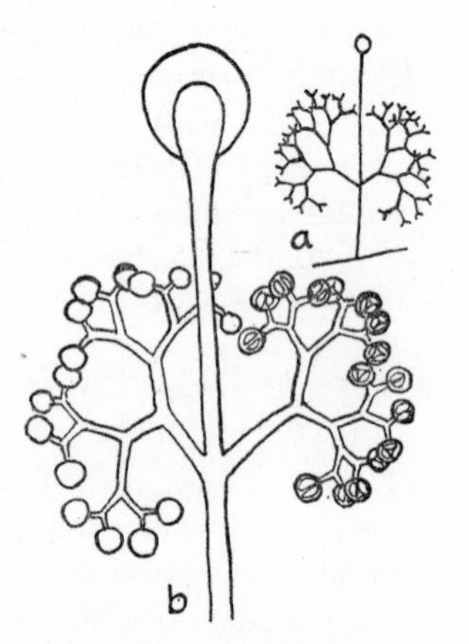

FIG. 8. **Thamnidium.** *a* habit; *b* sporangiophore (after Fischer).

A single species treated.

1. *Thamnidium elegans* Link

Turf 3 cm. high. Sporangiophore bearing a terminal sporangium, $100–200\mu$ in diameter, with a columella $50–70\mu$ wide $\times$ $62–90\mu$ long. The lateral branches divide in whorls and branch dichotomously. The length of the branch is diminished in proportion to its forking. The first arm, from the place of insertion on the principal filament to the first fork, 150–200μ long; the arm of the first order, 40–60μ, the last are 4–6μ long × 2μ in diameter. Sporangioles, very variable in size, up to 24μ in diameter. The smaller have not more than four, often only two or one spore. The spores are always the same size in all the sporangia, 6–8μ wide × 8–12μ long. Zygospores, according to Bainier, on the mycelium, round, black; exine verrucose, black; intine, yellow.

From soil: Czechoslovakia (97), England (37)

United States: Idaho (106), New York (67)

c. *PILOBOLACEAE*

Sporangia of a single sort, many-spored, with a membrane that is for the most part more firm, persistent and very dark, black, or inflated only toward the base. Some sporangia break away simply leaving a columella, some are thrown together with the columella and are open after the inflation of the membrane.

A single genus treated.

(1) **Pilaira** van Tieghem

Mycelium sunken in the substratum. Sporangiophores single, thread-like, nonseptate. Sporangia terminal; at first spherical; membrane cuticularized on the upper half; on the under half, delicate, thin; this latter becoming swollen upon the ripening of the spores, and evanescent. Columellae large, disc-like or spherical, persistent. Zygospores spherical, being borne on the ends of erect copulating branches which are somewhat twisted about each other.

A single species treated.

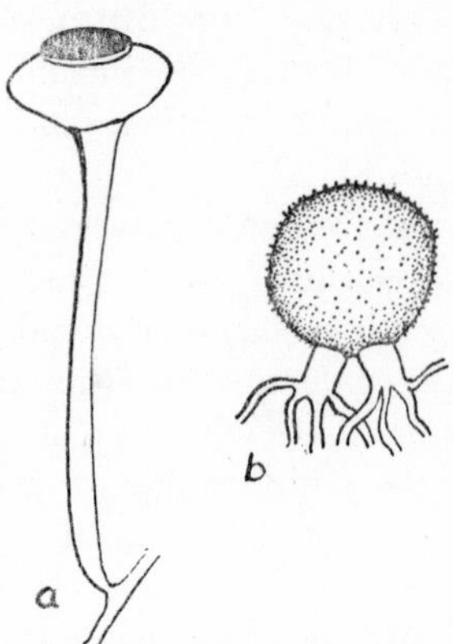

FIG. 9. **Pilaira.** *a* sporangiophore; *b* zygospore (after Fischer).

1. *Pilaira anomala* (Cesati) Schroeter

Sporangiophores only at first; at most 2 cm. high, (before they reach their full extension they appear like a Mucor turf), very soon collapsing and forming a high, loose, woolly, hyalin felt on which the black sporangia appear as black points. Sporangiophores extended, 10–12, even 20 cm. long, cylindric, 30–80μ thick, without basal and subsporangial swellings; with colorless, thinner, shallow wavy wall, entirely empty at the time of spore maturity. Sporangia at first white, then yellow; mature black, with colorless base; wet, globose, 100–250μ in diameter; dry, half-round;

many spored, ejaculated, sometimes nodding on the still upright sporangiophore. Columellae 100–150μ wide, 40–60μ high, flat, half-round, or knob-shaped, smooth, colorless. Spores long-oval 8–13μ long, 5–8μ wide, single, colorless; in mass yellow, with colorless thin membrane. Zygospores at maturity black, globose or slightly oval, 120μ long, 100μ wide, with smooth, thick, colorless intine and black, warty exine. Germination with a short sporangiophore.

From soil: Holland (100)

d. *MORTIERELLACEAE*

Sporangia without columellae, with diffluent wall, fugacious as in Mucoraceae. Zygospores enclosed in a felt of interwoven hyphae.

A single genus treated.

(1) **Mortierella** Coemans

Mycelium very thin and delicate; nutritive mycelium sunken, much branched, at times forming cysts; aerial mycelium creeping, many times anastomosing. Sporangiophores erect, at first short, at maturity rapidly becoming thread-like, with limited growth, simple or branched, very broad below, diminishing to the tip. Sporangia terminal, spherical, without columellae; membrane thin, diffluent. Spores spherical or ellipsoid. Zygospores spherical, covered by a thick case. Conidia formed on short side branches on the aerial mycelium, spherical, one-celled.

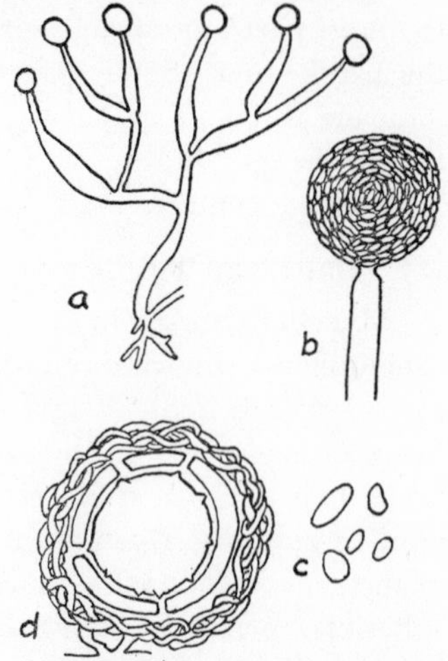

FIG. 10. **Mortierella.** *a* habit; *b* sporangium; *c* sporangiospores; *d* zygospore (after Fischer).

KEY TO THE SPECIES OF THE GENUS MORTIERELLA

a. Sporangia present
  b. Spores 1–4μ, seldom 5μ long
    c. Spores elongate — 1. *M. renispora*
    cc. Spores globose
      d. Spores 1.5–2μ long
        e. Turf white — 2. *M. turficola*
        ee. Turf wine-red — 3. *M. longicollis*
      dd. Spores 2–5μ long, sporangiophores 0.1–0.3 mm. high — 4. *M. pusilla*

e. Turf white — var. *typica*
ee. Turf smutty gray-brown — var. *isabellina*
eee. Turf russet vinaceous — var. *vinacea*
bb. Spores more than 5μ long
c. Spores 6–9μ, seldom 10μ long
d. Spores usually spherical
e. Sporangiophores unbranched
f. Sporangia with many spores — 5. *M. simplex*
ff. Sporangia with one spore — 6. *M. monospora*
ee. Sporangiophores branched
f. Branching sympodial
g. Side branches longer than the primary branch — 7. *M. candelabrum*
gg. Side branches equal to or shorter than the primary branch — 8. *M. spinosa*
ff. Branching racemose — 9. *M. marburgensis*
dd. Spores usually oval
e. Turf other colors than white
f. Turf brown — 10. *M. nigrescens*
ff. Turf yellow — 11. *M. elasson*
ee. Turf white
f. Sporangiophores branched
g. Sporangiophores cymosely branched — 12. *M. bainieri*
gg. Sporangiophore branching other than cymose
h. Sporangiophore branching racemose — 13. *M. raphani*
hh. Sporangiophore branching dichotomous — 14. *M. dichotoma*
ff. Sporangiophores unbranched with rhizoids — 15. *M. strangulata*
cc. Spores 10–25μ long
d. Sporangiophores up to 1 mm. high — 16. *M. polycephala*
dd. Sporangiophores 20–30 mm. high — 17. *M. tuberosa*
aa. Sporangia unknown
b. Stylospore wall reticulate — 18. *M. stylospora*
bb. Stylospore wall finely echinulate — 19. *M. humilis*

1. *Mortierella renispora* Dixon-Stewart

Mycelium always white; fine, sparse, cottony. Chlamydospores and swellings occur on the nutritive hyphae. Sporangiophores simple, coarser than the hyphae, with large basal foot formed from original hypha, 200μ long, tapering from 10μ to 3μ. Sporangia colorless, walls evanescent, 25μ in diameter, with basal collar. Spores 4 $\times$ 2μ, slightly kidney-shaped, hyalin. Stylospores absent. Zygospores bright brown, coat irregularly roughened, with few investing hyphae, 30μ in diameter without hyphal coat.

From soil: New Zealand (43)

2. *Mortierella turficola* Ling-Young

Differs from *M. pusilla* only by the smaller spores, 1.8–2μ.
From soil: France (78)

3. *Mortierella longicollis* Dixon-Stewart

Turf felt-like, at first white, later wine-red. Sporangiophores branched, 5–7μ thick. Sporangia brownish-purple, usually with base elongated into a neck, and thus pear-shaped, up to 80μ high, and maximum diameter of 30μ. Spores brown, rounded-cubical, only 1.5–2μ in diameter. Mycelial gemmae sparse, stipe gemmae and zygospores unknown.

From soil: New Zealand (43)

4. *Mortierella pusilla* Oudemans
Syn. *M. humicola* Oudemans

Tufts orbicular, always snow-white, woolly, composed of some stages sinuous or lobed, others less wide, which are higher. Hyphae creeping, hyalin, 2.5–10μ thick, forked, filled with dense protoplasm finely granular. Sporangiophores 4–6μ wide, larger at the base, diminishing at the tip, 120–170μ high, ending in a solitary sporangium. Sporangia globose, smooth, hyalin, 24–28μ in diameter, with hyalin membrane. Spores globose, smooth, hyalin, 2–2.5μ in diameter, without trace of nucleus or vacuoles. Differs from var. *isabellina* by the graded structure of the tufts and their unalterable white color; the content of the creeping hyphae; the form of the erect hyphae; the hyalin spores.

From soil: Austria (66) (131), Canada (15), Czechoslovakia (97), England (11), France (78), Germany (79) (97), Holland (100)

var. *isabellina* (Oudemans) Zycha
Syn. *Mucor mirus* Paine
Mortierella pusilla var. *ramifica* Dixon-Stewart

Tufts elliptic, zonate, at first snow-white, then pearl-gray, finally isabelline (Sacc. Chrom. No. 8). Hyphae creeping, branching in forks, continuous, filled with homogeneous protoplasm. Sporangiophores cylindric, slightly attenuated at the tip, continuous, 120–200μ high, hyalin, ending in a single sporangium. Sporangia globose, 12–25μ in diameter. with hyalin wall. Spores globose, smooth, nearly hyalin when single, pale yellowish-white in mass, 2–5μ in diameter. Chlamydospores submerged, globose or elliptic; smooth, hyalin, with thin membrane.

From soil: Canada (15), Czechoslovakia (97), England (11), Holland (100), New Zealand (43)

United States: Iowa (101), North Carolina (23)

var. *vinacea* (Dixon-Stewart) Zycha

Mycelium forming a thick cottony felt, becoming russet vinaceous as the sporangia mature. Chlamydospores formed, the hyphae irregularly swollen. Sporangiophores racemosely branched, 4μ in diameter. Sporangia brown, 20μ in diameter, no collar formed. Spores 3–5μ, irregularly angular, slightly colored. No stylospores or zygospores.

From soil: New Zealand (42)

5. *Mortierella simplex* van Tieghem and Le Monnier

Sporangiophores unbranched 0.7–1 mm. high, below up to 70μ, above about 15μ broad. Sporangia colorless, many spored. Spores globose, 10μ, with a bright oil-drop. Mycelial and stipe gemmae present, the latter colorless, with a warty surface and long unbranched small stipe. Zygospores unknown.

In soil: France (78) (142), Germany (79)

6. *Mortierella monospora* Linnemann

Substrate mycelium in concentric zones, about 1 cm. wide, which are arranged in a stellate manner. Aerial mycelium white, thick, matted, up to 1.5 cm. high. The sparse unbranched very small sporangiophores scarcely visible in the thick tangle of hyphae. Sporangiophores erect, often opposite each other, 40–95μ long, 2–5.5μ wide at their base, to 0.2–2μ wide just below the sporangium. Sporangia with smooth, evanescent wall and a single colorless spore, 9–18μ in diameter. Chlamydospores numerous in the aerial mycelium, irregularly shaped, about as large as the spores.

On soil: Germany (79)

7. *Mortierella candelabrum* van Tieghem and Le Monnier

Syn. *M. minutissima* van Tieghem

Turf white, thick, yet delicate, many mm. high with strong garlic odor. Sporangiophores up to 1 mm. high, sympodially branched, so that the first branch is almost horizontal and the side-branches form a many branched candelabrum. Each sporangiophore and side branch is somewhat swollen at the base and narrowed toward the sporangium. Sporangia white, many-spored, 20–40μ broad, very easily caducous. Spores thin-walled, usually 7μ, seldom 9μ in diameter, globose, seldom short oval. Mycelial gemmae numerous, 20–40μ. Stipe gemmae and zygospores unknown.

In soil: Greenland (96), Germany (79), U. S. S. R. (108)

8. *Mortierella spinosa* Linnemann

Aerial mycelium up to 0.5 cm. high, white, floccose, stellately arranged. Sporangiophores up to 500μ high, from 10–18μ broad to 2–6μ at tip. Branching racemose with repeated side branching forming a candelabrum with the heads of almost equal height as the primary sporangium. Sporangia hyalin about 35μ in diameter. Wall delicate, evanescent, or fragile, often leaving a collarette. Columellae arching into the sporangium, almost globose, and frequently furnished with a spine. Spores irregular in size, globose to elongate, about 6–12μ, without a central oil-drop. Chlamydospores globose, about 20μ in diameter, infrequent. Zygospores unknown.

From soil: Germany (79)

9. *Mortierella marburgensis* Linnemann

Substrate mycelium forming a rosette. Aerial mycelium white, 1–2 mm. high. The small sporangiophores singly or in groups on the prostrate, anastomosing, aerial mycelium. Sporangiophore from 3.6–6 to 1–1.2μ at the sporangium, 30–180μ (usually 60–80μ) long. Branching irregularly bush-like, with three to five shorter side branches often in a whorl. Sporangiophores not septate. Sporangia hyalin with few (sometimes but one) spores, 10–12μ in diameter. Wall evanescent, smooth. Spores globose or slightly elongate, 6.3–10μ in diameter, usually 7μ. Chlamydospores and zygospores unknown.

From soil: Germany (79)

10. *Mortierella nigrescens* van Tieghem

Turf white, thick, many mm. high, in older cultures yellowish and finally dark-brown. Sporangiophores 1–1.5 mm. high, without rhizoids, below 50μ, above 8μ broad. Sporangia many-spored, yellow. Spores elliptic, cylindric or kidney-shaped, 3–4 × 6–8μ. Gemmae unknown. Zygospores homothallic, without case 100–125μ.

In soil: Jugoslavia (103)

11. *Mortierella elasson* Sideris and Paxton

Turf sulphur-yellow. Sporangiophores unbranched, 0.2–0.5 mm. high, at the base 5–10μ, at the tip 3–6μ in diameter. Sporangia colorless, 10–24μ. Spores globose or oval, 3–6 × 5–10μ. Mycelial gemmae formed on malt peptone agar.

From soil: Canada (15)

12. *Mortierella bainieri* Costantin

Sporangiophores single, erect, branched, without cross-walls and rhizoids, 2–3 mm. high, at base 18–20μ and at the point 8μ broad. Sporangia many-spored, white, with basal collarette, about 50μ broad. Spores elliptic, irregular, 4–6 × 6–10μ.

In soil: Germany (79), France (78)

13. *Mortierella raphani* Dauphin

Colony zonate, zones about 0.5 cm. broad. Aerial mycelium sparse, 1–2 cm. high, always white, loose. Stylosporangia, appearing first, numerous in an irregular bushy arrangement, later the sporangiophores arise. Sporangiophores single to many, racemosely branched, side branches 30–40μ long, often in whorls. Sporangiophores tapering from 16μ at the base to 4–6μ at the tip. Sporangia 20–40μ, containing about 20 spores. Spores irregularly globose, sometimes almost elongate, with oil-drop, 10–12μ in diameter. Stylospores broader than long, echinulate, up to 20μ in diameter. Zygospores unknown.

In soil: Germany (79)

14. *Mortierella dichotoma* Linnemann

Colony neither zonate nor stellate. Aerial mycelium, white, matted, 0.5 cm. high. Sporangiophores usually dichotomous branches of the aerial hyphae. Sporangiophore walls almost parallel, slightly smaller below the sporangium, from 6–10 to 4–6μ in diameter, length irregular, up to 1 mm. Sporangia globose 20–40μ in diameter, with smooth evanescent wall and more than 20 spores. Spores, various in size and shape, mostly oval, sometimes cylindric and even globose, 2.7–5 to 4.5–8μ, or 2.5–5.4μ in diameter. Chlamydospores on aerial hyphae, cylindric, 4.2–8.1 to 7–14.4μ. Zygospores unknown.

From soil: Czechoslovakia (97), Jugoslavia (103)

15. *Mortierella strangulata* van Tieghem

Turf white, heavy. Sporangiophores occurring on projecting hyphae, about 25μ in diameter and constricted just below the sporangium. The sporangiophores form a felt of rhizoids at their base, which may occur as a thick brown sheath, up to a quarter of their length. Sporangia white, 50–150μ in diameter, with an easily diffluent wall, which leaves a basal collarette behind. Spores elliptic, 6 × 9μ. Mycelial gemmae present; stipitate gemmae globose, echinulate, 18–20μ, occur on unbranched sporophores, up to 20μ long.

16. *Mortierella polycephala* Coemans
Syn. *M. rhizogena* Daszewska
M. van tieghemi Bachmann

Turf formed by white, floccose aerial mycelium, wide-spread and creeping up the wall of the culture dish. Sporangiophores with characteristic bushy branching, 0.3–0.6 mm. high. Sporangia 30–70μ, white, evanescent, with four to fifty, usually globose or oval spores, which vary greatly in size. Average size 10–14μ, extremes of 3 and 25μ, with a large oil-drop in their content. Characteristic mycelial or sporophore gemmae occur singly or in twos, echinulate or warty, about 20μ in diameter. Monoecious zygospores, formed on the aerial mycelium, covered with a thick hyphal felt, at first white, later brown or black. This body may reach a diameter of 1 mm.

From soil: Czechoslovakia (97), France (78), Germany (104) (169), Switzerland (38)

17. *Mortierella tuberosa* van Tieghem

Turf white. Sporangiophores erect, single, unbranched, slightly thickened at the base and covered with a large globose vesicle, 20–30 mm. high. Sporangia white, many-spored, leaving behind after dispersal a collarette and a small smooth head in the middle of the sporangium. Spores elliptic, irregular in shape and size, usually 6–8 × 11–16μ. Mycelial gemmae present, stipitate gemmae echinulate, 20–25μ, single on unbranched carriers.

From soil: France (78)

18. *Mortierella stylospora* Dixon-Stewart

Turf delicate, white. Mycelial gemmae at the top of the hyphae frequent. Stipitate gemmae on thin stalks, 40μ high, with reticulate thickened walls about 18μ in diameter. Zygospore-like structures were observed. Sporangia unknown.

In soil: New Zealand (43)

19. *Mortierella humilis* Linnemann

Colony forming a rosette. Aerial mycelium white, matted, reaching 0.5 cm. in height. Stylosporophores arise erect from the aerial mycelium, simple or branched or in groups. Branching irregular, cymose or bushy, 46–80μ long, 2–4μ wide at the base. All branches end in stylospores. Aerial mycelium delicate and easily collapsing. Stylospores globose with a finely echinulate wall, 8–15μ in diameter. Sporangia, chlamydospores and zygospores unknown.

From soil: Germany (79)

Species of uncertain position:

Mortierella hygrophylla Linnemann

Linnemann (79) reported from Germany; it was not sufficiently distinguished from *M. polycephala* to place it among the forms here described.

e. *CHOANEPHORACEAE*

Sporangioles or pseudoconidia covering vesicular enlargements of the sporophores. Sporangia typical of Mucoraceae are present in some genera, absent in others.

KEY TO THE GENERA OF THE CHOANEPHORACEAE

a. Pseudoconidia longitudinally striate, in sporangia and sporangioles — (1) **Blakeslea**

aa. Pseudoconidia, single on vesicles — (2) **Cunninghamella**

(1) **Blakeslea** Thaxter

The genus is characterized by the formation of mucor-like terminal sporangia and sporangioles beside each other, resembling somewhat the conidial stages of Cunninghamella. Below the large globose sporangia with a columella there occur—as well as terminally on unbranched sporangiophores—smaller sporangia with a reduced columella. Single sporangiophores divide at their tip into many dichotomous branches and carry, on globose terminal vesicles, sporangioles with usually three, sometimes as many as six spores. The spores from the sporangia and sporangioles differ somewhat in size but both show a typical longitudinal striation and thread-like appendages on their ends. Mycelial gemmae occur on pincer-like suspensors and have a smooth exine.

Fig. 11. **Blakeslea.** *a* sporangiophore; *b* sporangioles; *c* sporangiospore; *c* zygospore (after Zycha).

A single species treated.

1. *Blakeslea trispora* Thaxter

Turf low, orange-yellow, woolly. Large sporangia 40–50μ, small sporangia about 16μ. At the tips of many times dichotomously branched tree-like sporangiophores occur up to thirty-two heads, each with thirty to forty sporangioles. The latter, sitting on rounded (about 3μ–long) spicules, are oval, 11–14μ in size, and contain three, sometimes up to six spores. The spores are red-brown, longitudinally striate and possess fine thread-like appendages on both ends; 5–8 × 10–15μ. Mycelial gemmae of various shapes occur, 24μ in diameter. The fungus is dioecious and forms globose zygospores, 40–60μ in diameter, on pincer-like suspensors.

From soil: Louisiana (23)

(2) **Cunninghamella** Matruchot

Mycelium white, floccose, slightly thickened, 3–6μ, continuous when young, later becoming septate, septa disposed here and there without order. Rhizoids very tenuous. Conidiophores straight, branched. The main axis, as well as the side branches, little or not septate, terminating in spherical heads, furnished with small swellings which are the points of insertion for the conidia. Conidia spherical or oval, often with an irregular outline, the external membrane spiny with needle crystals. Chlamydospores globose, intercalary in the mycelium.

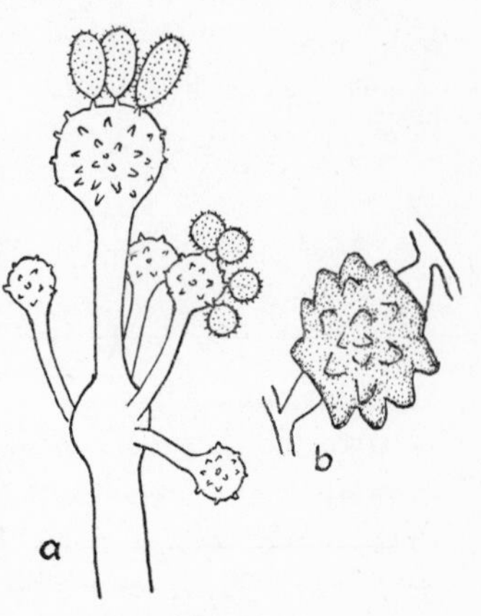

FIG. 12. **Cunninghamella.** *a* sporangiophore; *b* zygospore (after Lendner).

KEY TO THE SPECIES OF THE GENUS CUNNINGHAMELLA

a. Terminal vesicles more than 50μ in diameter
  b. Sporangiophores dichotomously branched; lateral branches more than 30μ long — 1. *C. elegans*
  bb. Sporangiophores not dichotomously branched; lateral branches less than 30μ long — 2. *C. verticillata*
aa. Terminal vesicles less than 50μ in diameter
  b. Lateral sporangia smaller than terminal — 3. *C. bertholletiae*
  bb. Lateral sporangia of same size as the terminal sporangia — 4. *C. echinulata*

1. *Cunninghamella elegans* Lendner

Turf white to silver, spreading. Filaments firm and interwoven, 7–13μ wide, with abundance of oil; circinate portions typical. Conidiophores erect, multi-branched; terminal vesicles 27–35μ in diameter, spherical, smooth; lateral branches lacking or up to three whorled, place of attachment to conidiophores swollen; subterminal whorl 38μ long, vesicles spherical, 16–28μ in diameter, smooth; intermediate whorl 24μ ong, vesicles spherical 14–16μ in diameter, smooth; basal whorl of pyriform branches, 14μ wide × 26μ in length, smooth; super-branches, arising from terminal head, of varying lengths. Vesicles spherical. Terminal conidia lemon-shaped, bearing spicules after separation from vesicle, 12μ long × 9μ in width, very finely echinulate; lateral conidia ovate in varying degrees; 6μ wide × 10μ in length; nonspiculate, very finely echinulate.

From soil: Austria (66), Canada (16), Czechoslovakia (96), France (77), Germany (169), Jugoslavia (102)

United States: Idaho (7), Illinois (128), North Carolina (23), Texas (91)

2. *Cunninghamella verticillata* Paine

Turf white to silvery, loose, erect, 2–4 cm. in height. Conidiophores long, 2 cm. or more, nonseptate; terminal vesicles globose to oval, about 50μ in diameter; lateral branches numerous, not exceeding 30μ long, subterminal, whorled. Conidiophores swollen at point of attachment of lateral branches. Lateral vesicles pyriform to oval, not over 16μ in diameter. Terminal conidia ellipsoid, pointed at the attached end, 10 × 13–15μ, very finely echinulate. Lateral conidia oval, bluntly pointed at the attached end, 8–12μ in diameter, very finely echinulate.

From soil: China (81), India (21)

United States: Colorado (74) (76), Iowa (101), Texas (92)

3. *Cunninghamella bertholletiae* Stadel

Turf gray, filaments firm and interwoven. Conidiophores erect, irregularly-cymosely branched, nonseptate. Terminal vesicles ovate, about 25μ wide × 33μ long. Lateral branches variable, alternately arranged in groups, numerous, 22–55μ long. Lateral vesicles round, about 23μ in diameter. Terminal conidia ovate, smooth, 5 × 9μ; lateral conidia similar to terminal but slightly smaller.

From soil: Idaho (7)

4. *Cunninghamella echinulata* Thaxter

Turf white becoming yellowish with age; filaments interwoven. Conidiophores erect, more or less irregularly and indefinitely branched. Terminal vesicles very variable in size, areolate, nearly spherical to obovoid, 28 × 35μ average, 45 × 65μ maximum; lateral branches similar to terminal but smaller. All conidia oval to elliptical; finely echinulate; 10 × 12μ average, 18 × 25μ maximum.

From soil: Egypt (115), Jugoslavia (103)

United States: North Carolina (23), South Carolina (23), Texas (92)

Species of uncertain position:
Cunninghamella dalmatia Pispek
Cunninghamella polymorpha Pispek
Cunninghamella ramosa Pispek
From soil: Jugoslavia (103)

f. *CEPHALIDACEAE*

Spores in chains, arising by the simultaneous division of the content of elongated cells, mero-sporangia, since the latter in turn, usually are part of a fruiting head. The mero-sporangia may be formed directly on the sporangiophore or on basal cells which in turn make up the head.

KEY TO THE GENERA OF THE CEPHALIDACEAE

a. Basal cells lacking; conidial chains numerous on spherical vesicles (1) **Syncephalastrum**
aa. Basal cells present
b. Basal cells on globose sporophores (2) **Syncephalis**
bb. Basal cells on boat-shaped sporophores (3) **Coemansia**

(1) **Syncephalastrum** Schroeter

Mycelium like that of species of Mucor. Sporangiophores with rich sympodial branching, often with somewhat curved side-branches, but lacking the cross-wall which in Mucor is observed above the division of the lateral branches. Sporangia-bearing heads globose, separated from the sporangiophore by a wall. Mero-sporangia many spored, without a basal-cell.

A single species treated.

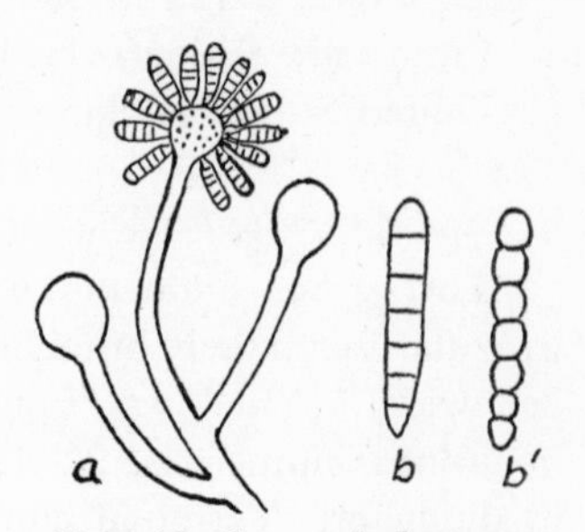

Fig. 13. **Syncephalastrum.** *a* sporangiophore; *b*, *b′* sporangiospores.

1. *Syncephalastrum racemosum* (Cohn) Schroeter

Turf at first white, later gray, about 6 mm. high. Mycelium with pseudo-holdfasts which connect with short rhizoids that soon become overgrown. Conidiophores vigorous, at first unbranched, later richly branched with strongly curved laterals. Fruiting head globose or oval, 22–70μ wide, brown or gray, with numerous small warts to which the mero-sporangia are attached. In these latter are five to ten spores of irregular size, mostly globose 2.5–5μ.

From soil: Canada (15), Germany (79)

(2) **Syncephalis** van Tieghem and Le Monnier

Vegetative hyphae inconspicuous in the substrata. Sporophores (often designated conidiophores) usually unbranched, (in a single species dichotomous) up to 3 mm. high and attached at the base to the substrate by a few rhizoid-like hyphae. At their tips the sporophores enlarge into a vesicle, which carries on its upper surface the basal cells on which in turn are one to five many-spored mero-sporangia. Spores usually cylindrical to cask-shaped. Zygospores form as outgrowths of one of the gametangia.

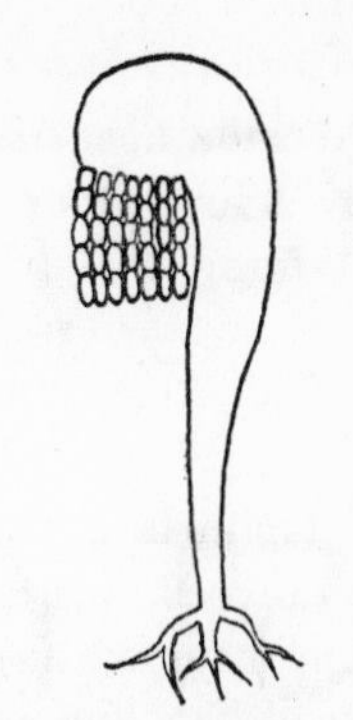

FIG. 14. **Syncephalis.** sporangiophore and spores.

KEY TO THE SPECIES OF THE GENUS SYNCEPHALIS

a.	Basal cells irregular, up to 12 spores	1. *S. depressa*
aa.	Basal cells cylindric, usually 5 spores	2. *S. sphaerica*

1. *Syncephalis depressa* van Tieghem and Le Monnier

Fruiting stage 400–700μ high, from 10–20μ in diameter. Heads 30–40μ in diameter, colorless, with twelve to fifteen irregular basal cells each with two to five papillae and a like number of mero-sporangia. The latter contain up to twelve cylindrical colorless spores, 2–3 $\times$ 5–7μ.

From soil: Holland (100)

2. *Syncephalis sphaerica* van Teighem

Sporophore 400–700μ high, 28μ broad at the base, lower half of head only 8μ, latter about 40μ in size, colorless. Basal cells cylindric, with each mero-sporangium containing usually five cylindric, smooth, colorless spores, 3–4 $\times$ 8–10.5μ.

From soil: Madeira (169)

(3) **Coemansia** van Tieghem and Le Monnier

Turf yellow, up to 6 mm. high. Mycelium creeping, branched and septate. Conidiophores unbranched or forked with many septa. Mero-sporangia, alternating, boat-shaped, quite broad, many-celled, and carrying on their inner surface a number of small basal cells which bear the spindle-shaped conidia.

A single species treated.

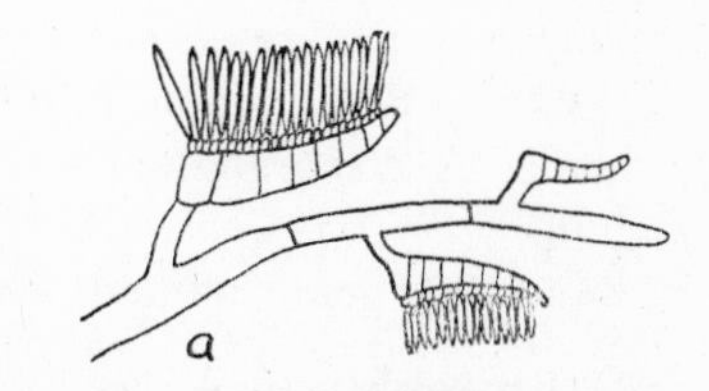

FIG. 15. **Coemansia.** *a* sporangiophore (after Zycha).

1. *Coemansia reversa* van Tieghem

Fertile heads sessile, sickle-shaped about 22μ long $\times$ 8μ broad, with only four septa. Basal cells, 2.8 $\times$ 3.1μ each with a spindle-shaped conidium, 2–3 $\times$ 7–8μ.

From soil: Austria (66)

B. MYCOCHYTRIDIALES

Thalli from the beginning surrounded by a membrane and divided into mycelium, reproductive organs, sporangia and resting spores, which often occur extramatrically. Mycelium usually slightly developed, slender, thread-like and extended, either branching root-like, or more seldom unbranched or scattered from the base or the entire surface of the sporangium.

Sporangia arising either in their entirety from greatly enlarged spores or as lateral growths of these, or from terminal or intercalary mycelial swellings; of very different form, globose, elliptic, pear-shaped, cylindric, star-shaped or irregular. Zoospores usually globose with an oil-drop and one flagellum, hopping, and escaping by one or many openings in the sporangial wall; at times by means of a lid (operculum). Resting spores like sporangia; in a few cases by conjugation or union with mycelial-swellings; sometimes with appendages.

a. *RHIZIDIACEAE*

Thalli predominantly interbiotic, monocentric, eucarpic, consisting of a well-developed usually extensive rhizoidal system, the tips of which at least are endobiotic and a reproductive rudiment which is converted into a sporangium, prosporangium, gametangium, or resting spore. Sporangium inoperculate. Zoospores posteriorly uniflagellate, with a single globule. Resting spore asexually formed or sexually by fusion of iso- or anisogamous aplanogametes which are never liberated into the outside medium, upon germination functioning as a sporangium or a prosporangium.

A single genus treated.

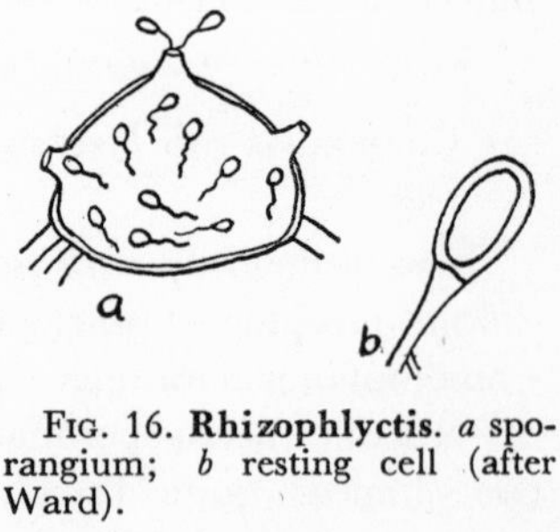

FIG. 16. **Rhizophlyctis.** *a* sporangium; *b* resting cell (after Ward).

(1) **Rhizophlyctis** A. Fischer

Sporangium free, saprobic or with only its mycelium penetrating into a host-cell; globose or elliptic. Zoospores spherical or elongate, with one long flagellum, escaping singly from the sporangium through the mouth.

Resting sporangia free, like the zoosporangia. Wall thick, smooth.

A single species treated.

1. *Rhizophlyctis rosea* (de Bary and Woronin) A. Fischer[1]

Thalli monocentric. Each thallus consisting at maturity either of a zoosporangium with an extensively developed rhizoidal system, or of a resting body with rhizoids. Zoosporangia rose-colored, smooth-walled, spherical or irregular in shape, quite variable in size, up to 130μ wide. Exit papillae one to several, of varying length and diameter, the ends being filled with a gelatinous plug. Rhizoids arising from one to several places on the zoosporangium, stout, up to 11.66μ in thickness near the point of origin, extensive, much branched, sometimes 650μ or more in length. Zoospores emerging through the papillae, about spherical, mostly 4μ thick, with a small refractive globule and a very long posterior flagellum. Resting spores smooth, relatively thick-walled, spherical, oval or irregular, contents olive-brown to orange-brown in color, granular and containing numerous refractive bodies. Germination not observed. Saprobic or weakly parasitic (according to Cornu).

From soil: United States: North Carolina (150)

b. *MEGACHYTRIACEAE*

Thalli epi- and endobiotic, or free in the medium, only the tips entering the substrate, polycentric, either rhizoidal, extensive, and much branched, with intercalary swellings, or broadly tubular, hypha-like, not distinctly rhizoidal, with swellings. Zoosporangia operculate, formed from terminal or intercalary swellings. Zoospores posteriorly uniflagellate. Resting spores thick-walled, apparently asexually formed, upon germination producing zoospores.

A single genus treated.

(1) **Nowakowskiella** Schroeter

Mycelium much branched, widely extended, irregular, often considerable swellings here and there at the points of branching. Sporangia usually formed on the surface of the substrate or in it, terminal or intercalary, ellipsoid or pear-shaped with

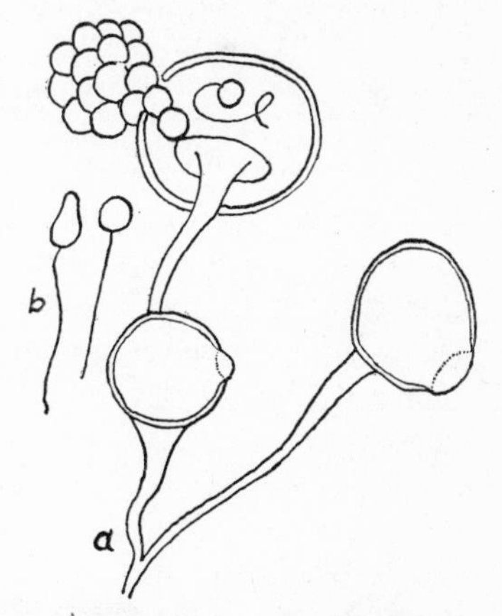

Fig. 17. **Nowakowskiella.** *a* sporangiophores; *b* zoospores (after Whiffen).

[1] Recently, Johanson (Amer. Jour. Bot. 31: 397-404, 1944) has transferred this species to *Karlingia rosea* because of the finding of opercula at the bases of the papillae.

a smooth membrane, opening at the tip or at some other points with a circular pore, covered by a lid. Swarmspores formed simultaneously or gathered before the mouth in a spherical mass, spherical, with a single oil-drop and one long flagellum, relatively large. Secondary sporangia often protruding through the primary sporangia. Resting-spores known from but one species, perhaps really absent in the other species, spherical with a smooth thick wall, usually massed next one another forming a pseudoparenchyma by the budding or division of the hyphae which bear them; at maturity becoming empty, as turbinate cells.

A single species treated.

1. *Nowakowskiella delica* Whiffen

Rhizomycelium, intra- and extramatrical, much branched with numerous elongated swellings, which give rise to zoosporangia or resting bodies. Zoosporangia, terminal or intercalary, 12.5 × 15.0μ to 22.5 × 28.8μ, spherical, ovoid, or pyriform; with a single exit pore, usually laterally but often apically placed, operculate; apophysis variable in shape, 2.5 × 15.0μ to 11.5 × 17.5μ. Zoospores hyalin, spherical, 5.7–7.5μ, posteriorly uniflagellate, flagellum 30μ long, uniguttulate. Resting bodies terminal or intercalary, usually intramatrical, ovoid to spherical, 10.0 × 15.0μ to 26.2 × 29.9μ, apophysate, hyalin, containing one to many oil globules. Germination of resting bodies not observed.

From soil: United States: Texas (154)

C. BLASTOCLADIALES

Microscopic fresh-water or terricolous saprobic fungi. The thallus coenocytic, eucarpic, sometimes with pseudosepta, true cross walls formed only to delimit the reproductive organs, composed of a system of branched rhizoids that anchor it to the substratum and either a single reproductive rudiment or, more commonly, a basal cell which bears one or more reproductive structures directly on its surface or on lobes or on extensive nonseptate or pseudoseptate dichotomously, sympodially or umbellately branched hyphae, sterile setae present or absent, the walls sometimes giving a reaction for chitin, never for cellulose. Protoplasm variable in aspect, frequently alveolately or reticulately vacuolate. The fungus sometimes differentiated into similar sporophyte and gametophyte phases; the asexual plant bearing one or more inoperculate, uni- or multiporous thin-walled zoosporangia and thick-walled, punctate, generally brownish resting spores borne singly within and completely filling the terminal segments of the hyphae, the sexual plant monoecious or dioecious, bearing one or more thin-walled inoperculate uni- or multiporous gametangia. Zoospores posteriorly uniflagellate, with a conspicuous subtriangular nuclear cap, monoplanetic, movement swimming or amoeboid, germinating directly to form the asexual plant. Resting spores upon germination producing either (a) posteriorly uniflagellate planonts which give rise directly to new asexual plants or to sexual plants, or (b) posteriorly biflagellate planonts which immediately encyst, each of the cysts forming endogenously four uniflagellate planonts that, after emerging from the cyst through a pore and swarming, germinate to form asexual plants. Gametes posteriorly uniflagellate, isogamous or anisogamous, if anisogamous the smaller gamete always containing carotinoid pigment, fusing in pairs to form a biflagellate planozygote which, without a period of rest, germinates to form an asexual plant.

A single family treated.

a. *BLASTOCLADIACEAE*

Characters the same as those of the order.

KEY TO THE GENERA OF THE BLASTOCLADIACEAE

a. Basal cell lacking, thallus a single reproductive rudiment — (1) **Sphaerocladia**

aa. Basal cell present, anchored by rhizoids

  b. Basal cell of thallus bearing a single sporangium, a single gametangium or a single resting spore — (2) **Blastocladiella**

  bb. Basal cell of the thallus bearing an indeterminate number of reproductive organs

    c. Thallus with an unlobed or unbranched basal cell; zoosporangia with more than one discharge papilla; gametophytes known — (3) **Allomyces**

    cc. Thallus with a simple lobed or branched basal cell; zoosporangia with a single apical discharge papilla; gametophytes unknown — (4) **Blastocladia**

(1) **Sphaerocladia** Stueben

Thalli one-celled, consisting of a simple spherical swelling from the surface of which numerous delicate much-branched rhizoids arise, the expanded part converted as a whole into a single reproductive rudiment. Sporophyte plants forming from the rudiment either a thin-walled zoosporangium with from one to several short discharge tubes through which the posteriorly uniflagellate zoospores escape, or a thick, dark-walled resting spore which upon germination produces posteriorly uniflagellate planohaplonts that give rise to gametophytes. Gametophytes similar to the zoosporangial plant but smaller, dioecious, forming either + or − isogamous gametes that fuse in pairs to form biflagellate zygotes, which germinate at once to produce either a zoosporangium-bearing or a resting-spore-bearing sporophyte.

A single species treated.

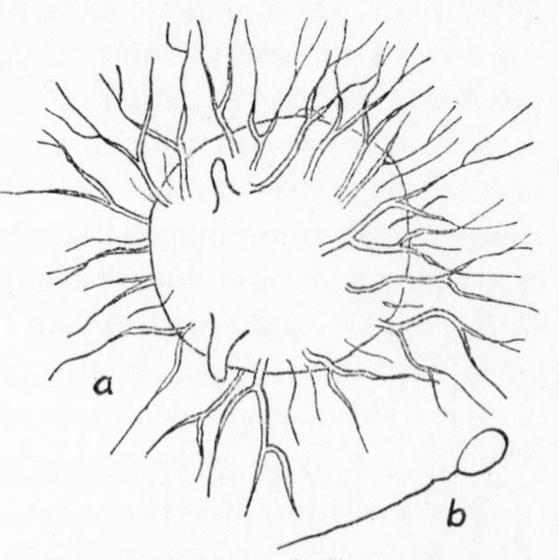

FIG. 18. **Sphaerocladia.** *a* sporangium; *b* zoospore (after Sparrow).

1. *Sphaerocladia variabilis* Stueben

Sporangia with a thin smooth wall, spherical or more or less ovoid or ellipsoidal, averaging 60μ in diameter on flies, 140μ on peptone agar. Rhizoids profusely developed. Zoospores emerging through one or several papillae, ellipsoidal, 3.5 × 4.8μ, the contents bearing in addition to a "food body" a lateral tear-shaped structure. Resting spores either ovoid and 13 × 17μ, the brownish wall 0.5μ thick, or spherical, 110μ in diameter, the wall 2.5μ thick. Planohaplonts one-half the size of the zoospores, emerging through a prominent papilla which protrudes through a crack in the resting-spore wall. Gametophytes similar to the zoosporangial plants but smaller, averaging 106μ on peptone agar, + or − gametes similar to the zoospores but smaller. Zygote biflagellate and germinating at once to form a sporophyte.

In soil: Mexico (127)

(2) **Blastocladiella** Matthews

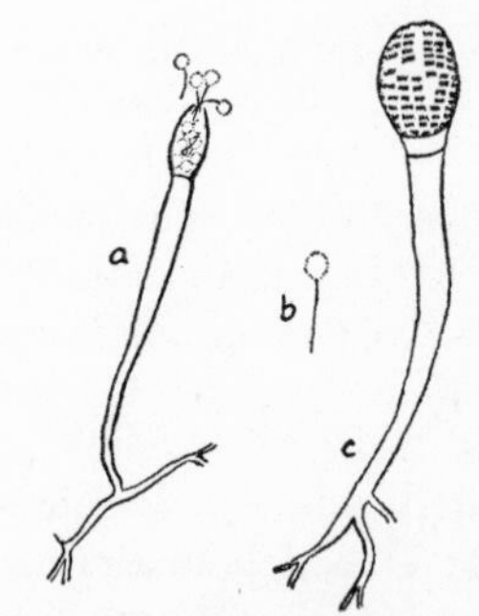

FIG. 19. **Blastocladiella,** *a* plant bearing zoosporangium; *b* zoospore; *c* plant bearing resting spore (after Sparrow).

Thalli microscopic, unicellular, consisting of a single extramatrical, tubular, pyriform or globose part anchored to the substratum by a branched rhizoidal system; extramatrical part giving rise to the stalked or sessile, thin-walled zoosporangium, resting sporangium, or gametangium. Thin-walled zoosporangia usually present, globose to cylindrical with one to several discharge papillae. Zoospores posteriorly uniflagellate, with a nuclear cap over the anterior part of the nucleus and with several laterally placed fat or lipoid globules. Thick-walled resting sporangia usually present, enclosed in the old thallus membrane; wall of resting sporangium composed of two distinct layers: the outer brownish, thick, usually sculptured layer, and the inner, hyalin, thin, smooth layer; outer layer coarsely or minutely areolate, warted or smooth or smooth except for one or several ridges. Resting sporangia germinating by the cracking of the exine into two or several irregular pieces after which the hyalin inner part sends out one or more papillae through which zoospores emerge to form zoosporangia and/or resting sporangia, or to grow into thallose gametangia, or to encyst, the cysts acting as gametes which fuse in pairs to form a zygote. Cell walls not giving a cellulose reaction with chloriodide of zinc.

KEY TO THE SPECIES OF THE GENUS BLASTOCLADIELLA

a. Short life cycle, only one generation
  b. Both zoosporangia and resting sporangia present
    c. Thalli more than 500μ tall; resting sporangia distinctly areolate — 1. *B. simplex*
    cc. Thalli less than 500μ tall
      d. Resting sporangia warted — 2. *B. asperosperma*
      dd. Resting sporangia smooth — 3. *B. laevisperma*
  bb. Only resting sporangia and cysts which act as gametangia formed — 4. *B. cystogena*
aa. Long life cycle, two equal generations, asexual and sexual. Sporangia and gametangia usually stalked; male gametangium orange — 5. *B. variabilis*

1. *Blastocladiella simplex* Matthews

Basal cell cylindrical, holdfasts delicate, branched, wall thin, smooth. Whole thallus, including rudiment of reproductive structure, 30–1,005μ long $\times$ 8–40μ in diameter. Sporangium cylindrical to globose, 15–105μ in diameter, wall thin, smooth, with from one to three discharge papillae. Zoospores ovoid to ellipsoidal, 5.5–7μ long $\times$ 3–4μ wide, with a long posterior flagellum and an anterior ring of refractive globules, emerging in a quickly evanescent vesicle. Resting spore borne like a sporangium, 15–180μ in diameter, surrounded by the thin wall of the container, clavate, with a rounded apex and a truncate base, wall dark brown, thick, coarsely and irregularly reticulate, upon germination forming posteriorly uniflagellate planonts which give rise to zoospore- or resting-spore-bearing thalli. Sexuality not known.

From soil: United States: Texas (127)

2. *Blastocladiella asperosperma* Couch and Whiffen

Thalli consisting of extramatrical stalked or sessile zoosporangia and resting sporangia anchored to the substratum by rhizoids; single or caespitose. Zoosporangial thalli spherical, ovoid, pyriform or nearly cylindrical; sporangia spherical, pyriform, ovoid, subcylindrical or irregular; 25–100μ thick when spherical, up to 88 $\times$ 155μ when ovoid or pyriform; stalk when present varying from a small basal part to a long tube 30 $\times$ 150μ, usually thickest just below the sporangium, frequently with one to several rings just below sporangium. Sporangia with one to several (up to eleven) emergence papillae. Zoospores usually becoming active in sporangium before emergence. The first spores emerging in a spherical vesicle formed from the gelatinized tip. After the vesicle bursts the spores remaining in the sporangium emerge slowly and swim away immediately upon reaching the outside. Zoospores 3.6–4.6 $\times$ 6–7μ. Resting sporangial thalli similar in shape, size and early development to the zoosporangial thalli; resting sporangia not completely filling the case in which they are formed; spherical, ovoid, pyriform, or subcylindrical; outer wall yellow-brown and distinctly warted; 14–80μ thick on grass when spherical, somewhat larger on hemp seed and spherical to oval in shape. Germinating after several weeks' rest to form zoospores as in zoosporangia.

From soil: United States: South Carolina (34)

3. *Blastocladiella laevisperma* Couch and Whiffen

Zoosporangial and resting sporangial thalli much as in *B. asperosperma* except resting sporangia usually fill the case and have a smooth wall except for a few ridges. Zoosporangia usually sessile on leaves and stalked on hemp seed; 25–117μ when spherical; stalk up to 26 $\times$ 105μ. Zoospores 3.8–4.6 $\times$ 6–6.5μ when active, 5–5.4μ when rounded up. Resting sporangial case as a rule stalked on hemp seed but usually sessile on leaves; rarely two resting sporangia may be formed on the same stalk; spherical, ovoid, pyriform or subcylindrical, sometimes constricted in the middle, at times lobed; 25–140μ thick when spherical, up to 63 $\times$ 95μ when ovoid; stalk 30 $\times$ 155μ, broadest at the top. Resting sporangia usually completely filling the case and conforming to its shape, up to 126μ thick when spherical; wall 1.5–2.8μ thick, reddish-brown or pale, dull yellow, smooth except for one or several conspicuous ridges. Germinating after a few weeks' rest to form zoospores which are similar to those from zoosporangia.

Distinguished from *B. asperosperma* by differences in the resting sporangia. In *B. laevisperma* the resting sporangia are reddish-brown, smooth walled and usually fill the case, while in *B. asperosperma* the resting sporangia are yellowish-brown, with a coarsely warted wall and only partly fill the case.

From soil: United States: South Carolina (34)

4. *Blastocladiella cystogena* Couch and Whiffen

Zoosporangial thalli lacking in life cycle. Resting sporangial thalli when on leaves usually extramatrical, spherical or pyriform and sessile or with a very small and inconspicuous stalk, frequently intramatrical on leaves and then subspherical ovoid or lobed; on hemp seed usually tubular, the case that encloses the resting sporangium globose to subcylindrical. Rhizoids well developed, attached basally if sporangium is extramatrical, or attached to several places on the sides if intramatrical. Resting sporangia when spherical 16–342μ thick, when subspherical 37–142 $\times$ 35–209μ; stalk when present very variable in size, 30–116 $\times$ 39–564μ; usually filling the case in which it is formed; wall two-layered, very faintly areolate, 1.5–6.1μ thick, nearly hyalin to dull yellow or orange-brown. Germination as in the genus. Zoospores uniflagellate when discharged, encysting almost immediately near the sporangial mouth in an irregular mass. Encysted spores (gametangia) 8.2–10.2μ thick, spherical. After one to two hours each cyst gives rise to four uniflagellated gametes which emerge through a minute pore to fuse in pairs near the empty cysts; gametes 4.1 $\times$ 4.9μ. Zygotes biflagellated, 6 $\times$ 8μ, swimming for an hour or more before germinating to form resting sporangia.

From soil: United States: Texas (34)

5. *Blastocladiella variabilis* Harder and Soergel

Basal cell cylindrical, holdfasts delicate, much branched, wall thin, smooth. Sporangia clavate, colorless; wall chitinous, thin, smooth, with several discharge papillae. Zoospores ovoid, with a refractive saddle-shaped "food body" and a long posterior flagellum. Resting spores borne like the sporangium, spherical, with a thick dark-brown, several-layered wall (sculptured?), cracking upon germination and allowing papillae to protrude, which upon dissolution form pores for the escape of numerous posteriorly uniflagellated planonts. Planonts upon germination forming thalli like the zoosporangial plants, on each of which is produced either a colorless or an orange colored clavate gametangium (+ or −). Gametes isogamous, + or −, fusing in pairs to form a zygote, which at once produces a sporophyte plant.

From soil: Dominican Republic (127)

(3) **Allomyces** Butler

Plant small, slender, the short or long stalk not conspicuously differentiated; branches usually dichotomous, often verticillate in groups of three to five, separated from the nodes by distinct and complete septa, not constricted at intervals; in vigorous cultures repeating the branching in the same way to form a complex plant. Sporangia oval, terminal, sympodially arranged, not rarely in chains of several, often clustered by the shortening of the branches, which continue the stem by one or more lateral buds beneath. Spores biflagellate at times, but the two flagella so closely approximated or fused as usually to appear as one. Resting bodies borne in the same way as the sporangia and of the same size and shape, at maturity enclosed in a thin, hyalin sheath out of which they finally fall through an apical slit; the wall brown and conspicuously pitted as in Blastocladia, the whole representing a thin-walled oogonium completely filled with a thick-walled parthenogenetic egg, or a resting sporangium as thought by Barrett.

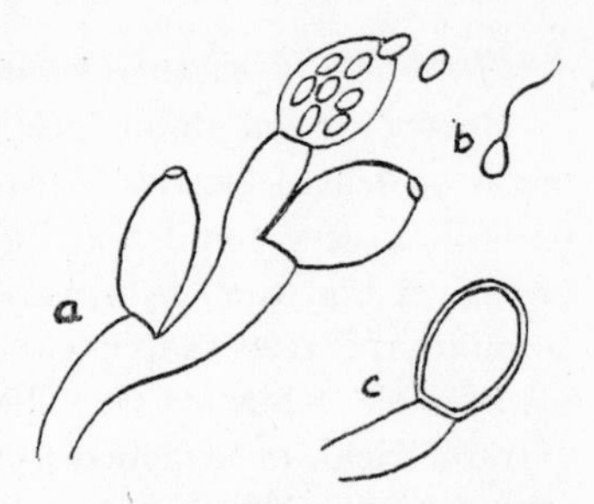

FIG. 20. **Allomyces.** *a* sporangiophore; *b* zoospore; *c* resting sporangium (after Kniep).

KEY TO THE SPECIES OF THE GENUS ALLOMYCES

a. Primary sporangia oval
  b. Gametangia, when first formed in pairs, with female terminal — 1. *A. arbusculus*
  bb. Gametangia, when first formed in pairs, with male terminal — 2. *A. javanicus*
aa. Primary sporangia cylindrical, 2.5–4.5 times as long as broad — 3. *A. moniliformis*

1. *Allomyces arbusculus* Butler

Characters of the genus. Threads extending about 3 mm. from the substratum on a termite ant, about 10–37μ thick, growing gradually more slender distally at each joint, basal joints 35–130μ long, those of central region up to about 675μ long; tips blunt, hyalin. Sporangia oval, 28–46 × 55–76μ; spores escaping singly or at times, according to Barrett, in a vesicle that soon bursts, emerging through one or two usually apical holes or short papillae, biflagellate (or uniflagellate by fusion of the two flagella?), oval when swimming, with the flagella apical, monoplanetic, amoeboid before encysting, 10μ thick when at rest; sprouting by a slender thread. Resting bodies appearing later than the sporangia but of the same shape, 25–39.2 × 36.3–49.2μ, the conspicuous pits apparently sunken from the outside in regular fashion, at maturity slipping from the thin, clasping sheath; sprouting into zoospores after a rest. The thick wall is divided into two parts, an outer layer (pitted) about 1.8μ thick and a homogeneous inner one about 1μ thick.

From soil: India (62), Mexico (155)

United States: Kentucky (61), Mississippi (61), New York (61), North Carolina (26) (61) (149), Oklahoma (61), Wisconsin (61)

2. *Allomyces javanicus* Kniep

Young mycelium shows large basal cell with strongly developed rhizoid system; on a rich nutritive substance it shows a typical dichotomously branched mycelium, whose thick hyphae contain very numerous, proportionally large nuclei. At the base of the forked branches are the characteristic crosswalls (pseudosepta). They are pierced by large pores, which allow free passage of the plasma and also the nuclei. With cultures on insects it is easy to obtain zoospores. The zoosporangia originate in such a way that the tips of the hyphae are cut off below the swelling by a closed crosswall. They appear in rather long rows behind one another and have mostly oval form. The dichotomous growth then changes mostly into a sympodial growth. Very young germ tubes can be induced to form zoosporangia by transference to water. The size of the zoospornagia is $60–80 \times 27–50\mu$.

The content of the zoosporangia is cleft into uniflagellate zoospores, after whose maturity the container opens through one or more pores The openings are recognizable at an early age by a raised membrane papilla. The naked zoospores force themselves one after the other through the opening; the flagellum is drawn behind in this process. The form of the zoospores in the swimming state is oval to elliptical; they are $11–12.5\mu$ long and $8–10\mu$ wide.

In addition to the zoosporangia the fungus also forms resting cells: several nucleate structures with three membrane layers, the middle one of which is the thickest and shows characteristic pores. They originate at the ends of the hyphae, in cells which are cut off by a separating wall. The entire plasma is spent in the formation of resting cells. Upon maturity the resting cells come forth from the mother cells. Their coloring is golden-brown.

Further, typical sexual organs may appear. They arise like the zoosporangia at the ends of the hyphae, and are mostly in series with the male gametangium terminal and under it one or several female gametangia. In longer chains of gametangia one observes intercalary antheridia also; likewise laterally stationed ones appear. The dichotomous growth of the fungus is then transformed into a sympodial one, as was also observed with the formation of the sporangia and resting cells. Several antheridia standing in a chain behind one another occur proportionately rarely.

The male gametangia are distinguished from the female at an early age by the smaller size and the appearance of a striking orange to vermillion color. The proportion of the long diameter to the largest cross diameter amounts to an average of 34.5–23.5μ for the male gametangia against 51–33μ for the female gametangia. Structure and opening mechanisms agree with those of the zoosporangia above. The uniflagellate gametes which originate in the male gametangia are very significantly smaller than the female gametes which they are like in structure otherwise: female gametes 9–11.5 $\times$ 7.5–8.5μ; male gametes 4.8–6.3 $\times$ 3.4–4.4μ.

From soil: Mexico (155)

United States: Arizona (156), California (156), Florida (158), Texas (156), Utah (156)

3. *Allomyces moniliformis* Coker and Braxton

Growth, dense, reaching a length of 7 mm. on a boiled hemp seed. Hyphae about 10–48μ thick, growing gradually more slender at each joint, basal joints from 45–150μ long, those of central region up to about 655μ long; tips blunt, hyalin. Primary sporangia cylindrical from 2.5–4.5 times as long as broad, 23–32 $\times$ 62–135μ; secondary sporangia mostly formed beneath the primary ones, forming long, usually much branched chains in old cultures, up to about twenty-eight in a single row, with the younger ones gradually becoming smaller and more nearly spherical, the basal ones as small as 20μ in diameter. The protoplasm in the sporangia becomes pink about the time the outlines of the spores appear, and gradually becomes browner as the spores develop. Spores 10–16.8μ in diameter, average 13–15μ, monoplanetic, escaping singly through one to four holes or short papillae, oval when swimming, amoeboid before encysting. Resting bodies appearing after two and a half or three days, of the same shape as the primary sporangia, except shorter, 21–35 $\times$ 43–66μ, with conspicuous pits as in *A. arbusculus*, at maturity slipping from the thin, clasping sheath.

From soil: Mexico (155)

United States: North Carolina (27)

(4) **Blastocladia** Reinsch

Mycelium unicellular, rather abundantly branched, separated into a main axis and secondary axes and sometimes having sterile, thin threads of unknown function. Sporangia usually ellipsoid to cylindrical in shape, often clearly forming sympodia through the shortening of the threads upon which they grow, but also thickly crowded together, seldom growing through each other. Zoospores ellipsoid to egg-shaped with broad blunt ends and one flagellum on the broader, colorless end, the other end containing small granules; emerging with force through an apical opening in the sporangium and swimming away, then, after being surrounded by a membrane, sprouting; while entering into the resting condition there is a moeboid motion. Resting stages usually broadly elliptical cells, which in occurrence and position resemble the sporangia but possess two membranes, of which the outer is smooth and colorless, the inner with fine regular dots. At maturity these cells are either shed as a whole, or the outer membrane splits and allows the escape of the inner membrane and its contents. Germination occurs after a rest-period by the breaking of the thick, brown wall, to allow the protrusion of a delicate bladder in which the spores develop.

A single species treated.

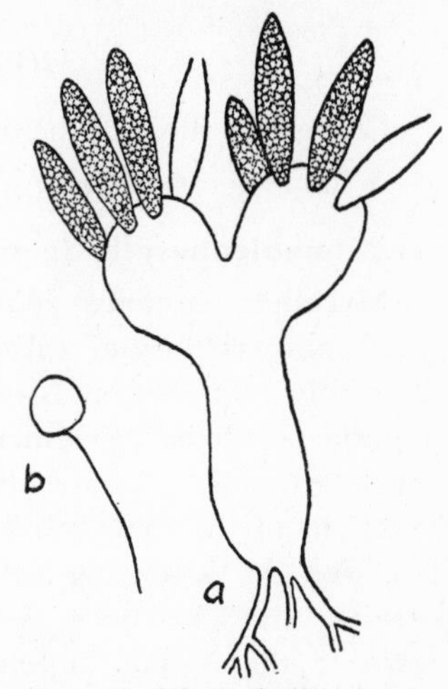

FIG. 21. **Blastocladia.** *a* sporangiophores; *b* zoospores (after Indoh).

1. *Blastocladia parva* Whiffen

Plant body cylindric, slender, 32–170μ high, 12.5–50.0μ wide; branching subdichotomous to dichotomous. Zoosporangia, irregular in shape, 40.0 × 65.6μ to 41.3 × 90.2μ, with one to six exit papillae. Zoospores from germinated resting bodies, posteriorly uniflagellate, with several small oil globules and a conspicuous food body, 5.6–6.1 × 2.8–3.2μ. Resting bodies spherical, elliptic, ovoid, 36.8 × 41.0 to 35.2 × 77.1μ; wall smooth, pale yellow in color; germination by cracking of outer wall and discharge of zoospores through one or two exit papillae.

From soil: United States: Texas (154)

D. MONOBLEPHARIDALES

Eucarpic fungi with a well developed filamentous mycelium, protoplast reticulate or foamy, coenocytic with septa delimiting only reproductive organs or with pseudo-septa; non-sexual reproduction by posteriorly uniflagellate zoospores borne in sporangia; zoospores usually with an anterior group of refractive granules; sexual reproduction where known, oogamous, by means of posteriorly uniflagellate antherozoids borne in antheridia and non-flagellate oospheres, borne singly in each oogonium; the fertilized oosphere becoming a thick-walled oospore, upon germination forming a hypha.

A single family treated.

a. *MONOBLEPHARIDACEAE*

Characters those of the order.

A single genus treated.

(1) **Monoblepharella** Sparrow

Mycelium, contents, zoosporangia, oogonia, antheridia, and antherozoids as in the family. Differs from Monoblepharis in that the fertilized egg emerges from the oogonium and, by means of the persistent flagellum of the male gamete, undergoes a period of swarming, after which it encysts and becomes a thick-walled oospore; the oospore upon germination forming the vegetative mycelium.

A single species treated.

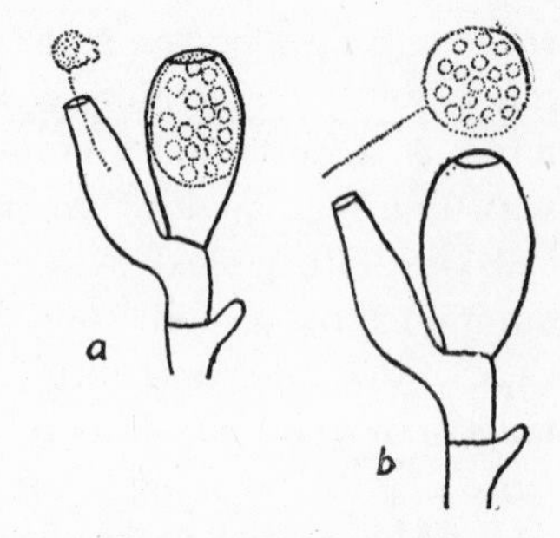

Fig. 22. **Monoblepharella.** *a* antheridium and oogonium; *b* emergence of fertilized egg (after Sparrow).

1. *Monoblepharella taylori* Sparrow

Mycelium well developed, consisting of tenuous flexuous branched hyphae, 2–5μ in diameter; sporangia narrowly siliquiform, with a tenuous wall, variable in size, 35–65μ long $\times$ 5–9μ in diameter, with a very narrow (2.5–4μ) base, occurring singly or in pairs at the tips of the hyphae or after sympodial branching of a hypha appearing laterally. Zoospores ovoid or somewhat cylindrical, 7–9 $\times$ 4.5–5μ, the posterior flagellum from two to three times the body-length. Oogonium at first terminal or after sympodial branching of the supporting hypha often appearing lateral, clavate or obpyriform, with rounded apex and narrow cylindric base, 15–17 $\times$ 8–10μ, tapering to 2–3μ at the base, at maturity containing one or occasionally up to four eggs with numerous large refractive globules. Antheridia hypogynous, often several developed in basipetal succession, consisting of a cylindric segment of the suboogonial hypha and a beak-like lateral outgrowth 8–10 $\times$ 4–5μ. Antherozoids, two to five, strongly amoeboid, posteriorly uniflagellate, ovoid when swimming, 3–5μ, escaping through a pore formed at the top of the beak. Zygote broadly ovoid to nearly spherical 10–13 $\times$ 8–10μ, posteriorly uniflagellate, free-swimming, the contents bearing numerous large refractive globules. Oospore formed free in the water, spherical, 8–10μ in diameter, with a slightly thickened light brown smooth wall, contents bearing globules, forming a mycelium upon germination.

In soil: British West Indies, Panama, Nicaragua (126)

E. SAPROLEGNIALES

Saprobic or parasitic, aquatic or terricolous fungi; the thallus commonly partly within and partly outside the substrate; holocarpic or eucarpic; when eucarpic the hyphae without constrictions and of unlimited growth, septa formed only in eucarpic species where they delimit reproductive organs; walls turning blue with chloriodide of zinc; contents granular, refractive only in growing tips; gemmae present or absent. Zoospores formed in sporangia, which in certain genera may be proliferous; zoospores biflagellate (flagella lacking in Geolegnia), mono- or diplanetic, if diplanetic the primary zoospore somewhat pyriform or pip-shaped, with two anterior flagella, the secondary zoospore reniform or grape-seed-like with two lateral or subapical oppositely directed flagella, capable in some individuals of repeated encystments and emergences before germination. Sexual reproduction oogamous, plants homo- or heterothallic, gametes never flagellate or set free in the medium. Oogonia producing one or more eggs without periplasm. Antheridia (occasionally functional or lacking) usually forming a fertilization tube. Oospore thick-walled, characteristically with a large reserve globule (partly or completely surrounded by one or more layers of minute globules) and a lateral bright spot, upon germination forming a mycelium or a short hypha terminated by a zoosporangium.

A single family treated.

a. *SAPROLEGNIACEAE*

Thallus eucarpic, mycelial, without constrictions, of unlimited growth, bearing numerous reproductive organs, homo- or heterothallic. Zoosporangia varied in character, usually terminal. Zoospores mono- or diplanetic, the secondary zoospores sometimes capable of repeated emergence. Antheridia from one to many, androgynous or diclinous, sometimes non-functional or entirely lacking. Oogonia with smooth or pitted wall, eggs from one to many, formed from the entire contents of the oogonium. Oospores sexually or apogamously formed, thick-walled, partly or nearly completely filling the oogonium, upon germination forming hyphae or a hyphal stalk bearing a zoosporangium

KEY TO THE GENERA OF THE SAPROLEGNIACEAE

a. Sporangia rare or absent; oogonia with very thick pitted walls, the antheridia arising from immediately below them and running up their sides — (1) **Aplanes**

aa. Not as above

  b. Spores normally leaving the sporangium by a common mouth

    c. Spores all (normally) swarming on escaping from the sporangium

      d. Sporangia not thicker than the vegetative hyphae; zoospores in a single row — (2) **Leptolegnia**

      dd. Sporangia usually thicker than the vegetative hyphae; zoospores not in a single row

        e. New sporangia formed within the empty ones — (3) **Saprolegnia**

        ee. New sporangia formed in the greater part by cymose branching

          f. Antheridia on all oogonia, androgynous — (4) **Pythiopsis**

          ff. Antheridia absent, or on less than half the oogonia, diclinous — (5) **Isoachlya**

    cc. Spores all collecting in a hollow sphere or an irregular group at the mouth of the sporangia on escaping

      d. Sporangia usually thicker than the vegetative hyphae; zoospores not in a single row — (6) **Achlya**

      dd. Sporangia not thicker than the vegetative hyphae; zoospores in a single row — (7) **Aphanomyces**

  bb. Spores not leaving the sporangium by a common mouth, but encysting within the sporangium

    c. Oogonia with more than one egg

      d. Eggs centric or subcentric with many peripheral oil drops; sporangium dehiscing by a cap — (8) **Calyptralegnia**

      dd. Eggs eccentric; spores set free by the breaking up of the sporangial wall — (9) **Thraustotheca**

    cc. Oogonia with a single egg

      d. Mycelium not depauperate but of the usual water mold type. Spores not conspicuously of different shape — (10) **Dictyuchus**

      dd. Mycelium depauperate, dense and opaque

        e. Sporangial wall evanescent; spores after encystment more or less spreading and separating; spore size variable in same plant, with many small subspherical ones — (11) **Brevilegnia** ee. Sporangial wall retained and obvious; spores large, subspherical, always non-motile and in a single row (12) **Geolegnia**

(1) **Aplanes** de Bary

Mycelium as in Achyla. Sporangia extremely scarce, often entirely absent for long periods in culture, cylindrical, renewed as in Saprolegnia and perhaps also as in Achlya. Spores at times escaping, at times retained in the sporangium and sprouting there, their behavior not well known. Oogonia abundant, in chains or single and terminal, barrel-shaped, spherical or pyriform, their walls very thick (more so than in other water molds) and heavily pitted. Antheridial branches arising from immediately below the oogonia, or when the oogonia are in chains arising from the top of one oogonium and attached to the next above, simple or branched, the antheridia with their sides attached to the oogonia. Eggs centric, at times elliptic from pressure.

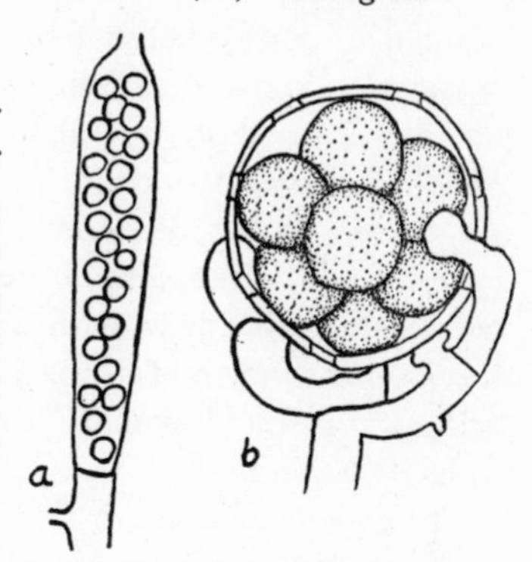

FIG. 23. **Aplanes.** *a* sporangium; *b* oogonium (after Coker).

KEY TO THE SPECIES OF THE GENUS APLANES

a. Oogonia strongly papillate
  b. Eggs one to twenty-five, 28–36μ in diameter 1. *A. treleaseanus*
  bb. Eggs two to forty, 22-34μ in diameter 2. *A. braunii*
aa. Oogonia usually not papillate; eggs 20–26μ in diameter 3. *A. turfosus*

1. *Aplanes treleaseanus* (Humphrey) Coker

Syn. *Achlya treleaseana* (Humphrey) Kauffman

Hyphae moderately stout, majority 28–48μ thick, sparingly branched. Growth slow but forming a heavy ring growth about 2 cm. in diameter on hemp seed in distilled water in three weeks. Sporangia rare, lacking in ordinary cultures, sometimes formed in old cultures on branches arising from gemmae or directly from gemmae, very variable in size, 12–30 $\times$ 100–500μ. Spores in a single row, or in several, spherical to cylindrical, 12–14μ thick and up to 16.8μ long, the majority of cases sprouting *in situ*, also escaping as in Achlya or rarely as in Dictyuchus. Gemmae very abundant, rod-shaped to oval in long chains, giving off numerous branches in old cultures, these branches usually cutting off one or more gemmae or rarely forming intercalary or apical sporangia. Oogonia abundant when grown on hemp seed, varying greatly in shape from spherical to cylindrical, wall hyalin, pitted, 2–4μ thick, usually with numerous papillae 2–24μ or more long, occasionally smooth, spherical and oval ones 30–88μ in diameter, usually about 84μ, other oogonia 50–72 $\times$ 132–577μ, usually borne terminally on main hyphae or terminating a chain of gemmae, not rarely on short lateral branches; intercalary ones plentiful, also occasionally borne two or three in a chain. Eggs one to twenty-five, usually six to ten to an oogonium, 28.8–36μ in diameter, majority 31–36μ, spherical or often elliptic or block-shaped from pressure, filling the oogonium, subcentric with a sheath of small oil drops surrounding all but a small portion of the protoplasm; wall thin. Antheridia androgynous, arising from the oogonial stalk, often with the antheridial cell not cut off as in *A. turfosus*.

From soil: Denmark (80)

United States: North Carolina (26) (30) (149)

2. *Aplanes braunii* de Bary

Syn. *A. androgynus* (Archer) Humphrey

Turf thick, up to 1.5 cm. broad, with long extensive hyphae, 20–60μ in diameter, which, branching irregularly, often bear many very fine lateral branches, with pointed tips. Sporangia terminal, very sparse, cylindric. Spores free within them, but irregularly many seriate, germinating by laterally erumpent threads. Oogonia numerous, terminal or intercalary, two to five in a chain or separated by hyphal fragments; in form very variable, club-, spindle-, or keg-shaped, when terminal often with snout-like elongated tip; wall heavy with many very real pits. Oogonia of various sizes, 470μ long $\times$ 60μ wide or 130μ long $\times$ 50μ wide or 80μ long $\times$ 30μ wide. Antheridia on androgynous, short, thin, slightly or unbranched secondary hyphae, which arise from the primary branch near or also above the hyphae which carry the oogonia and which often coil about them; small irregularly cylindrical to clubbed, lying at the side of the oogonia. When the oogonia occur in chains the antheridia arise from the top of one oogonium and become attached to the next. Oospores numerous, from two to forty, usually thickly crowded, globose, centric, germination by short tubes, whose content separates into uniseriated spores, which germinate by laterally erumpent threads; 22–34μ in diameter.

From soil: Germany (112)

3. *Aplanes turfosus* (Minden) Coker

Growth moderately stout, threads about 15–25μ thick; sporangia very scarce, usually entirely absent, cylindrical, rounded at the tip, proliferating internally in the few seen; spores about 11μ in diameter. Gemmae fairly plentiful (not nearly so abundant as in *Saprolegnia litoralis*), the great majority rod-shaped and in chains exactly as in *Achlya debaryana*, only here and there one fusiform or oval, etc. Oogonia spherical (without a neck), or rarely oblong or pyriform, smooth or at times papillate-warted, 27–90μ in diameter, nearly always racemosely borne on short stalks (no intercalary or cylindrical ones seen); wall hyalin, varying in thickness with the size of the oogonia; in small ones as little as 1.2μ thick, in larger ones thicker than in any species of Saprolegnia or Achlya, and reaching up to 4μ; pits numerous and very conspicuous (few and less conspicuous in small oogonia). Eggs one to thirty, mostly six to twenty in an oogonium, 20–26μ thick, often elliptic or block-shaped from pressure and usually well filling or even crowding the oogonium, centric; wall thick. Antheridia on all the oogonia, very peculiar, arising from short stalks which spring laterally from immediately beneath the oogonia, an antheridial cell not being cut off from the oogonial stalk except in very few cases; tubes from the partition wall into the oogonium lacking. The whole of the antheridial thread may be cut off as an antheridium or only a part of it; at times also no wall cutting off an antheridium can be seen.

From soil: United States: North Carolina (26)

(2) **Leptolegnia** de Bary

Hyphae long and delicate, sparingly branched. Sporangia long, apical, cylindrical, of the same size as the hyphae, at times multiplied by growth through empty ones, rarely branched. Spores formed in a single row, elongated on emerging, then changing their form to pip-shaped and swarming with two apical flagella, encysting and swarming again as in Saprolegnia. Gemmae absent. Oogonia borne on short lateral branches, small, smooth, subspherical, unpitted. Antheridia pyriform, diclinous or androgynous. Eggs single, completely filling the oogonium, the protoplasm nearly surrounded by a surface layer of small droplets.

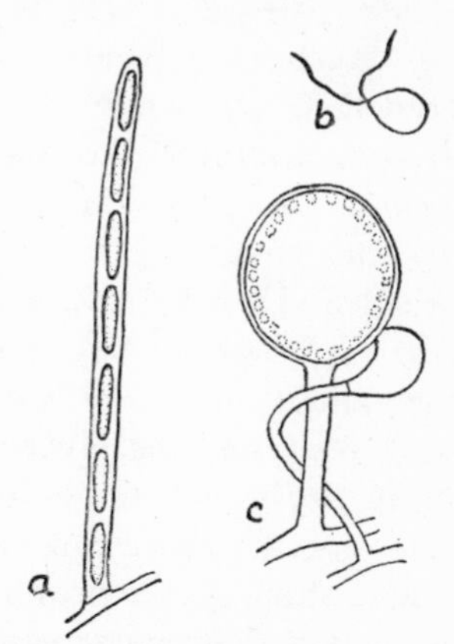

FIG. 24. **Leptolegnia.** *a* sporangium; *b* zoospore. *c* oogonium (after Coker).

KEY TO THE SPECIES OF THE GENUS LEPTOLEGNIA

a. Antheridia androgynous — 1. *L. eccentrica*
aa. Antheridia lacking — 2. *L. subterranea*

1. *Leptolegnia eccentrica* Coker

Hyphae very small, 4.8–7.2μ thick, sparingly branched, forming a ring growth of 2 cm. on hemp seed in water in ten days. Sporangia long, filamentous, same size as the hyphae, first ones usually unbranched, later ones may be branched, the spores escaping at the apex of any branch. Spores in a single row, spherical oval or elongated in the same sporangium, usually about 6μ thick and up to 16.8μ long, swimming away upon escaping, then encysting with a diameter of 7.2–9.6μ, the majority about 7.2μ, and swimming again. Oogonia abundant, spherical to oval with numerous short irregular projections, including these projections 19.2–36μ in diameter, most from 24–28μ, borne on lateral branches; wall hyalin, very thin. Eggs one to an oogonium slightly irregular, spherical to oval, 14.4–27.6μ thick, most 22–26μ, eccentric with a large oil drop on one side; wall extremely thick, up to 5μ or more, consisting of a dark outer portion, lighter irregular central portion and a clear inner portion. Antheridia androgynous arising from the oogonial stalk, often from immediately below the oogonium.

From soil: United States: North Carolina (26) (30)

2. *Leptolegnia subterranea* Coker and Harvey

Mycelium on hemp seed up to 15 mm. in five days. Hyphae sparingly branched, just enough for the plant to appear as a delicately matted growth, with the threads distinguishable practically throughout their length, 9.4–11.8μ thick. Sporangia filamentous, of the same size as the hyphae, often long but not so long as in Aphanomyces, about 11.8μ broad and up to 785μ long or longer, sometimes branched, in which case all the spores may emerge through a common mouth at the apex of the main trunk or any branch. Spores typically in a single row, spherical or oval to elongated, there being about as many of one form as another, 11.8μ broad and up to 16.4μ long; diplanetic, swimming away upon emerging, after encysting measuring up to 14.1μ thick. Oogonia abundant, spherical or subspherical to irregular, usually with low or high, blunt or pointed and irregular protrusions, which may reach a length of 9.5μ, oogonia without the protrusions 40–51.7μ; wall about 1.3μ thick. Eggs one to an oogonium and filling it completely, the mature egg with a cap of oil globules on one side; the wall extremely thick, about 3.7–5μ, rarely 6.5μ. Antheridia lacking.

From soil: United States: Mississippi (61), New York (32), North Carolina (30) (58) (61)

(3) **Saprolegnia** Nees von Esenbeck

Saprobic on animal or plant remains, or parasitic in some species on aquatic animals as fish, frog eggs, etc. Exposed hyphae branched or more or less simple, straight or crooked, usually tapering gradually outward, more or less pointed, springing from an intricately branched, in part rhizoid-like mycelium within the substratum; all vegetative parts colorless in transmitted light, white in reflected light, the threads not septate or constricted until the approach of reproductive stages. Sporangia at first terminal on main threads, typically long-clavate and thicker toward the distal end, or at times slender-fusiform, often irregular and polymorphic in older cultures; at maturity opening typically by an apical mouth, the spores emerging rapidly one by one through pressure from within; typically proliferating within the older ones in a "nested" fashion, but often also as in Achlya. Spores pip-shaped, with two apical flagella, swimming away as soon as discharged, soon coming to rest and encysting in spherical form; after a few hours emerging again through a minute opening in the cyst and swimming again more actively in a somewhat kidney-shaped form with two lateral flagella, finally coming to rest on a nutrient substratum (if such is available) and sending into it a slender tube which grows and branches into the extensive mycelium within. Resting bodies, called gemmae or chlamydospores, of very variable shape and size formed in greater or less number; often in chains like beads; after resting for a few days the contents producing spores of the usual type which emerge by a variously formed mouth. Oogonia terminal on main threads or in long or short lateral branches, or in some species intercalary singly or in chains; shape spherical or oval or pyriform or when intercalary sometimes fusiform; wall smooth or papillate, often pitted. Eggs one or many in an oogonium, formed of all its contents, but never completely filling it, smooth, the protoplasm entirely surrounded by one or two layers of fatty food material (centric or subcentric); undergoing a resting period before sprouting. Antheridia present or absent, of various

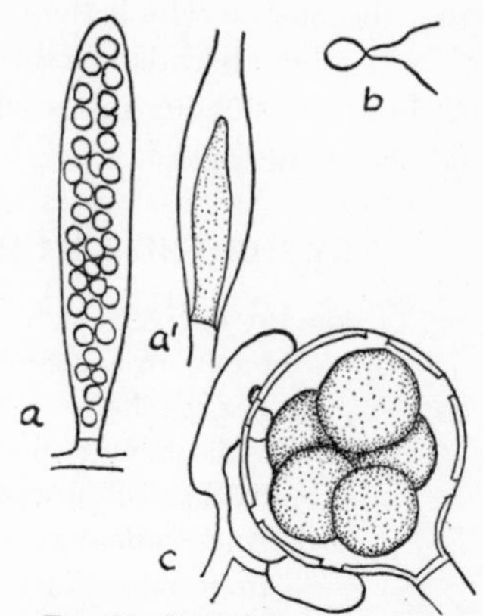

FIG. 25. **Saprolegnia.** *a*, *a'* sporangia; *b* zoospore; *c* oogonium (after Coker).

origin and appearance, usually terminating slender antheridial branches which are short or long, simple or branched, and originating from the same threads on which the oogonia they reach are borne (androgynous), or from other threads (diclinous); antheridia when present often forming one or more slender tubes which enter the oogonia through thin places and reach the eggs.

KEY TO THE SPECIES OF THE GENUS SAPROLEGNIA

a. Oogonium covered with blunt papillae; eggs only one to three in an oogonium — 1. *S. asterophora*

aa. Oogonia not papillate
- b. Antheridia present on all or nearly all oogonia
  - c. Antheridia all or mostly all androgynous
    - d. Antheridial branches present and not arising from immediately below the oogonia
      - e. First oogonia mostly terminal on main hyphae — 2. *S. litoralis*
      - ee. Oogonia on short lateral branches
        - f. Oogonia elongate or flask-shaped — 3. *S. bernardensis*
        - ff. Oogonia globose
          - g. Eggs very large, 30–52μ in diameter — 4. *S. megasperma*
          - gg. Eggs not more than 27μ in diameter
            - h. Eggs usually five or more in an oogonium, 16–22μ in diameter — 5. *S. monoica*
            - hh. Eggs usually less than five in an oogonium, 25–27μ in diameter — 6. *S. glomerata*
    - dd. Antheridial branches usually arising immediately below the oogonia — 7. *S. terrestris*
  - cc. Antheridia all or mostly diclinous
    - d. Spores of a single size
      - e. Oogonial wall without pits, eggs 20–26μ — 8. *S. diclina*
      - ee. Oogonial wall with a few pits
        - f. Eggs mostly 25–27μ in diameter — 9. *S. delica*
        - ff. Eggs 12–22μ in diameter
          - g. Oogonial wall colorless — 10. *S. pseudocrustosa*
          - gg. Oogonial wall yellowish — 11. *S. crustosa* var. I
    - dd. Spores of two sizes, large and small — 12. *S. anisospora*
- bb. Antheridia on part of the oogonia
  - c. Antheridia usually on not more than 15 per cent of the oogonia, often very few or none; eggs mostly about 26μ thick — 13. *S. ferax*
  - cc. Antheridia on about 75 per cent of the oogonia — 14. *S. floccosa*

1. *Saprolegnia asterophora* de Bary

Mycelium extensive, but thin and delicate. Hyphae slender, uneven, much or little branched, about 5–11μ thick, rapidly thickening towards the sporangia which are typically very scarce and often entirely absent in cultures on insects. They are up to 40μ thick, subcylindrical to clavate, and proliferate from within, or rarely laterally from below as in Achlya. Spores 14–15μ in diameter, emerging and swimming slowly and aimlessly in the neighborhood for two or three minutes, then encysting and after a few hours emerging in the usual form and in a more active state. Gemmae not abundant, often absent on insects, peculiar, shaped like the sporangia or pear-shaped, tuberous, knotted, etc. Oogonia numerous, usually thickly set with blunt papillae which are usually 2–4μ, rarely up to 8μ, long; oogonia about 30–57μ thick, including the papillae, most about 37–45μ, borne on even more slender lateral branches of small ordinary hyphae (the stalks rather long), or occasionally intercalary or terminal; walls thin and unpitted. Eggs one or often two, rarely three (very rarely four or five—de Bary), 18–35μ in diameter, dark, often a large and a small one together; structure subcentric, i.e., with the protoplasm completely surrounded by small oil drops, which are in a double layer on one side and a single layer on the other. Antheridial branches varying greatly in abundance, often nearly absent at low temperature, appearing close to the oogonium and usually from its stalk, rarely from neighboring hyphae, often branched and several arising in a twiggy group, but only one or two becoming fully developed. Antheridia short-tuberous or pear-shaped; antheridial tubes not seen.

From soil: Wales (31)

United States: North Carolina (30)

2. *Saprolegnia litoralis* Coker

Growth about as in *S. ferax*; more vigorous, extensive and irregular than in *S. delica*, the hyphae reaching a length of 1–1.5 cm. on a mushroom grub. Sporangia not abundant, far less so than in *S. diclina*, early ones nearly cylindrical, or more often irregular in diameter, usually curved, repeatedly proliferating, later ones more irregular and often pointed. Spores 10–12μ in diameter. Gemmae very abundant, spherical, pyriform, clavate, etc., often in chains, the terminal one very often with an elongated papilla. Oogonia plentiful as a rule, about as much so as in *S. ferax*, but at times not found on grubs in distilled water (more scattered than in *S. delica* and *S. anisospora*), about 35–80μ thick, the larger number terminal on main hyphae, others (usually appearing later) on short lateral branches; shape spherical, or if borne on the ends of main threads usually oval, the latter frequently with a slender, more or less lengthy terminal extension, which when short may be included in the cavity of the oogonium, but which is often extended into a thread 2.8–3μ thick, thus making the oogonium intercalary; furnished with rather few, very conspicuous and usually large pits, up to 11μ across. Eggs centric, large and dark, one to twenty, mostly two to six in an oogonium, their diameter 20–40μ, most about 30–33μ, often elliptic from pressure. Antheridia on every oogonium (one to several), androgynous on short branches which usually arise very near the oogonium, frequently, when the oogonium is on a short stalk, arising from immediately below it. In addition to the androgynous antheridia a few diclinous ones may arise rarely from other nearby threads.

From soil: Denmark (80), Germany (112)

United States: North Carolina (30)

3. *Saprolegnia bernardensis* Harvey

Primary mycelium very lax and delicate, the hyphae transparent and very crooked, 12–15μ thick, giving rise within twenty-four hours to a few small sporangia. Sporangia generally more or less cylindrical to long barrel-shaped, straight or curved, seldom irregular, as a rule terminal; seldom more than 150μ long × 27.7μ thick. Secondary hyphae appearing on the second day and growing rapidly, reaching a colony diameter of 62 mm. on halved hemp seed in water within one week, more extensive than in *S. ferax*; hyphae delicate, as a rule less than 12μ in diameter. Secondary sporangia scarce, similar to the primary ones in appearance but larger, up to 335μ long, possibly longer. Internal proliferation of hyphae common. Spores at rest spherical, 10.5–12.6μ. Gemmae not abundant; spherical, elliptical, pyriform, clavate, or elongate, sometimes branched, not very dense; terminal or intercalary, single or in short chains; a few becoming transformed into sporangia. Oogonia very numerous within five days, giving the more or less transparent colony a speckled appearance; borne commonly on short lateral stalks, seldom terminating main threads, sometimes clustered as the result of branching of the parent hyphae or of oogonial stalks, rarely proliferated from another oogonium, at times intercalary; spherical, oval, elliptical, clavate, to flask-shaped, frequently narrowed basally or apically into a long filament of the same diameter as the parent hypha; commonly measuring more than 300μ in length and even not infrequently over one millimeter, in one case reaching 1,106μ; usually 41.4–97.2μ at greatest diameter (on corn-meal agar, 39.5–47.5μ); with greenish or yellowish tinged walls, 1.5–1.8μ thick, freely pitted, pits commonly about 9μ in diameter. Oospores one to twenty-five or more, commonly eight to twenty (one, two, or three on cornmeal agar), not always filling the oogonium, centric; spherical, oval, or often in a single row and very long and narrow when found in the filamentous portion of an oogonium, such as 9.0 × 75.6μ, spherical ones 20–29.7μ, averaging 22.5–27μ. Antheridia generally present, at times androgynous, but more commonly diclinous; as a rule one to three, seldom more per oogonium, not completely covering the oogonium. Empty antheridia seen, germ tubes not observed.

From soil: United States: California (62)

4. *Saprolegnia megasperma* Coker

Mycelium on grubs and vegetable media about as vigorous as in *S. ferax* or *S. litoralis*, threads on mushroom-grubs or termites, 9–35μ thick, most about 15–20μ thick, reaching a length of 0.5–0.7 cm; threads straight to wavy. Sporangia abundant, apical, 15–45 × 100–400μ, variable in shape, the first ones usually long and distinctly swollen at the distal end, later ones usually smaller and more or less irregular in outline; emptying normally for the genus, renewed by internal proliferation or rarely by cymose branching as in Pythiopsis or Achlya. Not rarely in cultures slightly infected with bacteria, the sporangia may break away from the threads as in Dictyuchus, such sporangia emptying normally after a long or short rest. Spores diplanetic, flagellate, 11μ thick when encysted. Gemmae abundant, round to oval or very irregular, emptying upon the addition of fresh water by one or more long papillae. Oogonia produced in fair abundance, inversely in proportion to the number of sporangia and gemmae, 40–100μ thick, wall smooth (rarely with a papilla), not thick, without pits or rarely with a few small ones; usually borne on short racemose branches which in length are as a rule less than the diameter of the oogonia; not rarely borne singly or in clusters of several on the ends of main threads in cultures in which sporangia are sparingly produced. Eggs one to ten, single in over 50 per cent of the oogonia in most cultures (not rarely running considerably above or below this per cent); 30–52μ thick, usually about 38μ thick; subcentric (one row of oil droplets on one side, two on the other), not filling the oogonia. Antheridia present on all oogonia, applied by their ends, seldom by their sides; antheridial walls thick, easily visible even in old cultures; antheridial branches usually of androgynous origin but quite often diclinous, usually simple and unbranched; antheridial tubes developed and easily visible.

From soil: Australia (31)

United States: North Carolina (25) (30)

5. *Saprolegnia monoica* Pringsheim

Main threads straight, tense. Primary sporangia slender, clavate-cylindric. Antheridial branches androgynous, forming antheridia on all the oogonia, almost always arising near, and spring from, the same stalks as the oogonia to which they are attached or from neighboring ones. Oogonia usually borne on racemosely arranged, bent or straight short branches which are about as long as the diameter of the oogonia; the main hyphae from which these spring ending in an oogonium, or a sporangium, or a sterile point. Oogonia spherical, smooth, with several large pits in the membrane. Oospores from one to over thirty, mostly five to ten in an oogonium, centric, 16–22μ thick. Antheridia bent-clavate, with the concave side applied to the oogonium.

From soil: Denmark (80), Germany (112), Wales (31)

6. *Saprolegnia monoica* var. *glomerata* Tiesenhausen

Growth moderately extensive, the hyphae not very robust. Sporangia abundant, cylindrical or long club-shaped, later ones more irregular, proliferating from within or not rarely from one side also, varying greatly in size, rarely so small as to have only a single row of spores. Spores 10–11μ in diameter. Gemmae abundant or few, often in moniliform chains, pear-shaped or irregularly club-shaped, often modulated or branched, quickly forming spores when brought into fresh water. Oogonia abundant, usually lateral on short stalks which are mostly a quarter to equally as long as the diameter of the oogonia, rarely intercalary, occasionally terminal and then usually cylindrical in old sporangia; wall colorless, moderately thick, the pits few or numerous in the same culture, and rather conspicuous, 5.5–7μ in diameter. Eggs centric, generally one, two, or four, occasionally six or eight, rarely twenty (or more?), diameter 24–31μ; usually about 25–27μ. Antheridial branches short, typically clustered and contorted, often branched, arising androgynously from the main branches near the oogonia or at times from the oogonial stalks, not rarely reaching also to nearby oogonia on other threads (diclinous). Antheridia pear-shaped or tuberous, one or more on every oogonium; antheridial tubes formed.

From soil: Germany (112)

7. *Saprolegnia terrestris* Cookson

Growth on hemp seed about 0.5–1 cm. long, mycelial mat thick. Hyphae slender, up to 48μ broad at base. Sporangia abundant, very variable in shape, typically cylindrical or clavate, 16–48μ broad, 60–400μ

long, frequently almost spherical, sometimes irregular and contorted, opening apically by a more or less prominent mouth. Primary sporangia terminal, secondary sporangia develop either by internal proliferation or by the delimitation of a segment behind a discharged sporangium and the outgrowth from this segment of a sporangium; occasionally secondary sporangia develop in a truly cymose manner. Spores diplanetic, about 0.5μ when encysted. Gemmae usually not abundant, cylindrical, pyriform or irregular in shape. Oogonia form terminally or laterally on straight stalks which are as long as, or considerably longer than, the diameter of the oogonium, sometimes intercalary; typically spherical, sometimes with a neck, or when developed within an empty sporangium, cylindrical, occasionally with a short apiculus, 30–87.5μ in diameter, average 61μ. Oogonial wall usually yellow, frequently unpitted but in some cultures with well defined though not conspicuous pits. An upgrowth from the basal wall of the oogonium is frequently present, and an irregular internal thickening of the oogonial wall is sometimes met with in old hemp seed cultures. Oospores one to ten, usually two to six, 20–37μ in diameter, average 29μ, dark brown when immature, later becoming yellowish; eccentric in sense that the peripheral sheath of oil drops does not completely surround the protoplasm, or subcentric. Antheridia present on all oogonia, typically one, sometimes two or three; antheridial branches androgynous, usually simple, but sometimes slightly branched, typically arising from the oogonial stalk immediately behind the oogonium but occasionally developing from the same hypha as the oogonium; antheridia clavate attached by their sides to the oogonium, becoming inconspicuous; antheridial tubes large and conspicuous.

From soil: Australia (31)

8. *Saprolegnia diclina* Humphrey

Syn. *S. dioica* de Bary

Main hyphae of moderate size and length, little branched. Sporangia only slightly enlarged and broadest near the end, repeatedly proliferating inwardly, but also not rarely arising laterally from beneath the discharged ones, as in Achlya. Spores 11–11.5μ in diameter. Gemmae very abundant and variable in shape, long and pointed or stocky and knotted, the longer ones rather characteristic for this species, the other forms much as in *S. delica*, etc. By single spore cultures made several times the species has proved to be not dioecious (heterothallic) although highly diclinous, the mycelium from a single spore producing both oogonia and antheridia. Oogonia spherical or oval or pear-shaped, usually with a short neck, mostly terminating the main branches, but not rarely intercalary, occasionally two or three to five in a chain, rarely on short lateral branches or cylindrical in empty sporangia, very variable in size, even in the same cultures, 35–100μ in diameter; walls rather thin, without pits except where the antheridia touch. Antheridial branches arising diclinously from near or distant hyphae, branching, delicate, slender, and soon disappearing after the antheridia have been cut off. Antheridia on every oogonium, numerous, often completely covering the oogonium, usually slender and not much larger than the branches, occasionally somewhat swollen and tuberous, only moderately dense, but remaining visible for a long time after the anteridial branches have disappeared. Antheridial tubes nearly always invisible (if present). In only two cases were they seen. Eggs 20–26μ in diameter, most about 23–24μ, varying little in size in any one oogonium, one to twenty or more in an oogonium, usually six to twelve, centric.

From soil: Germany (112)

United States: California (48)

9. *Saprolegnia delica* Coker

Growth delicate and lax, but uniform and symmetrical, the hyphae straight and simple at first, then much branched. Sporangia long, nearly cylindrical or later irregular, abundant and symmetrical in most young cultures, the later ones often irregularly inflated or bent, repeatedly proliferating from within, and not rarely laterally from below; spores about 10.5–11.5μ in diameter; gemmae plentiful or few (not nearly so abundant as in *S. diclina*), spherical or pyriform to fusiform or clavate, often in moniliform chains. Oogonia typically spherical, abundant on most media, terminating the main branches and also racemosely borne throughout on rather long or rarely short lateral branches that are usually two or more times as long as the diameter of the oogonia; wall smooth, colorless, thin, about 1.8μ thick, furnished with rather few pits about 3.7–8.5μ in diameter which are not nearly so conspicuous as in *S. monoica*, or *S. ferax*; diameter of oogonia on termite about 40–63μ, averaging about 55μ, on fly 42–70μ, averaging about 60μ. Eggs mostly one to six, often eight. and very rarely up to sixteen (in abnormal cases when large oogonia are filled with very small eggs there may be up to forty), centric, quite dark when young (in transmitted light), lighter at full maturity, averaging about 25–27μ, with extremes of 14.8–33μ, the smallest often in oogonia of abundant size and not rarely mixed with the larger. Antheridial branches abundant, often long and rambling, the larger part diclinous, rather stout and persistent. Antheridia present and usually numerous on nearly all or all oogonia (95–100 per cent), each oogonium typically furnished with at least one diclinous antheridium, and at times with androgynous ones also, occasionally absent from oogonia that terminate long branches and are therefore removed some distance from the main mass; pear-shaped or irregularly tuber-shaped, well filled with protoplasm; antheridial tubes present and not inconspicuous.

From soil: Denmark (80)

United States: North Carolina (30)

10. *Saprolegnia pseudocrustosa* A. Lund

Hyphae slender and delicate. Sporangia cylindrical, the secondary ones formed by proliferation. Zoospores set free as usual in Saprolegnia. Gemmae mostly regular, pyriform to clavate or somewhat irregular. Oogonia spherical, 70–91μ in diameter, sometimes oval, terminal on main hyphae or on lateral branches that are one to four times as long as the diameter of the oogonia, rarely intercalary. Oogonial wall smooth, colorless, with few or numerous pits; sometimes an ingrowth into the oogonium is present. Oospores centric, 22.4–28μ, usually 25μ in diameter, four to twenty, often twelve to eighteen, in an oogonium. Antheridial branches diclinous, often winding about the oogonia.

From soil: Denmark (80)

11. *Saprolegnia crustosa* var. I Maurizio

Growth dense, 2 cm. long. Sporangia as usual. Oogonia racemose or clustered, also intercalary; and also in simple sympodia of two or four; spherical or elongated at times, if intercalary; stalk short, straight, of moderate thickness; membrane somewhat yellowish, of medium thickness, with pits of medium size that are not numerous; frequently an upgrowth enters the oogonium from the wall below; diameter of oogonia 31.5–60μ or when elongated 33–41μ broad $\times$ 50–55μ long. Eggs four to twenty-five, various in size, mostly 19.5–22μ, smallest 12μ; wall moderately thick and yellowish. Antheridia mostly present, diclinous, coming from a distance, and plentifully enveloping the oogonia, not observed to branch. In this species, conidia are present, in rows or chains, also in complicated sympodia, rarely single, becoming either sporangia or oogonia, or resting. There are also present in the conidia the tubes from the cross wall below, as in the oogonia.

From soil: Denmark (80), Wales (64)

12. *Saprolegnia anisospora* de Bary

Main hyphae about 5–8 mm. long on a mushroom grub, of moderate size at base, but quickly becoming smaller, the culture appearing quite delicate in comparison with many other species; main hyphae from 40μ

in diameter below to 11μ or even less near the tip. Sporangia usually borne on larger branches than the oogonia (but a good many oogonia also borne on the larger branches), usually rather stocky and irregular and largest in the middle or near the base, sometimes regularly tapering towards the end, very variable in size in the same culture, about 8.6–15.2μ, rarely up to 16.6μ thick, usually thicker than the strand that bears them; often short and broad, proliferating as usual in Saprolegnia, or when in distilled water the greater part as in Achlya. Dictyosporangia have been several times observed. Spores remarkable in being of two kinds, large and small and often intermediate sizes, usually in separate sporangia without constant regard to the size of the latter, a single sporangium usually with spores of only one size, but occasionally they are mixed; the smallest spores about 8–9μ in diameter, others from 10.5–11.5μ, the large ones from 13.7–14.8μ; small and large spores similar in structure, but the small are greatly in excess of the large ones; in nearly all cultures there are formed in addition a few very large spores at least twice the bulk of the ordinary large spores, these appearing usually mixed with the latter. Oogonia numerous and formed in all ordinary culture media, borne usually on the tips of long slender branches which arise from near the substratum, often intercalary (very rarely two or three in a row), varying to laterally sessile or on short or rather long lateral branches; typically spherical with a short neck when apical, but at times oval to pear-shaped, and when intercalary oblong to flask or spindle-shaped with long necks, 33–92μ in diameter, most about 55–65μ; at maturity with moderately thick walls that appear unpitted except beneath each antheridium where there is always a distinct circular pit. Eggs one to twenty, mostly four to six, quite variable in size even in the same oogonium, 17–38μ in diameter, most about 21–27μ, centric. Antheridial branches quite slender and soon becoming very inconspicuous, arising from any of the main branches, usually, for the proximal half, and running to oogonia on other branches than the one from which they arise (diclinous); oogonial branches often give rise to antheridial branches lower down. Antheridia cylindrical to tuberous, present on all oogonia, usually several to many, when young well filled with protoplasm, in age apparently empty; antheridial tubes formed in most cases and remaining visible after the eggs are formed. Gemmae more or less numerous or rather few, usually spherical, sometimes pear-shaped or tuberous, and of other shapes, usually in short or long chains, easily becoming sporangia on change of conditions, emptying by a proliferating tube. Often there may be several proliferation tubes, but only one opens for the escape of the spores.

From soil: Australia (31)

13. *Saprolegnia ferax* (Gruithuisen) Thuret

Hyphae moderately stout and vigorous, irregular, sparingly branched below. Sporangia plentiful, only slightly enlarged, typically wavy and bent and of unequal diameter, often tapering upward, rarely almost cylindrical, often proliferating laterally from below old ones; zoospores about 9μ in diameter. Gemmae not very abundant, more or less elongated usually, but varying to bulbous, or pyriform and sometimes jointed. There is a strong tendency to the formation of long tapering tips on the ends of stout threads, the ends of which are later cut off as rejected tips of irregularly tapering gemmae below. In such cases the gemma opens later by basal protuberance as is also usually the case even when no tip is cut off. Oogonia numerous, varying in diameter from 37–97μ, the wall only about 1.3–1.6μ thick, but with numerous conspicuous pits, which are about 4.5–5.5μ in diameter; either lateral on stalks which are usually short and frequently curved, or terminal on the main branches, sometimes intercalary, but not in chains; spherical to slightly oval with a basal neck, which is often curved; not rarely formed inside of empty sporangia, and then cylindrical; thread-like extensions of the oogonia, containing a single row of elliptic eggs, are not rare. Eggs centric, one to twenty, mostly four to sixteen, the diameter 24–30.5μ (rarely as small as 14.8μ or as large as 33.8μ), the greater number about 26μ, extremes sometimes occurring in the same oogonium. Antheridial branches short, stout, mostly androgynous, present in nearly all the cultures in varying number, usually on about 10–15 per cent of the oogonia, but the number of oogonia furnished with them varies from none to 98 per cent, depending on the medium used. The antheridia arise as a rule from the same main strands that bear their oogonia, but often also from the oogonial stalks, and in the latter case usually applying themselves not to their own but to nearby oogonia, either from the same or another strand. Antheridia usually cut off, short and tuberous, not more dense than the threads; fertilizing tubes suppressed or very rare.

From soil: Germany (112)

United States: California (62), Florida (158), Georgia (61), North Carolina (27) (30) (58), Oklahoma (61), Wisconsin (60) (61)

14. *Saprolegnia floccosa* Maurizio

The slightly branched, irregularly curved, delicate hyphae are 60μ in diameter at their base and in the middle of their extent 12–25μ in diameter. They taper to their tips and end abruptly. Sporangia typical for Saprolegnia. Oogonia usually botryosely arranged, 42–67 μ in diameter, are characterized by their strongly angular shape and the large pits. The oogonia occur on robust stalks which vary in length but in general are less than three times the oogonial diameters, straight or irregular, sometimes sharply bent. Oospores, centric, 21–27 μ in diameter, two to fourteen in number, are in some cases lying loosely in the oogonium, in others so closely packed as to be deformed by pressure. Antheridia, present on about three-fourths of the oogonia, arise from secondary hyphae which in turn arise partly from below the oogonium and partly from the oogonial stalk. Antheridia lie beside the oogonium.

From soil: Germany (112)

(4) **Pythiopsis** de Bary

Hyphae slender, much or little branched. Sporangia typically short and plump, spherical, oval, pyriform with a distinct apical papilla, or varying to elongated and irregular, primarily borne at the tips of the hyphae and multiplied from lateral stalks below the old ones to form more or less dense clusters. Spores emerging and swimming as in Saprolegnia, pip-shaped with two apical flagella, sprouting after the first encystment (monoplanetic). Gemmae resembling the sporangia or oogonia, formed plentifully, often in chains, producing zoospores after a rest. Oogonia borne like the sporangia and gemmae and resembling them in youth, typically spherical, oval or pyriform with unpitted walls, smooth or with a few blunt papillae. Antheridia short and thick, typically androgynous from the close neighborhood of the oogonia. Eggs one or few, eccentric, with a lunate cap of droplets on one side in *P. cymosa*.

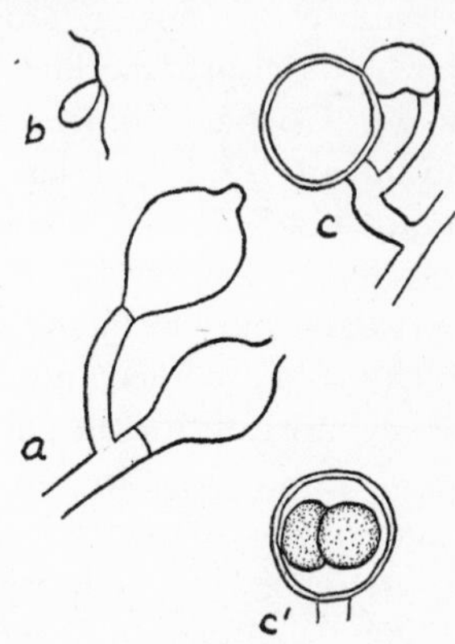

FIG. 26. **Pythiopsis.** *a* sporangia; *b* zoospore; *c, c'* oogonia (after Coker).

KEY TO THE SPECIES OF THE GENUS PYTHIOPSIS

a. Sporangia globular, or clavate, egg single, 14.8–18.5 μ in diameter — 1. *P. cymosa*

aa. Sporangia very variable, the majority elongated, irregular and tapering to a long crooked papilla, eggs usually one, 22–33 μ in diameter — 2. *P. intermedia*

1. *Pythiopsis cymosa* de Bary

Hyphae slender, 14.8–22.5μ in diameter at base, short or moderately long. Sporangia globular or clavate. Spores 8.6–10.8μ, most about 9μ; monoplanetic. Oogonia plentifully formed in old cultures, spherical to oblong or pear-shaped, unpitted, smooth, or sometimes with a few blunt outgrowths, terminal or rarely intercalary, 18–30μ in diameter, a few smaller. Eggs mostly 14.8–18.5μ in diameter, but sometimes up to 24μ, single (Humphrey says rarely two to an oogonium), eccentric, as described above. Antheridial branches short or none, usually arising from just below the basal walls of the oogonia, rarely diclinous. Antheridia one or two to each oogonium, clavate; antheridial tubes present, at times growing up through the basal wall of the oogonium. Gemmae of more or less globular or ovoid shape are formed in quantity and are often arranged in chains. After a rest these also form zoospores.

From soil: Germany (112), Wales (64)

United States: North Carolina (30)

2. *Pythiopsis intermedia* Coker and Harvey

Vegetative growth of long, slender, sparingly branched hyphae, 11–14μ thick which are stoutest in the neighborhood of the reproductive bodies, and which after maturity disorganize rather quickly. Sporangia very variable, the majority elongated, irregular and tapering to a long crooked papilla, varying to spherical, proliferating from below in a cymose manner; spores diplanetic, pear-shaped, 8–11.8μ long. Oogonia formed after two days, generally borne like the sporangia and not easily distinguished from certain forms of these when young, usually on short or long, irregular and often coiled lateral branches and often in groups by cymose branching, usually irregularly subspherical with a short basal neck, 35–50μ thick, usually about 35–45μ; wall wavy with low irregular protrusions. Eggs generally one, rarely two, three, or four, subcentric with two layers of oil droplets on one side and one on the other, as in *Saprolegnia asterophora*, etc., 23–33μ thick; wall about 3.5μ thick. Antheridia as in *Pythiopsis cymosa*, short-clavate and terminating a stalk that usually arises from immediately below the oogonium, rarely two; antheridial tube obvious, reaching and apparently fertilizing the egg. Gemmae resembling sporangia and oogonia abundant and forming spores after a rest.

From soil: Wales (64)

United States: North Carolina (27) (58) (61), Oklahoma (61)

(5) **Isoachlya** Kauffman

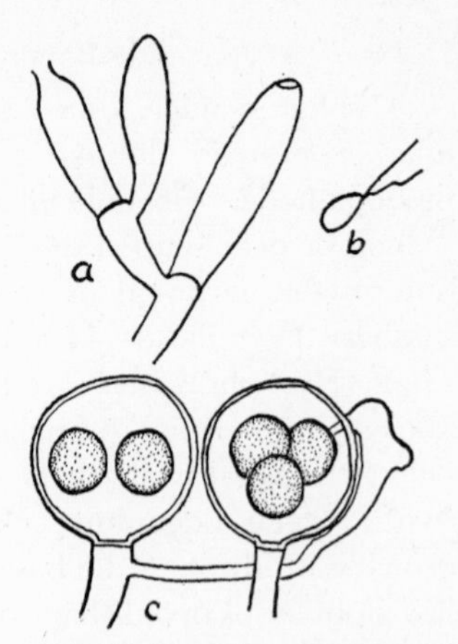

FIG. 27. **Isoachlya.** *a* sporangia; *b* zoospore; *c* oogonia (after Coker).

Hyphae rather stout or slender. Zoosporangia formed from their tips, oval, pyriform, ventricose-clavate, elongated pyriform to clavate or cylindrical-clavate; the later ones (secondary) arising either by cymose or pseudo-cymose arrangement, as in Achlya, or by internal proliferation as in Saprolegnia, both modes occurring earlier or later in the development of one and the same species, or frequently on the same main hypha. Zoospores diplanetic, as in Saprolegnia, escaping and swarming separately, and after encystment swarming the second time before the formation of a germ-tube. Oogonia terminal or torulose, occasionally intercalary. Oospores with centric or eccentric contents, the spores filling the oogonium incompletely. Antheridia present or few to none.

The genus is characterized and distinguished, in the main, by the presence of the cymose or Achlya mode of formation of secondary sporangia, coupled with diplanetic zoospores.

KEY TO THE SPECIES OF THE GENUS ISOACHLYA

a. Antheridia lacking
  b. Eggs usually one to two in an oogonium
    c. Eggs centric — 1. *I. unispora*
    cc. Eggs eccentric — 2. *I. eccentrica*
  bb. Eggs usually 2–6, centric — 3. *I. monilifera*
aa. Antheridia present, strongly androgynous
  b. Antheridial branches arising near oogonium (not farther than diameter of oogonium) — 4. *I. subterranea*
  bb. Antheridial branches farther from oogonium than oogonial diameter — 5. *I. terrestris*

1. *Isoachlya unispora* Coker and Couch

Mycelium vigorous, more extensive than in most species of Saprolegnia; hyphae irregular, not straight or cylindrical, normally little branched, about 10–35μ thick, and usually largest toward the periphery. Sporangia typically scarce, frequently almost none, often quite irregular; primary ones elongated, varying from sub-cylindrical and slightly if at all thicker than the hyphae to shorter, thicker and more flask-shaped; secondary sporangia arising by cymose branching, and also not rarely growing through the empty ones, but in such cases the new sporangia forming entirely outside the mouth of the old ones. Spores diplanetic, 9.3–13.7μ, most about 10.5–11.5μ thick at rest; emptying as in Saprolegnia and swimming rather sluggishly and aimlessly, some coming quickly to rest: on emerging from the cysts they swim longer and more actively. Not rarely the spores remain in the sporangia and sprout there, as in most other species at times. Gemmae plentiful or few, typically spherical, with or without a neck, usually in chains, the distal member of which is not rarely an oogonium; emptying on changed conditions by an elongated papilla. Oogonia abundant, mostly spherical, rarely pyriform, usually with a distinct neck and borne on lateral branches, quite often terminal on small hyphae, and in strong cultures frequently in clusters, with the arrangement of a scorpioid cyme, not rarely intercalary or in chains of two or three, sometimes cylindrical inside old sporangia; diameter 24–75μ, most about 50μ, wall clear at first, distinctly yellowish in age, about 2.8μ thick, with few visible (usually two or three) very large and conspicuous pits. Eggs few, usually one, often two, rarely three and very rarely four, 18.5–43μ in diameter, most about 32–35μ when two in one oogonium or 40–45μ when only one; centric. Antheridia never developed.

From soil: Wales (64)

2. *Isoachlya eccentrica* Coker

Threads long, slender, little branched, 6–18.5μ thick, most about 12–15μ, growth vigorous on termite ants, bits of boiled corn grain and on agar plates; length on corn grain up to 8 mm., on a termite ant up to 4 mm., tips pointed and clear. On corn grain many or most of the tips become sporangia of a regular cylindrical shape, with a distinct papilla, 30–45 × 142–400μ, at times broader in the middle, again near the tips. On termite ants the sporangia are less regular, with several apertures in most cases. Proliferation of sporangia not common, when present never internal as in Saprolegnia, but irregularly from below, as in Achlya and Pythiopsis. Dictyosporangia observed a few times. Spores diplanetic, 10–11μ thick when at rest, emerging rather slowly with the flagella directed backward, then reversing and swimming sluggishly for a very short time, many coming to rest in the immediate neighborhood of the sporangium; shape and contents as in Saprolegnia. Gemmae plentiful, following the sporangia, very irregular in shape and size, after a time forming spores. Oogonia spherical as a rule, seldom oval, 15–40μ, most about 30–35μ thick, usually single, at times in chains of four or five; commonly borne on short lateral stalks which are from one-half to twice as long as the diameter of the oogonia; often as well on the tips of threads which have proliferated through empty sporangia, and in many such cases not rarely formed inside the sporangia; sections of old threads may also become oogonia. Wall colorless, smooth, many without pits, some with a few large, inconspicuous ones. Eggs usually one, often two, rarely three or four; 12–31μ thick, most about 20–25μ, eccentric, with a single large oil drop at maturity. Antheridia none.

From soil: Wales (64)

United States: North Carolina (30) (58) (61)

3. *Isoachlya monilifera* (de Bary) Kauffman

Vegetative growth short; main hyphae 13–22μ in diameter near base. Sporangia scarce, often entirely absent, short or moderately long, usually largest near the tip; in older cultures the sporangia often proliferating laterally below as in Achlya. Spores 11–11.8μ in diameter. Oogonia abundantly produced in a very dense zone immediately surrounding the substratum, appearing before the sporangia, mostly in chains, the lower elements of the chain usually smaller and sometimes remaining as gemmae, commonly spherical with or without a basal neck, rarely elongated inside old sporangia; diameter about 40–93μ, most about 50–65μ, a large part of them breaking off more or less completely from the hyphae and from each other after the maturity of the eggs; walls yellowish-brown when old, smooth, slightly or not at all pitted. Eggs one to twelve in an oogonium, mostly two to six, 17.7–33.5μ in diameter, average about 23–25μ, extremes sometimes occurring in the same oogonium, yellowish-brown, centric, with two rows of small droplets all the way around or subcentric, with one row on one side and two on the other. Gemmae abundant in all cultures, spherical, pear- or club-shaped, very often borne in chains; upon change of condition they may become sporangia, discharging their spores through a lateral papilla. Antheridia not developed.

From soil: Wales (64)

United States: North Carolina (30)

4. *Isoachlya subterranea* Dissmann

Mycelium flaccid, up to 15μ in diameter, hyphal tip abruptly tapered. Sporangia long cylindrical, in the lower third only slightly broader than the sporangiophores and on the average about ten times longer than broad. Exit-papillae almost always present. Sporangia rarely proliferating through the empty case, more often by lateral branching. Zoospores sluggish, often gregarious, of great variability, and after a very short motile period encysting in the immediate neighborhood of the sporangium. Secondary zoospores extruded after many hours. Gemmae usually spherical, clustered on the hyphal tips, about 45μ in diameter, or in short chains on long club-shaped hyphae up to 170μ. Germination by tube. Oogonia 50–55μ, almost always spherical, single in clusters on the primary hypha or at the tips of lateral hyphae, never in chains. Oogonial wall, pronounced brown-yellow, surface irregularly striate, finely pitted, staining brownish-red with chloro-zinc iodide. Pits (dots) abound. Antheridial hyphae always present on androgynous secondary branches arising from the stalk of the oogonium, seldom farther from the oogonium than its diameter. Antheridia club-shaped, seldom branched, usually attaching themselves to the base of the oogonium. Oospores spherical, centric, when fully mature with a 2–3μ-thick hyalin wall, in diameter 25–45μ (usually 39–42μ), single or very rarely two in an oogonium. Germination not observed.

From soil: Germany (40)

5. *Isoachlya terrestris* Richter

The long straight extensive, delicate hyphae measure at their base 29–42μ, in their intermediate course, 18–21μ and at their tips, about 10μ in diameter. They taper throughout with rounded ends that show a hyalin zone. The protoplast in young hyphae dark brown, in older hyphae lighter. Relatively few lateral branches are long, thin and at right angles, more often they become curved. In addition short thorn-like branches occur infrequently. Terminal sporangia 293–543 $\times$ 25–34μ. They occur singly or in groups and proliferate in a characteristic manner for Isoachlya. Zoosporangial walls visibly pitted. One or many lateral, more or less large (16–20μ) exit-tubes occur in addition to the terminal tube. Zoospores 8–10μ in diameter. Oogonial stalk, straight, seldom somewhat curved, 6–9μ in diameter, one-half to three times the diameter of the oogonium in length. Oogonia globose, seldom somewhat irregular, clustered, 32–44μ in diameter with a dark wall free from pits and in most cases smooth or with few irregular protrusions, 10–12μ long. Oospores one to five, usually two or three, globose, or flattened on the side by pressure, filling the oogonium entirely, 16–27μ, with a large eccentric fat globule. Antheridia occur on about half the oogonia, often appearing late, 8μ in diameter and lying along the side of the oogonium or surrounding it thickly, in many coils. They are terminal on delicate, branched secondary hyphae which arise at some distance from the oogonium, but from the same hyphae as the oogonial stalk, or, in some cases, diclinous.

From soil: Germany (112)

(6) **Achlya** Nees von Esenbeck

Resembling Saprolegnia essentially in size, growth and appearance of vegetative parts and as now constituted approaching that genus closely in some species. Sporangia typically (except in the *Racemose* group) broadest in the middle or towards the base, gradually pointed, not increased from within others but by lateral branching from below the older ones, at times in close clusters, again in more interrupted sympodial arrangement. Spores on leaving the sporangium coming to rest at once, or after a short period of slow rocking, in a hollow sphere or irregular cluster (in several species at least furnished with flagella during emergence), encysting there and after a few hours swimming again as in Saprolegnia. Oogonia borne variously as in Saprolegnia, with or without pits or papillae. Eggs formed of all the contents of the oogonia and not completely filling them, one to many; varying in structure with the different groups. Antheridia of near or distant origin, androgynous or diclinous, in a few species absent; fertilizing tubes usually present.

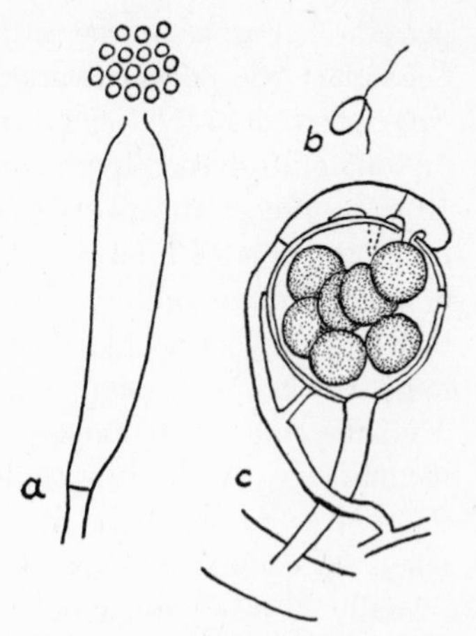

Fig. 28. **Achlya.** *a* sporangium; *b* zoospore; *c* oogonium. (after Coker).

KEY TO THE SPECIES OF THE GENUS ACHLYA

a. Oogonia all globose
  b. Oogonia without spines or papillate outgrowths or a few of them with such projections
    c. Antheridia arising from androgynous branches
      d. Eggs small, averaging less than 23μ in diameter
        e. Oogonial walls pitted; antheridial branches arising from the main hyphae between the near oogonial branches — 1. *A. americana*
        ee. Oogonial walls unpitted (except where antheridia touch)
          f. Antheridial branches arising from the oogonial branches — 2. *A. racemosa*
          ff. Antheridial branches arising from the primary hypha not from oogonial stalks — 3. *A. debaryana*
      dd. Eggs larger, averaging more than 23μ in diameter

e. Eggs eccentric — 4. *A. subterranea*
ee. Eggs truly centric
f. Oogonial branches about as long to twice as long as the diameter of the oogonia, or even longer; oogonial wall strongly pitted — 5. *A. conspicua*
ff. Oogonial branches much longer than the diameter of the oogonia; oogonial wall unpitted
g. Eggs about 27μ thick, five to twenty-five in an oogonium — 6. *A. polyandra*
gg. Eggs mostly 33–36μ thick, usually one to two in an oogonium — 7. *A. orion*
cc. Antheridia arising from diclinous or from both diclinous and androgynous branches
d. Oogonial wall unpitted
e. Oospores 42-52μ in diameter — 8. *A. megasperma*
ee. Oospores 20-30μ in diameter — 20. *A. rodriguenziana*
dd. Oogonial wall pitted
e. Antheridia arising from both diclinous and androgynous hyphae
f. Antheridia arising more often from diclinous than androgynous branches
g. Antheridial branches winding themselves about main hyphae — 9. *A. proliferoides*
gg. Antheridial branches not winding about hyphae and never arising from oogonial stalk
h. Homothallic — 10. *A. flagellata*
hh. Heterothallic
i. Oospores 24–30μ in diameter — 11. *A. bisexualis*
ii. Oospores 18–21.6μ in diameter — 12. *A. heteromorpha*
ff. Antheridia about equally distributed on diclinous and androgynous branches — 13. *A. imperfecta*
ee. Antheridia always arising from diclinous branches — 14. *A. prolifera*
bb. Oogonia mostly with spines or papillate outgrowths
c. Oogonia without antheridia
d. Eggs small 18–23μ — 15. *A. caroliniana*
dd. Eggs larger 24–64μ — 16. *A. abortiva*
cc. Oogonia (at least in part) with antheridia
d. Antheridia always on diclinous branches — 17. *A. inflata*
dd. Antheridia on both androgynous and diclinous branches
e. Oogonia with a single apiculus — 18. *A. apiculata*

ee. Oogonia with numerous papillae — 19. *A. recurva*

ddd. Antheridia all or almost all on androgynous branches

e. Antheridia, often arising from a hypogynal cell; eggs commonly three to five in an oogonium — 21. *A. hypogyna*

ee. Antheridia not formed as above

f. Eggs commonly one or rarely two or three in an oogonium

g. Oogonia covered with sharp spines; eggs commonly 34–36.5μ thick — 22. *A. radiosa*

gg. Oogonia covered with blunt warts; eggs averaging 20μ thick — 23. *A. glomerata*

ff. Eggs commonly one to four in an oogonium — 24. *A. colorata*

aa. Oogonia occasionally lobed — 25. *A. pinnulata*

1. *Achlya americana* Humphrey

Growth not dense, consisting of stout hyphae with more slender ones intermingled, the largest up to 100μ thick at base, the tips pointed. Sporangia long, slender, usually more or less fusiform (one of about average size measured 22 $\times$ 370μ); emptying normally, the spores furnished with flagella as they emerge (Humphrey), 10.5μ thick. Gemmae very few, not peculiar, elongated and formed by segmenting hyphae, single or two or three in a row. Oogonia numerous, racemosely borne from the base to the tip of main hyphae on short stalks which are usually straight and much shorter than the diameter of the oogonia (rarely oogonial stalks may be several times longer than diameter of oogonia); not rarely apical on main threads, no intercalary ones seen (rarely intercalary, Humphrey); spherical, rarely distorted, 40–90μ, most about 50–60μ thick; walls hyalin, rather thin, pits numerous and obvious. Eggs varying little in size, 18.5–25μ, the great majority about 22μ thick, rarely a very small one about half size occurs with the normal ones, three to thirty or even more, usually six to twelve, in an oogonium, eccentric. Antheridial branches androgynous, occasionally one from an adjoining strand, one or two, seldom more on each oogonium; arising from the main hyphae near the oogonia or rarely from the oogonial stalk. Antheridia elongated and closely applied to the oogonia, antheridial tubes developed and clearly visible.

From soil: United States: Florida (158), North Carolina (30) (149)

2. *Achlya racemosa* Hildebrand

Hyphae stout, usually 25–36μ thick at base. Sporangia long, almost cylindrical, rounded or tapering at the tips, about the size of the hyphae bearing them or sometimes slightly larger, sometimes twisted like a corkscrew. Spores 9–11μ in diameter; on emerging forming an irregular cluster or imperfect sphere which slowly expands as if embedded in jelly so that the spores become more or less separated singly or in groups. Gemmae usually few, formed by the distal parts of hyphae becoming divided into joints after being densely filled with protoplasm. Oogonia racemosely borne on short lateral branches, rarely intercalary, plentifully developed in all cultures, rather small, 40–70μ in diameter; wall distinctly yellowish at maturity, smooth and unpitted except where antheridia touch. Eggs variable in size, 16.6–27.7μ in diameter, most about 22μ, centric, one to eight in an oogonium (Humphrey says one to ten), in most cases two to five, centric, the wall thick (about 3.5μ). Antheridial branches short, arising from oogonial branches near the basal walls of the oogonia, or as often from the neck-shaped base of the oogonium or even from its curved surface, rarely from the main hyphae. Antheridia one or two, sometimes more, to each oogonium, short-clavate, usually bent and applied by their tips to the oogonia.

From soil: Australia (31), Germany (112)

United States: North Carolina (30) (58) (149)

3. *Achlya debaryana* Humphrey

Turf up to 1.5 cm. broad, of coarse stiff hyphae, 100–150μ in diameter, ending vegetatively or by sporangia. Sporangia terminal, usually cylindric to spindle-shaped, generally of the same diameter as the sporangiophores, sometimes up to 300μ long and about 50μ in diameter. Secondary sporangia produced by lateral budding. Oogonia on delicate, 8–14μ in diameter, botryose lateral hyphae, about one to three times the length of the diameter of the oogonia. Oogonia globose, often with a short cylindric protrusion in the stalk, and not pitted, smooth, or according to de Bary infrequently with some wart-like papillae and some thinner, circular places on the wall; 45–65μ in diameter. Antheridia always present, relatively large, usually many on each oogonium, cylindric to clubbed, on thin secondary branches which arise in groups of one to four in the neighborhood of the oogonium from the primary hypha, according to de Bary not from the oogonial stalk. Secondary branches much curved, usually richly branched, often bent toward the oogonium and intermingled with the antheridia; one to two fertilization tubes from each antheridium. Oospores usually numerous, three to ten and more, seldom only one to two, globose, eccentric; germinating by tube or as sporangia, 18–25μ in diameter.

From soil: Germany (112)

4. *Achlya subterranea* Coker and Braxton

Growth fairly dense on a hemp seed or bit of boiled corn grain, reaching a length of 9 mm. Main hyphae up to 92.5μ at base, gradually tapering toward the end, generally about 18–35μ on hemp seed; growing tips smooth, hyalin. Sporangia abundant, primary ones long, up to about 625μ, only slightly thicker than the hyphae bearing them. Spores generally discharged through one or more openings in the sporangium, biflagellate on emerging and often as many as half of them swimming away immediately and behaving as in Saprolegnia, the remaining ones coming to rest at once, forming an irregular, loose, spherical cluster at the mouth of the sporangium and behaving as in Achlya, but often several spores, occasionally as many as half, are left in the sporangium; encysted spores 10.5–13μ in diameter, generally from 11–12μ; the sporangial wall collapses somewhat immediately after the spores escape. Gemmae very abundant in all cultures after a few days, formed by partitions in the thread below the sporangia, oval to elongate or clavate, up to about sixteen in a straight row, often partially or completely breaking apart from each other at maturity, part of them soon forming and liberating spores as in Achlya (biflagellate during emergence), or more often producing one or two long slender hyphae which bear small sporangia at the end; the spores escaping mostly as in Achlya, though very often as in Dictyuchus and sometimes as in Thraustotheca. Oogonia abundant in all cultures after about two weeks, spherical or sometimes irregular, borne on short lateral stalks from the main hyphae, the stalks usually one-half to three-fourths as long as the diameter of the oogonium, rarely intercalary, from 53–82μ in diameter, generally about 58–68μ; the fairly thick walls producing from two to several very short, blunt outgrowths up to 5.5μ long, which occur mostly under the antheridia; walls unpitted except under the antheridia, an outgrowth under each antheridium. Eggs brown when mature, two to eight to the oogonium, generally four to six, spherical, sometimes elliptic from pressure, almost or completely filling the oogonium, from 26–33μ in diameter, usually 28–31μ, eccentric, with a single, large oil drop on one side. Antheridia one to several on all oogonia, small and apically attached, androgynous or diclinous, probably more androgynous, with the antheridial branch borne near or somewhat distant from the oogonium with several finger-like antheridia. Antheridial tubes sometimes visible, though indistinct.

From soil: United States: North Carolina (27) (30), Oklahoma (61)

5. *Achlya conspicua* Coker

Hyphae long and more stout than in most Achlyas, up to 166μ thick near the base or some as small as 30μ, the tips often withering and the hyphae extended from a bud below as in *A. imperfecta*, etc. Sporangia abundant, secondary ones plentiful, varying from short and slender to very long and slender, or rarely stocky when short, 18–60 × 105–550μ. Spores emptying and behaving as typical in Achlya, 10.5μ thick. Oogonia not abundant, borne laterally from the main hyphae, their stalks of moderate length, varying from about as long to twice as long as the diameter of the oogonia, or not rarely even longer; oogonia spherical or rarely oval, 51–118μ thick most about 70μ, the walls yellowish, not thick, often strongly pitted, the pits varying much in number and about 5.5μ wide. Eggs three to thirty or more, usually four to ten, with a diameter of 22–29μ, most about 25μ, not filling the oogonium as a rule; rarely maturing and of obscure structure, apparently about like those of *A. apiculata* when in normal condition, but nearly always degenerating immediately and becoming irregularly filled with large oil drops. Antheridial branches androgynous or less often diclinous, usually simple, arising near the oogonia from the main hyphae or often from the oogonial stalks, usually one or two, rarely more, for each oogonium. Antheridia on all oogonia, cylindrical or long-tuberous, usually touching the oogonia by foot-like projections at times applied by the entire side; antheridial tubes obvious. Gemmae not peculiar, long, often in rows by the abstriction of the longer threads, frequently with prongs, emptying as sporangia under suitable conditions.

From soil: Germany (112), Mexico (155)

United States: Florida (158), North Carolina (30)

6. *Achlya polyandra* Hildebrand

Hyphae stout, long. Zoosporangia often not abundant, secondary ones rare, nearly cylindrical. Oogonial branches usually very long and often recurved at the tip, racemose. Oogonia terminal, globular, with smooth and unpitted walls. Antheridial branches arising chiefly from the oogonial branches not from the oogonia, often branched. Antheridia one to several on each oogonium, short-clavate. Oospores five to twenty-five, usually ten to fifteen, in an oogonium, centric, their average diameter 27μ.

From soil: Germany (112), Wales (64)

7. *Achlya orion* Coker and Couch

Hyphal threads long, reaching a length of 1.5 cm. on house flies, more slender than in most species of Achlya, from 10–40μ thick close to base, rarely up to 85μ thick, often wavy; usually little branched and pointed at tips when young; becoming considerably branched with age. Sporangia abundant, cylindrical, usually borne singly on the tips of the main hyphae in young cultures, renewed by cymose branching, often forming several clusters at regular intervals on the same hyphae, irregular and wavy in old cultures, 12–37 $\times$ 36–600μ (rarely to 900μ). Spores 9–10μ thick, emerging as usually in Achlya, but often falling to the bottom in an open group instead of forming a sphere at the sporangium mouth. Oogonia abundant on flies, grubs, and vegetable media, spread over the entire culture from the bases of hyphae to tips, giving the culture a lacy interwoven or network appearance; the diameter 30–60μ, commonly 32–48μ; usually borne singly on long, crooked, recurved stalks which arise racemosely from main hyphae and which vary in length from two to ten times the diameter of the oogonia; often oogonial stalks may branch bearing two oogonia, and rarely oogonia may be borne on a stalk which arises directly from another oogonial wall; very rarely intercalary; oogonial wall usually without pits (except where antheridial tubes enter) when grown on flies or grubs, but as a rule with pits when grown on boiled corn. Eggs one to eight, usually one or two in each oogonium, 25–45μ in diameter, most 33–36μ, eccentric when ripe, with one large oil drop; usually spherical, but often elliptical from pressure. Antheridial branches almost always androgynous, usually arising from the oogonial stalk itself, less often from the main hypha; rarely diclinous. Antheridia on about 75 per cent of the oogonia, one or two on an oogonium, tuberous; antheridial tubes obvious, penetrating the oogonia and reaching the eggs.

From soil: Germany (112)

United States: North Carolina (27) (30)

8. *Achlya megasperma* Humphrey

Mycelium more slender than in most species of Achlya. Sporangia abundant, of the typical Achlya type, borne singly or in clusters (often as many as eight) on the ends of hyphae, varying much in shape from the long, slender, tapering sporangia of *A. apiculata* to a club-shaped form swollen at the distal end; 100–1,000μ long, most between 300 and 400μ. Spores 11μ in diameter. Gemmae developed in considerable abundance, either single and shaped like a sporangium with pointed tip or very elaborately branched; when solitary often separating from the hyphae and falling to the bottom. Oogonia racemosely borne on branches which are about as long as or shorter than the thickness of the oogonia; rarely the oogonial branches may be longer. Oogonia without an apiculus, usually spherical, occasionally oblong, rarely cylindrical, 60–119μ thick, usually between 70–80μ, oogonial wall thickened and without pits except for thin places under the antheridia. Eggs one to ten or rarely more, usually two to five, almost or entirely filling the oogonium, often elliptic from pressure; 39–66μ thick, usually between 42 and 52μ (in an oogonium in which there were nine eggs the average size was 44.1μ); structure subcentric and exactly as in *A. apiculata*; walls 3–4.6μ thick. Antheridial branches diclinous or androgynous but never arising from the oogonial stalk, usually diclinous, often much branched and not applied to oogonia; long and very slender, becoming barely visible after the eggs are formed. Antheridia tuberous and fairly conspicuous; usually one or two on each oogonium; not rarely absent.

From soil: United States: North Carolina (30) (149)

9. *Achlya proliferoides* Coker

Growth moderately dense and strong, reaching a length of about 1 cm. on a mushroom grub. Hyphae moderately branched, variable in size, usually wavy and irregular, the tips hyalin and dying back here and there as in *A. imperfecta* and *A. flagellata*. Sporangia subcylindrical, usually bent, often with several openings; about 35–45μ thick as a rule, short or long, at times up to 1,425μ long. Spores 11–12μ thick, double ones not rare, often falling to the bottom in an open group on emerging. Oogonia abundant, spherical, smooth, 40–55μ in diameter, racemosely borne on stalks that are about one to one and two-thirds times as long as the diameter of the oogonia; wall hyalin, not thick; pits numerous (usually), but not very conspicuous. Eggs, eccentric, with a large oil drop, about 18–24μ in diameter, often elliptic, the great majority always going to pieces before maturity on ordinary media. Antheridial branches numerous, diclinous (mostly) or androgynous, usually long, contorted and much-branched, in many cases coiling themselves about certain selected hyphae which may or may not bear oogonia. Antheridia, one or several, on every oogonium, elongated, applying their sides to the oogonium or touching it by several blunt, foot-like processes.

From soil: United States: North Carolina (30) (117)

10. *Achlya flagellata* Coker

Growth stout and moderately dense, reaching a length of about 1 cm. on a mushroom grub or ant larva. Hyphae branching, tapering outward, up to 150μ thick near the base, more or less crowded and uneven, the tips hyalin and often dying and renewed from one side below as in all members of this group. Sporangia plentiful, subcylindrical, very variable in size, often bent and at times with more than one opening, scattered or clustered. Spores often falling to the bottom in an open cluster on emerging, about 11–11.5μ thick. Gemmae abundant, usually in rows from the segmentation of the distal parts of hyphae, short or long, usually more or less cylindrical, but often pear-shaped or tenpin-shaped or at times very irregular; usually becoming sporangia on change of medium and discharging through an elongated papilla at either end. Oogonia abundant, typically spherical, but not rarely irregular by abnormal growth on one side, and one or two papillate projections may be seen rarely; usually about 48–75μ thick, rarely up to 100μ, racemosely borne on short, slender stalks about as long usually as the diameter of the oogonia or a little shorter, rarely on longer stalks and quite rarely intercalary; wall hyalin, not thick (about 1.5μ); pits very variable, perhaps more often absent, but again numerous and rather easily seen, about 5.5μ wide. Eggs spherical, eccentric with a large oil drop, one to ten (rarely twenty) in an oogonium, mostly two to six, diameter 26–35μ, most about 28μ, rarely small ones down to 18μ may be mixed with the others. Antheridial branches abundant, usually much branched and irregular, often so much so as to make an intricate network like a group of rhizoids, originating laterally and apically from hyphae which may or may not bear oogonia and applying themselves to oogonia on the same or on other threads or to both; more often diclinous than androgynous, perhaps about three times as often usually, but varying in this respect; the antheridial branches never arising from the stalks of the oogonia. Antheridia on nearly all oogonia, one or several, elongated with the side on the oogonium, frequently touching the oogonium with foot-like projections; antheridial tubes easily observed.

From soil: Wales (64)

United States: Florida (158), Mississippi (61), New York (61), North Carolina (27) (30) (58) (61) (149), Oklahoma (61), Wisconsin (61)

11. *Achlya bisexualis* Coker

Plant heterothallic; growth vigorous on ordinary media, reaching a centimeter or more in a week on hemp seed or corn grain. Hyphae rather stout, about as in *A. flagellata*, primary sporangia long, pointed, commonly about 30–60 × 300–950μ; spores 9.6–10.8μ, behavior as usual for the genus. Gemmae typically pyriform to flask-shaped, often nearly spherical, some elongated, most 40–100μ thick; after a rest sprouting a tube (often long) and liberating spores directly or forming sporangia at the ends of the tubes. Oogonia borne on rather long (rarely short) laterals or apically on main threads, prevailingly spherical, not rarely oblong, about 50–80μ thick; the wall thin and apparently pitted only where the antheridia touch. Eggs two to ten, about 24–30μ thick, eccentric with one large oil drop, only rarely maturing. Antheridial branches long, much branched, abundant, producing a mass of filamentous, branched antheridia, which as a rule almost or quite cover the oogonial surface; no fertilization tubes yet seen.

From soil: United States: North Carolina (26) (30)

12. *Achlya heteromorpha* Harvey

Strain "A": Mycelium up to 35–37 mm. (diameter growth) within ten days; hyphae prominent to the naked eye, dense and white; generally about 50μ thick, although they may reach 100μ or more, and of practically the same diameter throughout; straight or crooked, branched. Sporangia appearing on the third day and very abundant within one week, terminating practically all hyphae, also subterminal or intercalary; more or less cylindrical, or broader at the middle or below the middle, and gracefully tapered distally; seldom with lateral branches, such branches when present arising commonly from subterminal or intercalary sporangia; up to 960μ long, and 56–104μ at the greatest diameter; hyphae renewed by sympodial branching, often in such a manner as to produce clusters of three or four sporangia. Spores discharged through apical or lateral papillae, forming a hollow cluster at the point of discharge, these clusters seldom becoming detached even in old cultures except when disturbed, and the discharged spores uncommonly disintegrating in place; at rest 10.8–12.6μ. Gemmae plentiful within ten days; for the most part long and more or less cylindrical, or they may be somewhat tenpin- or flask-shaped, and in some cultures spherical, single or in filaments up to twelve for the elongated ones, and the spherical ones single or in chains of two or three, dense in young cultures, freely breaking apart in old cultures; the more cylindrical ones measuring up to 360μ long and generally 50–60μ in diameter, the spherical ones about 200μ, suggesting immature oogonia. No oogonia produced to date, unless crossed with strain "B."

Strain "B": Mycelium up to 30 mm. (diameter growth) from hemp seed substratum within ten days; individual hyphae very large, white and prominent (more so than in strain "A"), 30–100μ in diameter, gracefully tapered and pointed at outer ends, and freely branched, especially in the outer half of the colony. Sporangia scarce, the terminal ones, when present, tapered distally, and intercalary ones long and slender, one of the latter type measuring 1,600 × 45μ. A few free spores seen, swimming or floating, and some germinating in water culture. Gemmae abundant, borne singly and terminally or produced in a row by hyphal segmentation, often tapered when terminal, otherwise more or less cylindrical, commonly with rounded ends, and frequently with laterally placed terminal, germ-tube-like appendages, or otherwise branched, often partially separating and maturing, black; some measuring more than 1,000μ in length, others much shorter. No sexual reproduction occurring to date, unless crossed with strain "A."

In mating "A" and "B" strains: Oogonia plentiful within three to five days on both corn meal agar and halved hemp seed in water; spherical, oval or pyriform, terminating lateral branches, or filamentous and intercalary; spherical oogonia commonly 54–68.4μ in diameter, seldom larger, intercalary ones up to 195μ in length and of the diameter of the hyphae bearing them, 18–30μ; walls unpitted, except where antheridia touch, smooth, about 1.3μ thick, and with a greenish tinge. Oogonial stalks generally longer than the diameter of the oogonia, though sometimes shorter, frequently reaching more than 1,000μ; as a rule very narrow basally, 9–18μ across, and commonly broadened distally, to 27–40μ, at point of attachment of oogonium. Oospores commonly three to nine, sometimes more, up to twenty-one in one case, spherical or sometimes oval to elongate when produced in a row in the filamentous oogonia, otherwise not completely filling the oogonia, eccentric; 18–23.4μ, seldom smaller or larger, one elongated oospore measuring 25.2 $\times$ 12.6μ; walls about 1μ thick. Antheridia very numerous on all oogonia, with their stalks frequently greatly coiled about oogonia and hyphae, there being many greatly branched and gnarled antheridial hyphae approaching each oogonium; fertilization tubes not seen.

From soil: United States: California (62)

13. *Achlya imperfecta* Coker

Growth dense or rather open, not very long, many stout hyphae with more slender branches, tips hyalin, often dying and then a new growing point produced below. Sporangia plentiful, subcylindrical, little larger than the hyphae that bear them, not very long as a rule, often irregular and twisted. Spores about 10–11.5μ thick, dark, emerging as usual but often falling to the bottom in an open group instead of forming a sphere at the sporangium mouth. Gemmae formed by the segmentation of the hyphae, therefore mostly subcylindrical and in rows, but often ovate and frequently with knobs or projections at one or both ends. Any part of the culture may be segmented into gemmae even to parts of the antheridial branches. They often become loosened from each other in part, rather rarely completely separating and falling singly to the bottom. Oogonia usually abundant, spherical, 37–60μ thick, most about 40–45μ, borne racemosely on short stalks about one-half to one and one-half times as long as the diameter of the oogonia; wall without pits, or with several to numerous small, inconspicuous ones; from the basal wall a protuberance of varying length is present in many cases, and there are rarely present one or two papillate protuberances. Eggs eccentric, with a large oil drop, two to eight in an oogonium, commonly four to six, diameter 17–23μ, most about 19.5–20μ, often elliptic from pressure. The great majority of the eggs go to pieces before maturity. Antheridial branches androgynous or diclinous, variable in origin and length, usually branched and irregular, arising from hyphae that also bear oogonia and then apply themselves to nearby oogonia or most often by extensive growth and then branch to more distant oogonia either on the same or other hyphae; or certain threads may give rise to antheridial branches only, which then seek out oogonia on other threads.

From soil: Denmark (80), Germany (112), Jamaica (32), Victoria (31), Wales (64)

United States: Kentucky (61), New York (32), North Carolina (30) (149)

14. *Achlya prolifera* (Nees) de Bary

Main threads stout, ending in primary sporangia, under which the secondary are formed sympodially. Oogonia racemosely arranged on short side branches of the main hyphae, as a rule terminal, globose, the wall with numerous, very sharply defined and obvious pits, 40–60μ in diameter. Eggs variable in number, usually six to ten, eccentric, 20–26μ in diameter. Antheridial branches diclinous, much twisted and branched, winding like a parasite about the oogonia and the threads that bear them; the oogonial walls thickly enwrapped and often completely covered by these branches which bear numerous, at times intercalary, antheridia, which lay their sides against the oogonium and send out fertilization tubes.

From soil: United States: North Carolina (149)

15. *Achlya caroliniana* Coker

Hyphae rather stout, about 48μ thick at the base and 20μ near the tip, in strong cultures reaching a length of 1.5 cm. Sporangia irregularly cylindrical, about 20–30μ in diameter, often discharging by several openings, sometimes remaining closed and emptying as in Dictyuchus, flagellated on emerging but behaving as in other species. Spores 11–12μ in diameter, most about 11.2μ. Oogonia abundant, very small, 24–55μ thick, most about 30–37μ, spherical when terminal; wall smooth, or not rarely with one or two papillae or angles, thin, not pitted, light yellow in age; terminating short or moderately long, slender branches, which are racemosely borne on the strong main hyphae, or rather rarely intercalary and elongated, at times filiform with several elongated eggs in a row. Oogonial branches generally simple, but often giving off near the base, or sometimes near the oogonia, one or two branches which also terminate in oogonia, and, as a rule, are curved downward. Eggs generally one to two, not rarely four, eccentric, with a large oil globule, 18.5–23μ in diameter, averaging about 22μ, often elongated by pressure. Antheridia absent. A papilla, thick-walled and soon empty, often grows into the oogonium through the basal partition exactly as in other members of the Prolifera group and in *A. hypogyna*.

From soil: Denmark (80), Jamaica (32), Wales (64)

United States: New York (32), Mississippi (61), North Carolina (27) (30) (58) (61)

16. *Achlya abortiva* Coker

Growth fairly dense on a bit of boiled corn, hemp seed, or jimson weed seed, reaching a length of about 1.3 cm. Main hyphae up to 128μ thick at base, gradually tapering toward the end, averaging about 18–30μ throughout the culture; growing tips smooth, hyalin. Sporangia generally abundant, long, up to about 525μ, slightly thicker than the hyphae bearing them, but usually tapering somewhat toward the tips, often bent, sometimes branched, many of them discharging spores through several openings. Spores 10.5–12μ in diameter, generally behaving as in Achlya, forming a sphere at the mouth of the sporangium or falling to the bottom of the culture in a group; several spores are sometimes left to encyst within the sporangium. Gemmae abundant in all cultures after a few days, generally oval to elongate or clavate, often spherical, or less often long rod-shaped, formed by the segmentation of the hyphae, usually borne several in a row, often up to twelve, up to about 75μ thick, and the long ones up to about 300μ long. Oogonia not formed except at low temperatures, fairly abundant in most cultures kept in an ice-box, appearing in a few days on mature cultures and after seven to eight days on young cultures, predominantly (about 75 per cent) oblong and usually with a long neck, walls unpitted, completely covered with many irregular, short or long, blunt outgrowths up to 23μ long; diameters, including the warts, 37–66 $\times$ 46–81μ (not including the neck); borne singly on the tips of slender branches from the main hyphae; branches from one to three times as long as the diameter of the oogonia and generally collapsing somewhat soon after the eggs are formed. Eggs one to three to the oogonium, generally single, 24–46.5μ in diameter, mostly 30–36μ, spherical to subspherical, often elliptic, not filling the oogonium, eccentric, with a single large oil drop at one side when mature (over 95 per cent go to pieces before maturity). No antheridia are produced.

From soil: United States: North Carolina (17) (27) (30)

16a. *Achlya abortiva* forma *normalis* Coker

Growth not very dense, reaching a length of 8 mm. on hemp seed or bits of boiled corn. Main hyphae slender, up to 115μ at the base, averaging 12–24μ throughout the central portion of growth. Sporangia terminating the main hyphae, not abundant, soon disappearing, sparingly reproduced by cymose branching, up to 720μ long, usually 270–550μ, slender, occasionally containing as few as two rows of spores, broadest near the distal end. Spores 11.5–12.5μ in diameter, behaving as usual for Achlya. Gemmae fairly abundant in old cultures, varying in shape and size, rod-shaped or clavate to oval, up to six or more in a row. Oogonia abundant in most cultures, especially on corn, terminating the main hyphae or borne on short lateral branches usually one-half to two times as long as the diameter of the oogonia, two or more often borne on a lateral stalk, usually spherical, often somewhat irregular, 30–98μ in diameter, mostly 55–75μ. Oogonial walls about 3μ thick, with numerous conspicuous pits, often giving the wall a wavy appearance. Eggs one to five to an oogonium, usually two to four, 21–45μ in diameter, averaging 31–36μ with walls 3–3.3μ thick, eccentric, with a large oil drop at one side of the egg, the great majority of eggs reaching normal maturity. Antheridial branches long and slender, always diclinous, soon disappearing; antheridia one or more on most of the oogonia, becoming less noticeable as the culture ages because of the early disintegration of the antheridial branches.

From soil: United States: North Carolina (26)

16b. *Achlya abortiva*, contorted form

Growth on hemp seed very short, dense, and slow, much like Geolegnia, reaching a length of 4 mm. Hyphae short and irregular, gnarled, much branched. No sporangia noticed. Gemmae rough or irregular, spherical or clavate. Oogonia rather numerous, borne on the tips of main hyphae or on short lateral branches, very irregular in shape, usually somewhat spherical, with numerous outgrowths; from 40–120μ in diameter, walls fairly thick, unpitted. Eggs one to five, usually one to three to the oogonium, spherical, often made somewhat irregular by oogonial pressure, 32–45μ in diameter, eccentric, many of them going to pieces before maturity. No normal antheridia seen.

From soil: United States: North Carolina (26)

17. *Achlya inflata* Coker

Vegetative growth fairly dense on boiled hemp seed or morning glory seed, reaching a length of 14 mm., growth very poor on corn. Main hyphae not large, up to 75μ at base, averaging 15–25μ throughout the culture; growing tips smooth, hyalin. Sporangia not very abundant, found rarely in old cultures, secondary ones not usually formed; about the same width as the hyphae bearing them and up to 275μ long. Spores 10.5–11.5μ in diameter, forming an irregular group at the mouth of the sporangia and behaving as a typical Achlya. Gemmae found in all older cultures, though not very abundant, generally long-clavate to spherical, often up to five or six in a row. Oogonia abundant in all cultures, appearing after two or three days, though the contents of many of them disintegrate before forming eggs; large globular, with an inflated appearance, 78–176μ in diameter, averaging 120–150μ, terminating long lateral stalks from the main hyphae, the stalks two to three times as long as the diameter of the large oogonia, sometimes shorter. Oogonial wall with several rather conspicuous pits, rarely with one or two blunt wart-like outgrowths near the base. Eggs rather numerous, from two to twenty to the oogonium, averaging seven to twelve, rarely filling even half the oogonium, 29–37μ in diameter, averaging 31–35μ, eccentric with a large oil drop at one side, the majority disorganizing before maturity. Antheridial branches long, slender, always diclinous, not traceable even remotely to hyphae-bearing oogonia, touching each oogonium with one to several somewhat tuberous antheridia, soon disappearing. Antheridial tubes sometimes visible.

From soil: United States: North Carolina (26) (30)

18. *Achlya apiculata* de Bary

Vegetative growth ample and abundant, but not so stout as in *A. oblongata* or in the Prolifera group. The main filaments mostly about 40–60μ thick, tips rounded; breaking up soon after maturity into segments with little or no change in the appearance of the threads, each segment becoming a gemma and resting indefinitely until the conditions change, then forming spores like sporangia. Sporangia moderately plentiful, long or short, usually somewhat larger than the threads and gradually pointed towards the end, emptying as usual for an Achlya, or often remaining closed and emptying as in Dictyuchus. Spores flagellated on emerging and capable of swimming under certain conditions, 12.5–14.5μ in diameter or at times larger. Oogonia not formed regularly or abundantly except at low temperatures, racemosely borne on the tips of short or rather long branches which are usually bent and sometimes make a complete turn, rarely intercalary, ovate, short pyriform or spherical, at low temperatures very rarely formed within empty sporangia (as in *Saprolegnia ferax*), typically with (but often without) a more or less prominent apiculus; 60–119μ thick, most about 80μ; walls thin, smooth, unpitted. Eggs few, large, very dark, subcentric, one to five usually two or three (rarely ten), 25–40μ thick, sometimes larger, average about 36μ. Antheridial branches usually androgynous, but often diclinous, arising from the main hyphae or from the oogonial branches, soon becoming inconspicuous. Antheridia small, tuberous, or cylindrical, usually one or more to each oogonium.

From soil: Australia (31), Germany (112), Wales (64)
United States: North Carolina (30) (149)

19. *Achlya recurva* Cornu

Growth extending about 1 cm. from the substratum, with strong main threads about 90μ thick at base. Sporangia long, cylindric or slightly spindleform; secondary ones few. Oogonia numerous, borne terminally on certain little-branched, often slender main threads or on more or less elongated, at times very long and branched, side branches which are always very slender and bent like a bow, and which spring from the sporangia-bearing main hyphae. Oogonia spherical, rarely elongated by an extension of the tip, and covered with many crowded, blunt, hollow projections; diameter of oogonia 50–90μ with the spines, the latter 7–11μ long. Antheridia cylindrical to clavate, small on slender branches, which are little or not at all branched and also not looped, but mostly only bent like a bow, and which are mostly only one to three to an oogonium, and are borne in part from the stalk of the oogonium or its main thread or in part from other threads. Eggs spherical, one to twenty-five, mostly about ten, filling the oogonium, 22–27μ thick.

From soil: United States: North Carolina (30) (149)

20. *Achlya rodrigueziana* F. T. Wolf

Growth on hemp seed rather dense, reaching a diameter of about 2–2.5 cm. Main hyphae about 40–50μ in width at the base. Sporangia abundant, renewed by cymose branching from below. Zoospores on discharge encysting to form a hollow sphere at the mouth of the sporangium; encysted zoospores about 10μ in diameter. Gemmae fairly abundant, rod-shaped, formed by segmentation of the hyphae. Plant homothallic. Oogonia spherical, abundant in older cultures, 30–50μ, averaging 42μ in diameter, borne on short lateral stalks from the main hyphae; wall of the oogonium smooth, hyalin, unpitted. Oospores one to four in an oogonium; about 50 per cent of the oogonia with a single oospore, 40 per cent with two oospores, 10 per cent with three oospores; four oospores very rare. Oospores 20–30μ in diameter, averaging 27μ, at maturity eccentric, with a single large oil droplet; oospore wall smooth, thick. Antheridia almost invariably diclinous in origin, very rarely androgynous; antheridial hyphae very slender and branching. Antheridia on a majority of the oogonia, one to three when present, rather long and tubular, irregularly swollen; antheridial tubes visible. Oospores in oogonia lacking antheridia maturing parthenogenetically.

From soil: Costa Rica (156)

21. *Achlya hypogyna* Coker and Pemberton

Hyphae slender, tapering gradually toward the apex, at base about 35μ in diameter, at or near tip about 8μ, in vigorous cultures reaching a length of 1 cm. Sporangia rather plentiful or few, nearly cylindrical, a little larger at the rounded and papillate distal end, usually curved, somewhat like those of *Protoachlya paradoxa*; dictyosporangia common, sometimes more abundant than the typical sort. Spores on emerging flagellated, a part usually dropping to the bottom and showing a little motion from the sluggish flagella. Gemmae at times abundant, again few, pyriform or flask-shaped, less often spherical, often in chains of two, three, or four; long, rod-shaped gemmae are also formed by segmentation of the hyphae. Oogonia generally borne on short branches, racemosely arranged on the main hyphae, but occasionally terminating a main hypha, and very rarely intercalary; globular or rarely oblong, the walls not pitted, more or less abundantly producing short or long rounded outgrowths, or a varying proportion smooth; yellow when old; diameter 26–83μ without the papillae which are up to 30μ long, the longest at times on the smallest oogonia. Eggs one to seven (commonly three to five), centric, diameter 20–36μ, averaging 27–28μ; not rarely elliptic and then up to 45 $\times$ 57μ. Antheridia cut off from oogonial branches just below the oogonia, very rarely absent; simple antheridial branches with one or more branched, tuberous antheridia also present at times and arising from the suboogonial cell or below it or even from the main hypha; in the latter case rarely diclinous. Fertilizing tubes arising through the common septa from the suboogonial cell and penetrating the oogonia from below (hypogynous), also from the other antheridia when present.

From soil: United States: North Carolina (58) (61)

22. *Achlya radiosa* Maurizio

Hyphae dense, about 1–1.5 cm. long, 14.5–49μ thick, thickened in places. Sporangia typical of the genus, cylindrical, sympodially arranged, often bearing an oogonial branch below them. Zoospores as usual. Oogonia typically racemosely borne, or also on main or secondary hyphae; spherical, with thorny, pointed warts over whole surface; wall yellowish, diameter without spines 31.5–46μ, with them 40–54.5μ. Spines 7–12μ long, 9.5–12μ thick at base. Eggs one, rarely two, or more rarely three in somewhat elongated oogonia, filling the oogonium. Egg membrane clear yellow; contents thick, with numerous large and small oil-drops; diameter 29–39μ, mostly 34–36.5μ. Antheridia on short bent stalks from the oogonial stalk or the main hyphae; club-shaped, present on most of the oogonia. Antheridial tubes nearly always present. In some cases sporangia halt in development and after a while drop off, and when brought into nourishing media sprout to hyphae. Typical gemmae not present.

From soil: Wales (64)

23. *Achlya glomerata* Coker

Hyphae rather stout, branched, not long. About 40–45μ in diameter at base and tapering to slender tips about 12μ in diameter. At maturity the main hyphae strongly incline to segment into elongated sections with dense protoplasm, but the slender apical section is apt to remain almost empty. Sporangia almost cylindrical, inclined to be somewhat irregular and often opening by a bent papilla. Oogonia abundant, approximately spherical, without pits; completely covered with short, blunt, irregular warts, 29–44μ thick, with the warts most about 33μ thick. Oogonia borne on the tips of very slender and delicate, but contorted lateral branches that are either simple, in which case there is but one oogonium, or more or less intricately branched, in which case there are a number of oogonia borne on the tips of the group of branches. Eggs single or very rarely two in an oogonium, eccentric, their diameter 15–23μ, averaging about 20μ. Antheridia absent from a good many oogonia, when present club-shaped; borne on the tips of branches from the same glomerulus and one or several on an oogonium.

From soil: United States: Mississippi (61), New York (32), North Carolina (30)

24. *Achlya colorata* Pringsheim

Hyphae stout, 25–50μ in diameter at base. Sporangia long, almost cylindrical, or slightly tapering toward the end, very little or not at all larger than the hyphae bearing them. Spores 11μ in diameter, emerging and behaving as in *A. racemosa*. In neither species is any spontaneous movement shown before encystment. Oogonia varying greatly in size, 41–90μ in diameter, rarely as much as 107μ, commonly 55–66μ, racemosely borne on short lateral branches and also at times on the tips of main branches; the yellow walls producing short, blunt outgrowths in varying number or rarely almost smooth. Eggs, mostly one to four, rather rarely five and very rarely six, 26–39μ in diameter, mostly about 30–37μ, centric, the wall very thick. Antheridial branches short, arising from the oogonial branches near the basal wall of the oogonium, and, as in the typical *A. racemosa*, often from the neck-shaped base of the oogonium itself, rarely from the main hyphae. Antheridia one to four on each oogonium, commonly two, short-clavate, usually bent and applying their tips to the oogonium. Gemmae formed at the maturity of the culture in large numbers. They are scarcely enlarged sections of hyphae arranged in rows of rarely over five, one end often projecting to one side below the partition and somewhat thickened. They do not form all the way to the substratum, but only near the ends of the hyphae. When brought into fresh water they sprout by tubes or become sporangia.

From soil: Germany (112)

United States: North Carolina (27) (30) (149)

25. *Achlya pinnulata* Harvey

Colony growth on halved hemp seed up to 20–22 mm. within a week and of limited growth thereafter, after one week becoming denser, matted, opaque, and pale, occasionally to almost black in old age. Sporangia more or less cylindrical, or broader basally or near the middle, early ones (after one day) commonly very small, 128 × 52μ, often quite long and narrow, 1,080 × 48μ in one case, thin-walled, and often greatly distorted and wrinkled after spores are released, not uncommonly emptying by several mouths or papillate branches, such branches often up to 160μ or more in length. Proliferation of new hyphae occurring as in other species, by sympodial (alternate) branching or very commonly by opposite or even occasionally by whorled branching, these in turn being likewise renewed; sporangia often produced in clusters. Spores dischargeable by one or several pores to congregate at points of discharge, many of them remaining permanently at those points while others may be diplanetic in their behavior, but frequently some or all spores remaining undischarged, these often germinating in place; discharged spores at rest, 10.8–12.6μ, undischarged ones occasionally up to 19.8μ. Gemmae formed by segmentation of main hyphae, at times not very numerous, similar to sporangia and generally in terminal position. Oogonia appearing by the second day, often so abundant within one week, along with the parent hyphae, antheridia, and antheridial stalks, as to obscure colony details; borne singly on lateral stalks or not infrequently two or three oogonia from the same stalk; for the most part spherical, often narrowed basally to diameter of the supporting stalk, or seldom comprising part of the main hypha, rarely irregularly lobed with spherical bulges in such a way as to suggest a cluster of grapefruit, with each lobe enclosing one or two oospores; spherical ones commonly 54–63μ, often larger; walls less than 2μ thick, seldom pitted except where antheridia touch. Oospores eccentric; numerous, four to six for the smaller oogonia and up to a dozen or more, seldom as many as thirty-five, for the larger oogonia; spherical or oval when compressed, mostly 21.6–23.6μ in diameter, sometimes larger. Antheridia commonly many to each oogonium, usually diclinous, but frequently androgynous from a distance, or antheridial stalks often arising from one oogonial stalk to be applied to an oogonium elsewhere, often at considerable distance.

From soil: United States: California (62)

(7) **Aphanomyces** de Bary

Hyphae very delicate, long, sparingly branched. Sporangia formed from unchanged hyphae, long to very long, not proliferating within old ones and rarely laterally from below. Spores borne in a single row, emerging apically in elongated form, then rounding up and encysting in a clump at the end of the sporangium as in Achlya, then emerging and swimming again as in that genus. Specialized gemmae absent. Oogonia terminal on short or long branches, smooth or warted, wall thin and unpitted. Antheridia diclinous or androgynous, not always present. Eggs single, not filling the oogonium, eccentric, with a single large oil drop in the protoplasm near one side or with a lunate disc of oil droplets on one side.

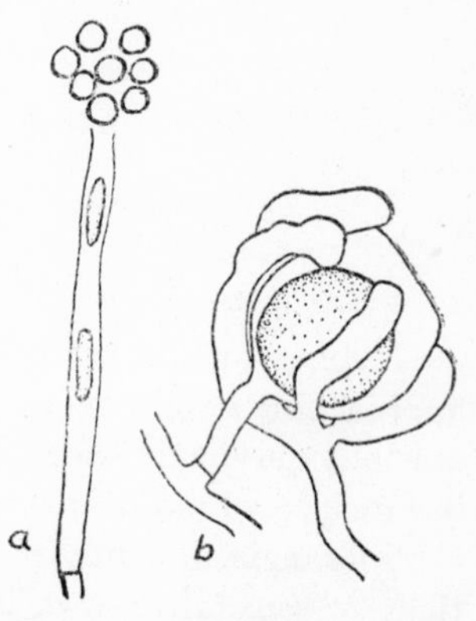

FIG. 29. **Aphanomyces.** *a* sporangium; *b* oogonium (after Coker).

KEY TO THE SPECIES OF THE GENUS APHANOMYCES

a. Oogonial walls smooth, not with spines or warts; eggs 16.5–26μ thick
  b. Oogonia with smooth thin walls — 1. *A. laevis*
  bb. Oogonia with thick wall, interiorly sinuous in contour — 2. *A. euteiches*
aa. Oogonial walls with distinct spines or papillae, oogonia 22–23μ (including papillae) — 3. *A. stellatus*
aaa. Oogonial walls uneven or tuberculate, not spiny; eggs 13–18.5μ — 4. *A. scaber*

1. *Aphanomyces laevis* de Bary

Hyphae saprobic or rarely parasitic on desmids and diatoms, slender, much branched, about 5–7.5μ thick. Sporangia long and of the same size as the hyphae, often extending to the substratum. Spores 7.3–11μ in diameter after emerging, most about 9–10μ, rod-shaped in the sporangium. Oogonia terminal on short lateral branches, globular or nearly so, with smooth thin walls without pits, 18–33μ in diameter. Eggs single, 16.5–26μ in diameter, mostly about 19–22μ, thick-walled, eccentric, with one very large oil drop enclosed in the protoplasm and very near the surface on one side. Antheridial branches very abundant, sometimes twining around the oogonial branches in a knot, androgynous or diclinous. Antheridia large, abundant on all oogonia and extensively wrapping them about; antheridial tubes developed and plainly visible.

From soil: Wales (64)

United States: Florida (158), New York (67), North Carolina (30) (58) (60)

2. *Aphanomyces euteiches* Drechsler

Hyphae hyalin, branching at moderate intervals (20–150μ) at angles approaching a right angle; 4–10μ in diameter, the individual filaments not abruptly varying in width; occurring in nature within cortical cells of host, in nutrient solutions as extensive nebulous translucent mycelia.

Sporangia in artificial culture arising by conversion of extensive portions of vegetative mycelium delimited by one or more septa; often including many ramifications; discharging through one or several (up to four) tapering branches, the distal portions of which measure usually approximately 4μ.

Zoospores cylindrical, in escaping from evacuation branches becoming attenuated to vermiform bodies, usually 3.5μ in diameter by 30–50μ in length; forming spherical cysts at mouth of sporangium, measuring usually 8–11μ in diameter, rarely up to 16μ; diplanetic, the empty spherical wall being distinguished by a protruding evacuation tube 1μ long by 2.5–3μ in diameter.

Oogonium generally, if not always, terminal on a short lateral branch, from which it is delimited by a partition sometimes present as a simple septum, at other times as a columella-like structure protruding into the oogonial cavity; subspherical, measuring usually 25–35μ in diameter; when mature exhibiting a heavy peripheral wall with smooth outer contour and sinuous inner contour, hence of irregular thickness, this dimension varying between 1–5μ, generally between 1–2.5μ.

Antheridia typically of diclinous origin, borne on a stalk frequently involved with the oogonial stalk, and often branching once or several times; measuring 8–10μ in diameter × 15–18μ in length, or when considerably larger often more conspicuously arched, somewhat lobulate, and becoming compound by the insertion of transverse septa.

Oospores subspherical or more rarely ellipsoidal owing to intruding columella-like septum; 18–25μ (generally 20–23μ) in diameter; provided with a wall of uniform thickness, this dimension varying between 1.2–1.8μ (generally 1.5μ); slightly eccentric in internal structure ("subcentric"); germinating without protracted resting period either directly by one to three germ hyphae or by production of a single unbranched sporangial filament usually 200–350μ in length, in the latter event producing generally thirteen to eighteen zoospores, approximately half of which are delimited within oospore wall.

From soil: United States: North Carolina (27) (30)

3. *Aphanomyces stellatus* de Bary

Hyphae straight, delicate, little branched, about 5.5–6.5μ in diameter, springing abundantly from the substratum, the tips rounded. Sporangia produced from the unchanged hyphae, very long, usually reaching to the substratum. Spores when in the sporangium irregularly rod-shaped with uneven ends, on escape becoming rounded and encysting in an irregular group at the mouth of the sporangium, diameter 8–8.5μ (at times a few larger double ones 11–12μ in diameter mixed with the others), emerging and swimming actively with the usual form, the large cysts giving rise to two spores of normal size. Oogonia subspherical, borne on rather long or short lateral branches, normally covered more or less densely with conspicuous blunt papillae up to 5.5μ long, diameter of the oogonia, including the papillae, about 22–33μ; walls rather thin, unpitted, cavity extending into the papillae. Eggs about 16–26μ thick, most about 18.5μ, single (rarely two-de Bary), contents eccentric when fully mature, with an inconspicuous lunate series of droplets on one side in optical section. Antheridial branches androgynous or also from neighboring threads, often branched. Antheridia short-tuberous, large, present on all or nearly all oogonia. Fertilization uncertain.

From soil: Germany (112)

United States: North Carolina (58) (61)

4. *Aphanomyces scaber* de Bary

Hyphae delicate, branching, about 5–7.5μ thick, rarely as small as 2.5μ. Sporangia like the hyphae, of indefinite length. Spores on encysting about 9.5μ thick; narrow and elongated in the sporangium. Oogonia terminal on short or moderately long branches, very small, 15–23.7μ in diameter, averaging about 21.5μ, surface uneven or varying to tuberculate, but projections never so prominent as in *A. stellatus*; wall thin, not pitted. Eggs single, 13-18.5μ in diameter, averaging about 13.3–15.5μ, eccentric, a single large oil drop near one side, protoplasm small in quantity and light in color, wall rather thick. Antheridia present on most of the oogonia, according to Fischer; not on all oogonia, according to Humphrey.

From soil: Wales (64)

United States: California (62), New York (32)

(8) **Calyptralegnia** Coker

Mycelium of the usual Achlya-like type. Spores encysting within the sporangium, therefore angular. Sporangium dehiscing by the breaking off of an apical segment, the spores escaping intermittently by the swelling of consecutive groups from above downward, afterwards emerging from their cysts and swimming as in Achlya. Eggs multiple, centric or subcentric with numerous small oil drops. Antheridia androgynous.

A single species treated.

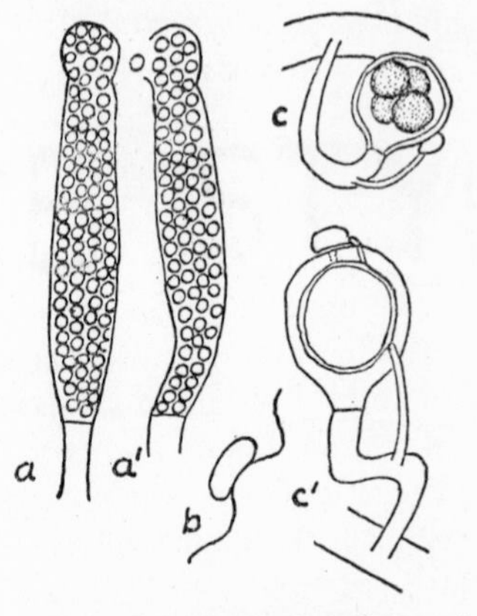

Fig. 30. **Calyptralegnia.** *a*, *a'* sporangia; *b* zoospore; *c*, *c'* oogonia (after Coker).

1. *Calytralegnia achlyoides* Coker and Couch
   Syn. *Thraustotheca achlyoides* Coker

Growth very vigorous but slow, the largest threads sometimes reaching a diameter of 150μ near the base, long, straight or sinuous, rarely or not at all branched. Sporangia formed as in Achlya or Saprolegnia, of equal or greater diameter than the threads which bear them, not tapering, but often of irregular thickness throughout their length, the tips rounded; the early sporangia straight or with slightly curved tips, the later ones almost invariably with recurved ends. Spores formed as in *Thraustotheca clavata*, Achlya, etc., but discharged by the breaking away of a considerable part of the end of the sporangium, caused by the swelling of an apical group of spores, after which the spores may emerge immediately, or may come to rest, to emerge several days later. Usually a few seconds after the cracking of the sporangium the spores of the tip ooze out in a group exactly as in Thraustotheca. The spores next below this apical group now swell, extending somewhat the truncated tip of the sporangium and after a few seconds begin to move out in their turn. This continues in a series of partial discharges involving a few layers of spores each time until in about five to ten minutes all the spores become loosened and most of them discharged from the sporangium tip where they are spread out in a loose irregular colony. A few spores are always left in the sporangia. The spores encyst in irregular, not spherical, forms before emerging, and are not connected by threads as in Achlya, but exhibit a distinct mutual attraction while emerging as shown for Dictyuchus. They slide over each other and shift their relative positions but always keep in contact with the emerging mass. Spores usually emerging from their cysts immediately upon discharge, some of them coming out of their cysts even while being pushed from the sporangium. The emergence from the cysts is much more rapid than in Achlya or Saprolegnia, occupying only about ten seconds. Gemmae not observed. Oogonia formed rarely under laboratory conditions, spherical or slightly oblong, 55–100μ thick with smooth walls, borne on lateral stalks which in length are from once to twice the diameter of the oogonia; oogonial stalks usually once coiled, not rarely straight. Eggs one to eight in an oogonium, 42–60μ thick, rarely up to 77μ thick, but when so large always single in the oogonium; often crowded and elliptical from pressure; structure as in *Achlya apiculata* with a central sphere of protoplasm surrounded by oil droplets; wall of the egg about 4μ thick. Antheridia apparently not always developed, but when visibly present quite often arising from the oogonial stalk, not rarely diclinous, one to several on an oogonium; antheridial tubes developed.

From soil: United States: North Carolina (26) (27) (28) (30)

(9) **Thraustotheca** Humphrey

Primary threads in greater part stout, branching. Sporangia clavate to subcylindrical, often irregular, proliferating from below as in Achlya. Spores always, or in the great majority, encysting within the sporangia when formed and later, in more or less angular form, swelling and escaping by the irregular rupture or disintegration of the sporangial wall; not escaping at once by an apical papilla except in the Achlya-like primary sporangia of one species. Oogonia with one to several eggs; antheridia present.

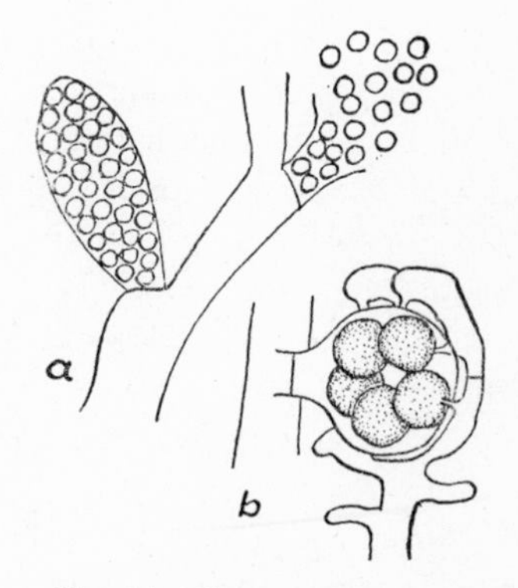

FIG. 31. **Thraustotheca.** *a* sporangia; *b* oogonia (after Coker).

KEY TO THE SPECIES OF THE GENUS THRAUSTOTHECA

a. Sporangia clavate, dehiscing laterally or apically, by fracture or disintegration of the sporangial wall — 1. *T. clavata*

aa. First sporangia shaped as in Achlya and spores escaping as in that genus; later sporangia long clavate to cylindrical and spores escaping by irregular rupture and disintegration of the sporangial wall

b. Oogonial wall showing projections — 2. *T. primoachlya*

bb. Oogonial wall smooth — 3. *T. irregularis*

1. *Thraustotheca clavata* (de Bary) Humphrey

Main hyphae stout, straight, reaching a length of 2 cm. in strong cultures and a thickness of 20–120μ, averaging about 37μ; profusely branching into secondary hyphae near their tips; secondary hyphae much curved and twisted, and often curiously knobbed and gnarled. Sporangia 37–85 $\times$ 66–370μ, terminal or rarely intercalary, proliferating as in Achlya, usually short, broad, and clavate, but often elongated somewhat as in Pythiopsis or even as in Saprolegnia, varying from nearly spherical to fusiform, differing from the sporangia of any other of the Saprolegniaceae. Spores about 12.5μ thick, encysting within the sporangium immediately after they are formed, and liberated passively and slowly by the gradual cracking and disintegration of the sporangium wall, which is probably due to internal pressure. They now emerge from their cysts and swim actively in a laterally biflagellate form, encyst again and sprout. In the sporangium they are polyhedral in shape, through pressure, each having a hyalin membrane of its own. Occasionally among the ordinary spores large irregular spore masses are liberated. These masses slowly round up somewhat and encyst, sprouting later without a swimming stage. Gemmae small, pyriform or rarely spherical, falling into spores in suitable environment. Oogonia borne singly on short, straight, perpendicular stalks from the secondary hyphae, rarely from the primaries; 30–70μ thick, spherical, smooth, and very slightly pitted, the pits appearing only after staining with chlor-zinc-iodide. Eggs one to ten or rarely eleven, usually four to six or eight, eccentric, with a single large peripheral oil globule; size very constant, the diameter about 18–22μ. Antheridial branches diclinous, arising from the secondary hyphae, very crooked, and quite stout; antheridia club-shaped, cut off by a wall; antheridial tubes obvious.

From soil: Australia (31), Germany (112), Jamaica (32), Wales (64)

United States: Kentucky (61), New York (32) (61), North Carolina (27) (30) (60), Wisconsin (61)

2. *Thraustotheca primoachlya* Coker and Couch

Growth fairly dense on mushroom grubs, termites and bits of boiled corn grain; hyphae 10–100μ thick near the base in the same culture, considerably branched, tips pointed and hyalin when growing. Sporangia produced in great abundance: the first ones borne on the ends of the main hyphae and usually of the Achlya type; subcylindrical, rather stout and regular to slender and irregular, the wall thin and delicate, and soon (in a day or two) disappearing in part or entirely after emptying; the spores discharging through an inconspicuous apical papilla and clustering at the tip, connected by threads as in Achlya while emerging; later sporangia borne singly or in large clusters (up to ten or more) on the ends of hyphae, irregular in form, usually bent-cylindrical, rarely short-clavate as in *T. clavata*, thickest at the distal end, more or less rounded at the tip or sides, or in both places, the spores swelling out by degrees and the sporangial wall in large part disappearing. Spores from the Achlya type of sporangia round when encysted, those from the Thraustotheca type angular; emerging and swimming with lateral flagella or sprouting in position. Gemmae not observed. Oogonia spherical or oval, borne on racemose branches, the length of which is from one to several times the diameter of the oogonia, 30–75μ thick, wall set with a few to a good many large, conspicuous, blunt projections 3–11μ high, and closed at the end by a very thin membrane. Eggs 16–23μ thick, one to sixteen in the oogonium, usually 4–8μ, eccentric, with a single, large, lateral oil-drop; germinating after about three weeks into small sporangia which usually grow out through the oogonial papillae. Antheridial branches androgynous, usually arising from the same stalk that bears the oogonia, often branched; antheridia small, inconspicuous, finger-like and laterally applied to the oogonia; soon entirely disappearing, as if dissolved away.

From soil: United States: North Carolina (29) (30), Oklahoma (26)

3. *Thraustotheca irregularis* Coker and Ward

Homothallic; growth dense and rapid on boiled hemp seed; hyphae often over 2 cm. long, 12–71μ thick near the base, rarely 128.5μ; mostly branching, seldom unbranched; tips mostly somewhat pointed and less granular than the remainder of the hyphae. Sporangia abundant; the first ones borne on the ends of the main hyphae, with few exceptions of the modified Achlya type, rather stout and regular to slender and irregular, subcylindrical or usually thicker near tip or center. Wall thin and delicate and eventually disappearing in part, more often entirely disappearing some time after emptying; the spores escaping through an inconspicuous apical papilla and encysting at the tip in the form of a hollow sphere, soon emerging and swimming away in the Achlya form. Encystment of most of the spores sometimes taking place within the sporangium, but a true dictyosporangium is never formed. Later sporangia of the Thraustotheca type borne singly or in large clusters up to four or five on the ends of hyphae, irregular in form, often curved or very crooked and very often forked; spore mass usually breaking away as a whole or in part; spores encysting within, spherical or subangular, the sporangial wall soon disappearing; spores leaving a false net after emerging. Gemmae usually abundant, spherical, pyriform, or flask-shaped, oval, clavate, or somewhat fusiform, intercalary or borne on the ends of hyphae, often in chains of two or three, sometimes as many as five; cylindrical gemmae also formed by the segmentation of the hyphae after sporangia have emptied, breaking off in some cases just as do the sporangia. Oogonia borne on short lateral stalks, the length of which may not equal the diameter of the oogonia, rarely on the ends of main hyphae or intercalary, smooth-walled, spherical to oval, very regular in size, 21.4–71.5μ, mostly about 50μ, wall rather thin, pitted or occasionally unpitted. Eggs 16.6–38.3μ thick; one to nine in an oogonium, usually three to six, eccentric with a single, large lateral oil drop when mature. Antheridia on all oogonia that reach maturity, diclinous, a large number of antheridial branches commonly found on each oogonium, twining about it; antheridia simple or branched and usually finger-like in shape, being applied to the oogonia by their sides or ends.

From soil: United States: North Carolina (149)

(10) **Dictyuchus** Leitgeb

Vegetative structure and appearance as in Achlya, but of much more tardy development in cultures (at least in *D. monosporus*); tips of hyphae rather blunt. Primary sporangia nearly cylindrical, blunt, borne typically in a zigzag sympodium with long internodes; later they are formed by the segmentation of the hyphae into long joints and in such case, after the spores are formed, rest like gemmae for a change of media before liberating the spores, and show a strong tendency to fall away from each other and from the hyphae and to lie free in the water. Spores not escaping from the sporangium as in other genera (except Aplanes), but remaining in the sporangium and forming there a network of walls from which they emerge, after a rest, by individual openings to

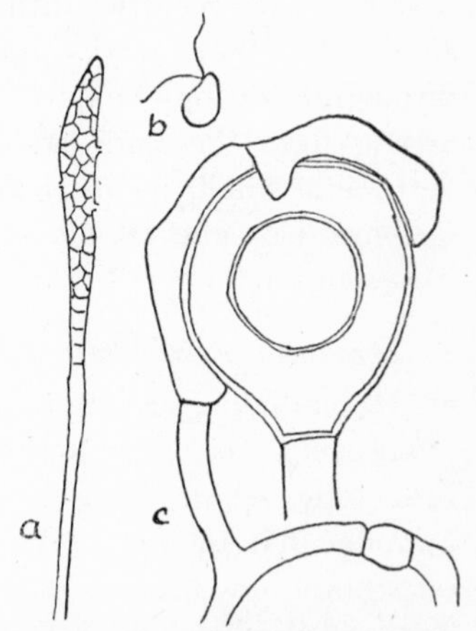

Fig. 32. **Dictyuchus.** *a* sporangium; *b* zoospore; *c* oogonium (after Coker).

the outside, where they swim by two flagella as in the form of the second swimming stage in Saprolegnia. They then, as a rule, sprout as usual in the family; some of them may again emerge and swim before sprouting. Gemmae not represented unless the resting sporangia with spores in them be considered such. Oogonia spherical, smooth, the wall unpitted, terminal on slender branches (absent (?) in *D. monosporus*). Antheridia much as in Saprolegnia and Achlya, diclinous or androgynous; fertilizing tubes observed. Eggs one to many, not filling the oogonium, at maturity containing one or a few large oil drops inside the protoplasm.

KEY TO THE SPECIES OF THE GENUS DICTYUCHUS

a. Sporangia of a true-net type (sporangial wall persisting) 1. *D. monosporus*
aa. Sporangia at some time of the false-net type
  b. Sporangia of two kinds, the first of the Achlya type, the later ones of the false-net type, oogonia papillate 2. *D. achlyoides*
  bb. Sporangia all of the false-net type; oogonia smooth
    c. Antheridia lacking 3. *D. missouriensis*
    cc. Antheridia usually present 4. *D. pseudodictyon:*

1. *Dictyuchus monosporus* Leitgeb
Syn. *D. sterile* Coker
D. magnusii Lindstedt

Vegetative growth moderately stout. Main hyphae branching, up to 55μ thick, mostly 30–45μ at base, very gradually tapering towards end, the larger up to 22–37μ near tip, many much smaller. Primary sporangia borne on the tips of hyphae, later ones formed by cymose branching, but usually separated from the earlier ones by some distance by the elongation of the threads. As the culture ages the arrangement becomes more irregular and complicated and most of the threads become segmented towards the periphery into numerous sporangia in rows or branched groups. They are usually a little larger in the distal half, often bent, sometimes branched, of various size, in old cultures often very long, not rarely thread-like with only a single row of spores. They usually break off from the hyphae about the time the outline of the spores becomes distinct and go into a resting state which may last a few days or many weeks depending on conditions. During this time the spores are separated by walls which in this condition are scarcely visible, the individuality of the spores being indicated by the usually conspicuous vacuole that each contains. On emerging the spores escape singly and swim as normal in the genus or they often sprout in position into slender hyphae. Spores 11.8–16.6μ in diameter before sprouting, with large conspicuous vacuole. Oogonia not developed.

From soil: United States: Florida (158)

2. *Dictyuchus achlyoides* Coker and Alexander

Growth moderately dense on boiled hemp seed, reaching a length of 6 mm. in five days. Main hyphae from 21–45μ thick at base, gradually tapering toward the end. Sporangia arising from tips of hyphae and by cymose branching; both Achlya and Dictyuchus types abundant, the Achlya type appearing first, the Dictyuchus ones later. In old cultures the latter strongly predominate. Sporangia frequently branched, occasionally borne in rows. Spores 10–16μ, mostly 12.5μ in diameter, the same size in both types of sporangia, occasionally sprouting in place. Oogonia plentiful, spherical, 25–37.5μ in diameter, borne singly on long, slender, usually curved stalks which arise from the main hyphae and vary in length from about two to four times the diameter of the oogonia. Oogonial walls unpitted, slightly yellow, provided with many papillae which vary in length from 3–11μ. Eggs eccentric, 20–27μ in diameter, mostly 25μ; one in each oogonium. Antheridia androgynous, usually arising from the oogonial stalk, occasionally from the main hyphae, appearing (singly) on about 65 per cent of the oogonia.

From soil: United States: North Carolina (26) (30)

3. *Dictyuchus missouriensis* Couch

Mycelium fairly vigorous, up to 2 cm. in diameter on hemp seed; hyphae up to 75μ thick near base, most about 50–60μ, more or less wavy throughout and zigzag in the distal half by formation of sporangia and renewed growth from beneath. Sporangia cylindric, thickest in the middle, 10–40 $\times$ 84–400μ, most 25–35 $\times$ 250–300μ, the wall thin, disappearing soon after the formation of the spores. Spores usually rounding up more or less before encysting within the sporangium, thus forming only an imperfect net, emerging from their cysts and swimming in laterally biflagellate form (by special treatment early formed sporangia may be induced to discharge as in Achlya), about 10μ thick when encysted. Gemmae very rare, spheric to elliptic. Oogonia abundant, spheric except for the basal elongation, 29–44μ in diameter, usually about 33μ, rarely 60μ, borne on rather thin lateral branches about two to three times as long as the diameter of the oogonium, the stalk usually bent at the base of the oogonium and the oogonium often joined to the stalk by a beak-like process from its base. Eggs single, 23–38μ in diameter, usually about 26μ, eccentric. Antheridia lacking.

From soil: United States: Missouri (30) (33)

4. *Dictyuchus pseudodictyon* Coker and Braxton

Vegetative growth moderately stout, reaching a length of 1 cm.;

hyphae up to 80μ thick at base, mostly 30–45μ, branching freely, the larger branches 18–30μ thick, many much smaller; primary sporangia borne on tips of the main hyphae, secondary ones formed by cymose branching, separated from earlier ones by some distance, in old cultures the arrangement becoming more irregular and complicated with many threads becoming segmented toward the periphery into sporangia in rows of two or three, or forming profusely branched groups. Sporangia usually a little larger in distal half, sometimes bent, very often branched, 12–30 $\times$ 100–830μ, the rows of encysted spores generally remaining attached to each other and to the hyphae until after the spores have escaped. Spores 12–15μ in diameter, mostly 13–14μ, with a large conspicuous vacuole, escaping as usual either before or after a rest depending on conditions, at times sprouting in position. Oogonia abundant, spherical or occasionally pyriform, 36–44μ in diameter, the pyriform ones up to 45.5 $\times$ 60μ, the wall smooth, rather thin, unpitted except where antheridia touch, borne on short lateral stalks from the main hyphae or from branches. Eggs one to an oogonium, not filling it, 29–34.5μ in diameter with a wall about 2.5μ thick, eccentric. Antheridia on 85–95 per cent of the oogonia, one to many to an oogonium, often almost entirely enwrapping it, finger-like or tuberiform, cut off by a cross-wall, borne on branches arising from tips of the main hyphae or from secondary branches, very irregular and tortuous, often much branched, of androgynous or diclinous origin.

From soil: United States: Missouri (33), North Carolina (30)

(11) **Brevilegnia** Coker and Couch

Mycelium depauperate, dense and opaque, never aerial. Sporangia in the great majority behaving about as in Thraustotheca, the wall soon disappearing (in some species sporangia of the Achlya type also occur.) Spores very variable in size and shape in the same culture, the larger ones multinucleate, encysting in position (except in the achlyoid sporangia) and only slowly separating after the disintegration of the sporangial wall; after encystment either emerging and swimming once or not swimming, depending on the species. Oogonia small, with a single eccentric egg. Antheridia present or wanting, prevailingly androgynous. Gemmae wanting in most species.

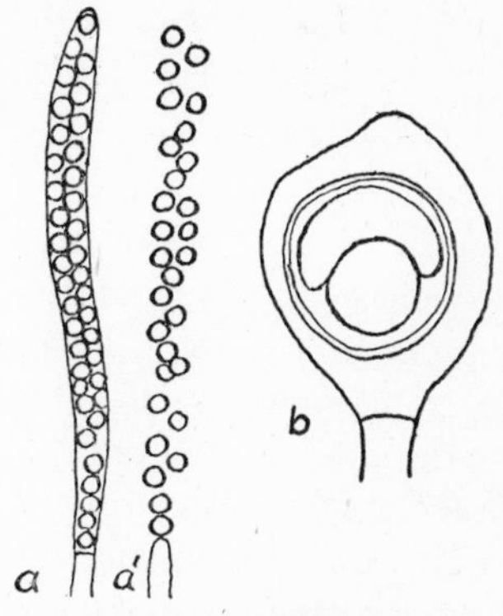

FIG. 33. **Brevilegnia.** *a, a'* sporangia; *b* oogonium (after Coker).

KEY TO THE SPECIES OF THE GENUS BREVILEGNIA

a. No sporangia achlyoid; gemmae lacking
  b. Spores normally swimming once
    c. Antheridia present — 1. *B. unisperma*
    cc. Antheridia lacking
      d. Oogonia borne on loose proliferating branches — 1a. *B. unisperma* var. *montana*
      dd. Oogonia usually borne singly on short slender branches from the main hyphae
        e. Sporangia long, slender; some spores with and many without a swimming stage
          f. Eggs 15–17.5μ in diameter — 1b. *B. unisperma* var. *delica*
          ff. Eggs 21–34μ in diameter — 2. *B. megasperma*
        ee. Sporangia shorter, entirely thraustothecoid; most of the spores with a swimming stage — 1c. *B. unisperma* var. *litoralis*
  bb. Spores normally not swimming
    c. Sporangia short, subclavate — 3. *B. subclavata*
    cc. Sporangia elongated
      d. Sporangia all with a single row of spores — 4. *B. linearis*
      dd. Sporangia with some but not the majority of spores in a single row — 5. *B. diclina*
aa. Primary sporangia achlyoid; gemmae present — 6. *B. bispora*

1. *Brevilegnia unisperma* Coker and Braxton
   Syn. *Thraustotheca unisperma* Coker
   Calytralegnia unisperma Coker

Growth fairly dense on boiled hemp seed, jimson weed seed, and mushroom grubs, reaching a length of 8 mm. under good conditions. Main hyphae from 10–68μ thick at base, freely branched into secondary branches, mostly from 18–33μ thick; growing tips pointed and hyalin. Sporangia abundant, long-cylindrical, a few short clavate ones appearing in young cultures, 18–30 $\times$ 55–370μ, generally 23–38 $\times$ 190–350μ; primary ones terminal on main branches, secondary ones formed by cymose branching; thicker in the distal half, with spores occasionally borne in a single row. Sporangial wall very thin, completely disappearing as the spores escape, except for a small basal cup that often remains. Spores 10–16.5μ in diameter, abnormal ones sometimes reaching 20μ; escaping by the rupture of the sporangial wall at any point, as base, side, or tip, or by the seemingly complete disappearance of the sporangial wall, leaving the encysted and somewhat angular spores free to exude in all directions; after a rest the spores escape and swim away as usual. Oogonia abundant in nearly all cultures, appearing when cultures are six to eight days old;

borne on long, slender, often branched, bent or sometimes once coiled branches of 4.5–6μ diameter. These oogonial stalks with their antheridia often form a very dense, untraceable, conglomerate growth. Oogonia 15–24μ, generally 19–22.5μ in diameter, abnormal ones up to 39μ, spherical to oval, often very irregular, and with outgrowths up to 6–7μ long; oogonial walls varying greatly in thickness in the same oogonium, from 1.6μ to almost no thickness. Eggs spherical, single in the oogonium and not filling it, from 12–19.5μ in diameter, generally 14–18μ, eccentric; with a single, large, round oil drop on one side. Antheridia androgynous, on from 25–65 per cent of the oogonia in most cultures, borne as a rule on the oogonial stalks generally profusely branched and irregular, with from one to many finger-like projections touching the oogonia. No antheridial tubes have yet been observed.

From soil: United States: North Carolina (26)

1a. *Brevilegnia unisperma* var. *montana* Coker

Growth fairly dense on boiled hemp seed, jimson weed seed, or corn, reaching a length of 5 mm. Main hyphae freely branched, up to 75μ thick at the base, averaging from 15–30μ throughout the cultures. Sporangia and spores, as in the species, usually densely grouped in old cultures. Oogonia plentiful in most old cultures, appearing after about ten days, borne as a rule on long, slender, loosely-branched hyphae about 6μ thick, which arise from the main hyphae. Oogonia obovate or oblong to subspherical, usually somewhat irregular or tuberculate and rarely prettily papillate, 17–26 $\times$ 18–32μ, averaging 20–23 $\times$ 25–30μ; oogonial wall thin, unpitted. Eggs spherical, sometimes elliptic, 12–24μ in diameter, averaging 17–19μ, usually single, rarely two to the oogonium, eccentric, with rather thick walls. No antheridia present.

From soil: United States: North Carolina (26)

1b. *Brevilegnia unisperma* var. *delica* Coker

Mycelium rather dense, reaching a length of 6 mm. on hemp and corn. Hyphae delicate, 15.2–21μ thick in main part of the culture. Sporangia extremely abundant, 94–682μ in length; many with only one row below and two above; primary sporangia borne on main hyphae, secondary ones by cymose branching and often from lateral branches of various lengths at some distance below. Spores more or less angular, 11.7–16.2μ, most about 12.5μ in diameter, both extremes occurring in the same sporangium; sporangial wall disappearing soon after maturity, but the spores, while spreading a little, remaining united for a long time as if in a jelly. Spores in the great mass without a swimming stage; a few, however, emerge about as in Dictyuchus and swim under normal laboratory conditions. Oogonia abundant throughout the culture, small, borne on long, slender, usually curved oogonial stalks which arise from main hyphae which in the great majority of cases bear only oogonia but may bear a sporangium at the tip. The oogonial stalks vary in length from about three to seven (usually three and one-half to four) times the diameter of the oogonia. Rarely an oogonial stalk may branch and bear two oogonia. Oogonia globular, with or without an apical projection, 16.5–21 μ, mostly 18 μ in diameter; wall smooth, unpitted. Eggs, one in each oogonium, 15–17 μ in diameter, eccentric with a single large oil drop. No antheridia present.

From soil: United States: North Carolina (26)

1c. *Brevilegnia unisperma* var. *litoralis* Coker and Braxton

Syn. *Thraustotheca unisperma* var. *litoralis* Coker

Growth short on boiled hemp seed, reaching a length of 3 mm., approaching Geolegnia in slow and limited growth, but not so dense. Main hyphae from 9–60 μ thick at base, mostly 15–25 μ thick, freely branched, branches often rough and gnarled near the tip; growing tip hyalin. Sporangia abundant, all primary ones small, short clavate, 21–31 × 31–67 μ, the smallest ones often containing only eight or ten spores; secondary ones subcylindrical, from 15–29 × 90–255 μ, generally about 140–220 μ long, generally borne in clusters of three to eight near or at the end of the hyphae, larger in their distal half, sporangial wall breaking at the distal end. Spores somewhat angular, from 11.5–15 μ in diameter, generally 12.5–14 μ, behaving as usually for Thraustotheca. Oogonia abundant in all cultures, appearing after five or six days, exactly similar in structure and measurements to *Brevilegnia unisperma*, borne on slender bent stalks from the main hyphae, but not forming such a conglomerate growth as *B. unisperma*. No antheridia present.

From soil: United States: North Carolina (27)

2. *Brevilegnia megasperma* Harvey

Mycelium in mature cultures very dense and opaque, prominently white, usually reaching a ring growth of 15 mm. within three weeks on hemp seed, sometimes greater, often less, the mycelium seemingly at times slightly shrunken, a few hyphae distinguishable at the outer edges. Hyphae slender, sparingly branched, practically the same width throughout; 8.3–25 μ broad, mostly 12–13 μ. Primary sporangia terminal on main hyphae, not much broader than the hyphae, often slightly swollen toward the outer end, rarely pronounced clavate; 19.7–235 μ long, 18.3–

25μ broad. Secondary sporangia similar to primary, formed at the tips of hyphal branches which arise by lateral proliferation from immediately below the primary sporangia. Spores usually about 11.7μ broad, rounded or slightly angular when compressed within the sporangium; formed in a single row, or rarely in more than three rows, varying with the thickness of the sporangia or of parts of the same sporangium. Spores encyst within the sporangium and escape after decay of the sporangial wall by floating away or very often by the spore protoplasts escaping by amoeboid movement, these assuming a reniform shape and acquiring two laterally placed flagella; the spores very often sprout in place, the delicate hyphae ramifying through the entire mycelium, adding to the opaqueness of the mycelium. Gemmae plentiful in some cultures after a few days, dark and very dense, rarely with a central clearer region; seldom varying from 50–60μ in length and 25–36μ in breadth; may give rise to sporangia, but more often sprout into hyphal filaments, which add to the density of the mycelial mat. Oogonia strikingly abundant, very often in compact masses; spherical, occasionally elongate or ovate to obovate, seldom with one or more papillae; borne for the most part at the tips of lateral branches arising from the main hyphae, though occasionally terminating the main hyphae, seldom intercalary; when on lateral branches there may be as many as five or six oogonia in a cluster, or other oogonia may arise at the tips of secondary lateral branches, which may be long and coiled; 30–55μ, mostly 31.7–35μ, in diameter. Oogonial wall about 1.5μ thick. Oogonial stalks up to 115μ long, being usually not over 7μ broad. Oospores, constantly one to each oogonium, large but not filling the oogonium, spherical; eccentric, possessing a large lateral oil drop; oospore protoplasm finely granular, continuous or discontinuous on the outer side of the oil globule; 21.6–33.3μ broad, mostly 25–26.6μ. The oil drop may reach the diameter of 20μ. The oospore wall usually about 1.3μ thick. Antheridia conspicuously absent, but sometimes present and then hypogynous; empty antheridia seldom noticed.

From soil: United States: California (62) (69), Kentucky (30) (61), New York (61)

3. *Brevilegnia subclavata* Couch

Mycelium forming a rather dense growth on hemp seed; reaching a maximum diameter of 0.9 cm. Hyphae short, usually not more than 3 mm. long; up to 62μ thick at the base; considerably branched. Sporangia formed in great abundance, 30–100 × 108–140μ, terminal, the renewal of the sporangia taking place by cymose branching as in Achlya or Thraustotheca and also by successive formation in basipetal series, as many as twelve sporangia sometimes being formed in a single row; usually short clavate, but varying from nearly spherical to long cylindrical with spores in a single row. Sporangia without any papillae of dehiscence, the spores escaping by swelling and bursting the sporangial wall, or not rarely remaining within the sporangium and sprouting *in situ*. Spores 10.8–19 × 10.8–28.8μ, varying from nearly spherical to cylindrical, but usually distinctly polyhedral in shape, with a distinct large vacuole; without a swimming stage. Sexual reproduction occurring in most but not all cultures, appearing after the culture is three or more days old. Oogonia 19–25μ, spherical, but more often slightly subspherical, formed on the ends of very delicate, usually long, stalks which arise racemosely from the much thicker, main branches. Eggs 15–19μ thick, spherical, single in the oogonium, eccentric. Antheridia on most oogonia, androgynous, usually arising from the oogonial stalk itself; becoming practically empty during the early development of the egg.

From soil: Mexico (155)

United States: New York (32)

4. *Brevilegnia linearis* Coker

Growth limited, dense and opaque as in other species of the genus, the hyphae reaching a length of 5 mm. on hemp seed or jimson weed seed. Hyphae slender, about 8–14μ thick throughout the culture, sparingly branched; growing tips even, hyalin. Primary sporangia terminating all the main hyphae, occasionally intercalary, long, slender, sometimes branched; secondary ones usually shorter, borne on the tips of lateral branches formed by cymose branching of the main hyphae. Sporangia of about the same thickness as the hyphae bearing them, with spores in a single row, up to sixty in a row, usually ten to twenty-four; the sporangial wall soon disappearing, but spores remaining held together for a long time by some invisible substance. Spores spherical to long rod-shaped, all with a central vacuole that varies with the spore size, primary ones mostly subspherical, those of secondary sporangia more rod-shaped, 8–20 × 15–60μ, mostly 8–14 × 15–25μ, never swimming but sprouting in position. Oogonia fairly abundant, borne singly on long, slender, often coiled lateral branches from the main hyphae; typically spherical, sometimes slightly irregular, 16–21μ in diameter; wall thin, smooth, colorless. Eggs one to each oogonium, spherical, 14.5–18μ in diameter, eccentric, with a large lateral oil drop. Antheridia on nearly all oogonia, single, large in proportion, short and tuberous, usually androgynous from near the oogonium, rarely diclinous. Antheridial contents often entering the oogonium before the egg is differentiated.

From soil: United States: North Carolina (26) (30)

5. *Brevilegnia diclina* Harvey

Mycelium dense and rather opaque, of fast but limited growth and having much the appearance of Geolegnia species. On boiled hemp seed in water reaching a diameter of 1 cm. in about five days and 15 mm. in two weeks. Hyphae straight, sparingly branched, at first resembling Leptolegnia, 5–28μ thick, usually 10–13μ. Primary sporangia formed within one day, singly or in clusters at the tips of practically all hyphae, sometimes many arising from the same point in dense sympodial groups; ovate to long club-shaped as in *Thraustotheca clavata* or sometimes long and slender with the spores in a single row throughout or in part. Secondary sporangia dense in sympodial clusters and also not rarely from segments cut off below the primary ones. Spores spherical or oval to elongated, frequently angular from crowding, most of the smaller ones about 10.3μ thick or if elongated 10.3–12.8 $\times$ 12.8–25.6μ, larger ones with a central clear spot as in *Geolegnia inflata*; escaping by dissolution of the sporangium wall, then sprouting with a germ tube, a swimming stage lacking. Oogonia abundant 21–33μ, mostly 21–25μ thick, borne singly and apically on branches smaller than the main hyphae, often as in *Achlya caroliniana*; formed usually after one week, often within three days; spherical to irregular and often with few to many projections which may reach 36μ in length, rarely over 5.5μ in breadth, wall about 1μ thick; earlier oogonia often proliferating, the contents forming a new oogonium at the tip, which may mature an egg. Eggs one to each oogonium, 18–25μ, mostly 18–21μ thick, spherical, usually filling the oogonium, though often not filling it, eccentric with a large lateral oil drop which is entirely surrounded by protoplasm; egg wall up to 2μ thick. Mature eggs never found in oogonia bearing projections, and mostly in the spherical ones. Antheridia more often absent, when present only one and usually diclinous as in *Thraustotheca clavata* and closely resembling that form in that the antheridial stalks are long, irregular and branched; rarely androgynous, in which case the antheridial stalks may be much coiled and irregularly wrapped about the oogonium. Fertilization apparently does not take place.

From soil: Mexico (155)

United States: Kentucky (61), Oklahoma (61), Mississippi (61), New York (61), Wisconsin (30) (59) (61)

6. *Brevilegnia bispora* Couch

Mycelium forming a dense growth on hemp seed; reaching a diameter of 3 cm. Hyphae up to 1.5 cm. long, not rarely as much as 65μ thick near the base; considerably branched. Sporangia very abundant, 21–45 $\times$ 125–400μ, terminal, renewed by cymose branching as in Achlya, long clavate, broadest near the middle. First sporangia resembling the Achlya type, many of which, however, lack a distinct papilla of dehiscence, a considerable part of the sporangial tip giving away for the exit of the spores; later sporangia of the same shape and size as the early ones but dehiscing by the swelling of the spores and consequent bursting of the sporangial wall much as in the later sporangia of *Thraustotheca primoachlya*; more numerous than the first type. Spores of the Achlya type mostly 10.8μ thick and spherical when encysted; spores of the Thraustotheca type up to 18μ thick and polyhedral when encysted, usually with a large conspicuous vacuole; spores of the Achlya type usually diplanetic but those of the Thraustotheca type remaining inactive or sprouting *in situ* if fresh water is added. Gemmae formed in considerable numbers and especially under unfavorable conditions as when the cultures become old or infected with bacteria; spherical, pyriform, or cylindrical, often in chains. Oogonia usually very abundant, but appearing only after the cultures are three or more days old; very small, 16–28μ thick, most about 21.6μ thick, spherical, smooth, formed usually singly but sometimes in pairs on the ends of very long, delicate stalks which arise from the main branches. Eggs 11.3–19.5μ, most about 18μ thick, spherical, eccentric when ripe. Antheridia one to several on most of the oogonia, of androgynous origin, often arising from the oogonial stalk itself. Antheridial tube developed.

From soil: United States: New York (32)

(12) **Geolegnia** Coker

Mycelium of very limited growth, forming a dense, opaque mat; hyphae very slender. Sporangia inflated at regular intervals or segmented into two or more compartments (unless very small); spores in a single row, very large, encysting within the sporangium with a thick wall and without any motile stage; escaping by the decay of the thin-walled sporangium and sprouting by a germ tube. Oogonia abundant, even, containing a single eccentric egg that does not fill the cavity. Antheridia always present and androgynous.

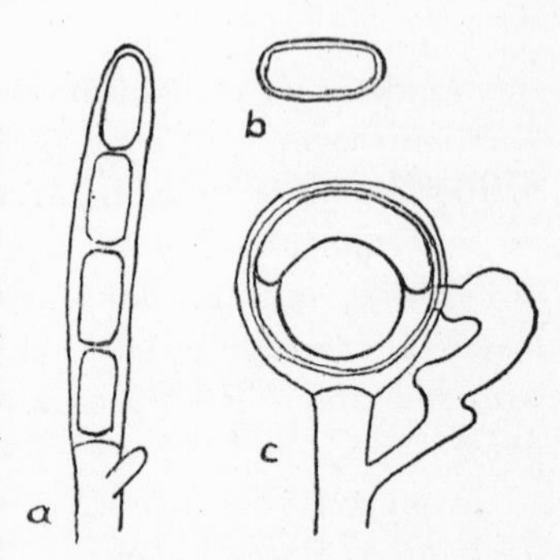

FIG. 34. **Geolegnia.** *a* sporangium; *b* zoospore; *c* oogonium (after Coker).

In regard to the sporangia and spores this is the most peculiar genus of the Saprolegniaceae and it must occupy a section of its own. The peculiar sporangia and the large motionless spores with thick walls separate at a considerable distance all the other genera.

KEY TO THE SPECIES OF THE GENUS GEOLEGNIA

a. Sporangia constricted at intervals, spores spherical to oval ... 1. *G. inflata*
aa. Sporangia not constricted, spores elongated ... 2. *G. septisporangia*

1. *Geolegnia inflata* Coker and Harvey

Mycelium of great density and slow and unlimited growth, as noted under the genus, forming after a few days a very dense, quite opaque, white mat, with individual threads distinguishable only on the margin. Growth on boiled hemp seed up to 2 mm. in six days, never reaching a length (ring growth) of over 3 mm. Hyphae straight, sparingly branched at first and then resembling Leptolegnia, 2.3–16μ thick. Primary sporangia formed from the straight ends of the larger hyphae, soon (usually before abstriction) becoming swollen at regular intervals; swellings 15–21μ thick; secondary sporangia, usually shorter, formed immediately below old ones, on the same thread or from lateral threads of irregular position. Spores very large and peculiar, spherical to oval, mostly spherical, rarely elongated, three to fifteen to a sporangium, usually four to six formed singly in each swelling, 14–21μ thick, encysting in position with a thicker wall than that of the sporangium (0.5μ) and without any motile stage, escaping from the sporangium by decay of the delicate walls, which soon occurs; sprouting promptly when brought into new media. Oogonia abundant, spherical, 15–19μ thick, with thick (2μ), smooth, unpitted walls, usually appearing later than the sporangia, though occasionally earlier, borne singly and apically on smaller and more irregular branches than the main hyphae. Eggs one to each oogonium, 13–15μ, eccentric, with one large lateral oil drop. Antheridia, short, swollen, tuberous, always present, borne on slender, irregular often contorted branches which are mostly androgynous from near the oogonia, rarely diclinous.

From soil: Germany (112)

United States: California (62), Kentucky (61), Mississippi (61), New York (32) (61), North Carolina (30) (58), Oklahoma (61), Wisconsin(61)

2. *Geolegnia septisporangia* Coker and Harvey

Mycelium exactly as in the preceding species except of even more limited growth, forming after a few days a dense, opaque mat with individual threads distinguishable only at the edge of the mat. Growth on boiled hemp seed up to 2 mm. in five days, but never reaching a length (ring growth) of 3 mm. Primary sporangia formed at the ends of practically all hyphae, these and later ones usually divided into several cells by cross partitions after being cut off and just before the spores are formed; swollen at places, but not so often nor so greatly and regularly as in the preceding species, 11.8–21.15μ thick and up to 136μ long. Secondary sporangia formed at the tips of lateral branches somewhat as in Achlya. Spores very large and peculiar, rarely spherical, mostly oval to ovate or elongated, formed in a single row, one to fifteen to a sporangium, usually two to five, 11.8–21 $\times$ 20.3–56.4μ, encysting within the sporangium with a rather thick wall (0.9μ) and never escaping from the sporangium except by the decay of the walls. On change of conditions, as removal to corn meal agar, sprouting at once by a germ tube; never forming swarm spores. Oogonia abundant, appearing very suddenly in young and old cultures alike and without apparent cause, borne singly and apically on smaller branches than the main hyphae; subspherical, 22–34μ thick; wall smooth, thin, colorless. Eggs one to each oogonium, spherical to slightly oval, 20–32μ thick, eccentric, with one large lateral oil drop; walls very thick, 2μ. Antheridia always present, elongated and apically attached to the side of the oogonium, in all cases observed borne on short, irregular androgynous branches from the oogonial stalk at little distance below the oogonium, one to four attached to each oogonium. Emptying of antheridial contents into egg observed.

From soil: United States: Mississippi (61), New York (32) (61), North Carolina (27) (30) (60), Oklahoma (61), Wisconsin (60) (61)

F. LEPTOMITALES

Thalli with or without a well-defined basal cell and holdfasts, the hyphae divided into constricted pseudocells by pseudosepta of cellulin, walls giving a cellulose reaction; reproductive organs consisting of segments of the hyphae or specialized pedicellate structures cut off by cross-walls and constrictions from the mycelium; zoosporangia forming mono- or diplanetic biflagellate zoospores; oogonia with or without periplasm, forming (except in Apodachlyella) a single egg; antheridium either a single segment or borne on specialized branches of mono- or diclinous origin, with or without a fertilization tube; oospores single (except in Apodachlyella), thick-walled, upon germination forming hyphae. A single family treated.

a. *LEPTOMITACEAE*

Filaments constricted at intervals to form a series of long or short segments; often showing conspicuous particles of material, supposed to be cellulin, which may entirely fill the constriction. Oogonia if present containing a single egg, which is surrounded by periplasm except in Apodachlya.

A single genus treated.

(1) **Apodachlya** Pringsheim

Hyphae constricted into segments of variable length, more slender than in Leptomitus, the branching taking place from any point in a segment, but usually near the distal end. Sporangia swollen, pyriform, oval or spherical. In three species spherical resting bodies are known with the contents entirely filling them, and these are regarded as true oogonia containing a single egg.

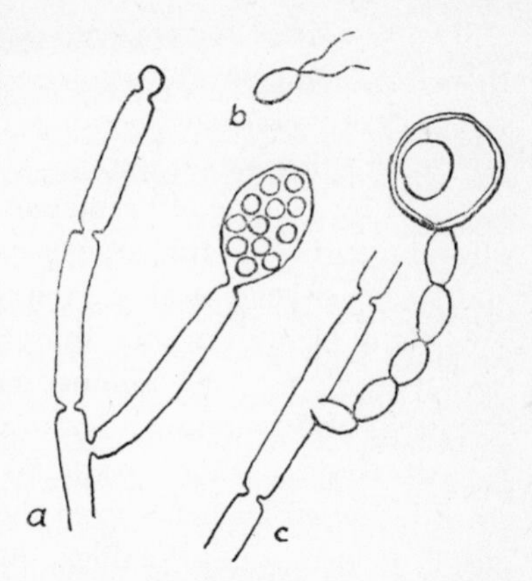

FIG. 35. **Apodachlya.** *a* sporangium; *b* zoospore; *c* oogonium (after Coker).

KEY TO THE SPECIES OF THE GENUS APODACHLYA

a. Oogonia borne on tips of main hyphae or on short lateral branches ... 1. *A. pyrifera*

aa. Oogonia borne on tips of lateral branches, made up of many short segments ... 2. *A. brachynema*

1. *Apodachlya pyrifera* Zopf

Syn. *Leptomitus pyriferus* Zopf

Mycelium composed of long, branched, segmented hyphae, the basal segments larger, becoming more slender toward the tips. Sporangia terminal, usually pyriform, more rarely oval or spindle-shaped, 12–22μ broad × 12–44μ long, sympodially arranged, at times with as many as twelve in such an arrangement. Spores encysting at the mouth of the sporangium immediately after emerging, about 11μ in diameter, after a rest emerging from the cysts as motile biflagellate spores. Oogonia spheric with a thick, colorless, double membrane, at maturity with colorless contents and a large oil-drop, usually terminal, rarely intercalary or on short lateral branches.

From soil: Mexico (155)

2. *Apodachlya brachynema* (Hildebrand) Coker and Matthews

Main hyphae slender, the segments about 4.5–8.5μ thick and 110–185μ long on termite ants, but 4–23.4 × 20–150μ on corn meal agar, becoming shorter near the sporangia as a rule, the protoplasm moderately dense and with small refractive drops here and there; branching rather sparsely from any point on the segments, but usually near the distal end. Sporangia terminal, single or rarely two or three in a row, swollen, pyriform or oval or spherical on termite ants, about 23–29μ thick and 23–46μ long, renewed by sympodial branching, opening by a distinct papilla formed a few minutes before discharge of spores; papilla usually apical in the longer sporangia, either apical or lateral in the short or spherical ones. Spores few, about eight to twenty, short-oval, nearly always swimming sluggishly and aimlessly for a few minutes with two apical flagella on emerging, then encysting and swimming again after a rest; diameter 8.5–10μ when encysted. Resting bodies (oogonia) formed plentifully on the tips of short, lateral, jointed branches from the main hyphae, spherical or very rarely short pyriform, 23.5–29μ thick, smooth, dense, at first nearly homogeneous, then forming a number of fat droplets and finally one eccentric, conspicuous droplet as in the eccentric-egged species of Achlya; wall unpitted, about 1.8μ thick; the suboogonial cell (antheridium) as a rule nearly spherical, at first denser than the other members of the chain, then discharging its contents into the oogonium and becoming quite empty before the maturation of the egg.

From soil: United States: Mississippi (61)

G. PERONOSPORALES

Mycelium well-developed, thread-like, much branched, without septa except in the formation of reproductive bodies. Fruit-bodies of two sorts, asexual and sexual. Asexual, deciduous sporangia formed on the ends of one-celled sporangiophores, germinating by zoospores or by tube. Sexual fruit in the substrate. Antheridia and oogonia morphologically different on the ends of lateral branches. Antheridia clavate, smaller than the oogonia, to which they become attached and pierce with a fertilization-tube. Oogonia large, sack-like or spherical, with a single egg. Oospores spherical with a many-layered wall, germinating by zoospores or a tube.

A single family treated.

a. *PYTHIACEAE*

Aquatic, amphibious or terrestrial saprobic or parasitic fungi, thallus a richly branched hyphal complex, septa in vigorously growing parts formed only to delimit reproductive organs, thick-walled chlamydospores and gemmae sometimes formed. Zoosporangia either undifferentiated portions of the mycelium, or an irregularly expanded complex of lobulate elements and an evacuation tube, or an ovoid, spherical, or bursiform structure with or without a more or less prolonged evacuation tube, formed singly or in catenulate series, sometimes internally proliferous. Zoospores of the reniform, laterally biflagellate type, either formed outside the sporangium in a vesicle or free in the water or produced within the sporangium, capable of repeated emergence. Oogonia terminal or intercalary, spherical or cylindrical, smooth or spiny-walled, usually containing a single egg, which is often differentiated into ooplasm and periplasm. Antheridia terminal or intercalary, rarely lacking, diclinous or androgynous, hypogynous or amphigynous, each forming a well defined fertilization tube. Oospore lying loosely in the oogonium or completely filling it, smooth or rough-walled, upon germination producing a germ-tube or zoosporangium.

KEY TO THE GENERA OF THE PYTHIACEAE

a. Zoospores formed in a vesicle produced at tip of evacuation tube 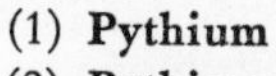 (1) **Pythium**

aa. Zoospores formed within the sporangium 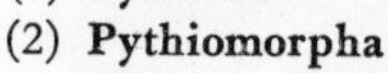 (2) **Pythiomorpha**

(1) **Pythium** Pringsheim

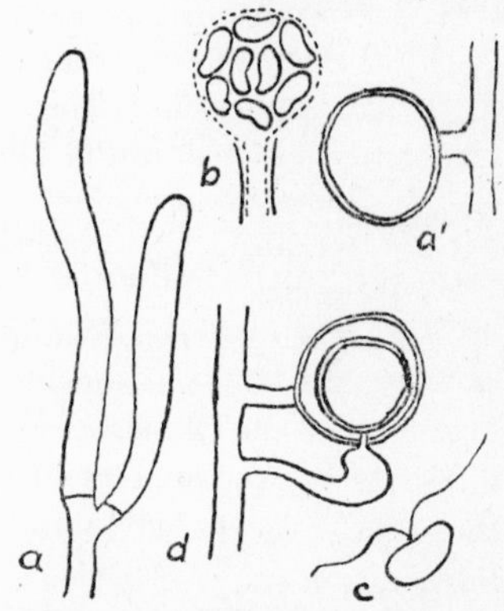

FIG. 36. **Pythium.** *a, a'* sporangia; *b* zoospore discharge; *c* zoospore; *d* oogonium (after Fischer).

Mycelium parasitic in living plants or saprobic on insects and plants rotting in or on water, with very thin, not more than 8μ thick, often much thinner, richly paniculately branched threads, at first always one-celled, in age often with some irregularly placed cross-walls, growing intra- or intercellular, always without special haustoria; in water often forming thin, Saprolegnia-like turf; colorless. Sporangia not on the special sporangiophores of other Peronosporales but partly at the end of hyphae, partly intercalary in or on the substrate, variously formed; partly thread-like, not thicker than the mycelial branches, partly spherical or lemon-shaped; the still undivided content empties into a bladder and breaks up here into zoospores, which become freed by the bursting of the bladder. In many species the sporangia remain sitting on the mycelium, in others only in water on submerged mycelium, while they may break off as conidia, spherical or lemon-shaped, germinating either as zoospores, or by a tube. Zoospores kidney-shaped with two flagella inserted at the side, monoplanetic, colorless, movement uniform. Sex organs partly in the interior of the substrate, partly on the hyphae growing out of it especially in water, numerous, always androgynously arranged. Oogonia small, spherical with colorless, unspotted, smooth, or warty-spiny membrane, one egg and little periplasm. Antheridia mostly club-shaped, on the end of short branches, sprouting from below the oogonium, curved secondary branches, seldom cylindrical and divided hypogynously as pieces of the oogonial-bearing threads; very rarely lacking also the antheridia. Oospores single in the oogonium, spherical with large central colorless fat-drop, colorless content, with thick smooth or spiny, yellowish or gray exine. Germination either by zoospores or tube.

KEY TO THE SPECIES OF THE GENUS PYTHIUM

a. Sporangia filamentous, resembling vegetative hyphae or composed of irregular inflated elements
  b. Oospores not filling the oogonium — 1. *P. gracile*
  bb. Oospores filling the oogonium
    c. Antheridia present
      d. Antheridia one to two to the oogonium — 2. *P. monospermum*
      dd. Antheridia one to six to the oogonium — 3. *P. graminicola*
    cc. Antheridia lacking — 4. *P. papillatum*

aa. Sporangia spherical to subspherical, cut off by septa
  b. Sporangia proliferous — 5. *P. proliferum*
  bb. Sporangia not proliferous, often transformed into conidia
    c. Sexual reproduction present
      d. Oogonia smooth
        e. Oospores smooth, filling the oogonium — 6. *P. rostratum*
        ee. Oospores smooth, not filling the oogonium
          f. Sporangia lacking or rare, conidia present
            g. Conidia irregular, not abundant — 7. *P. vexans*
            gg. Conidia subglobose, abundant
              h. Antheridia with short stalks on oogonial stalk — 8. *P. ultimum*
              hh. Antheridia usually single, one to six, with longer stalks — 9. *P. debaryanum*
          ff. Sporangia present; conidia also present, oogonia usually intercalary — 10. *P. pulchrum*
      dd. Oogonia with few to many spines, rarely smooth
        e. Oogonia with few spines, 14–46μ — 11. *P. irregulare*

ee. Oogonia with numerous spines
f. Antheridia androgynous — 12. *P. mamillatum*
ff. Antheridia usually hypogynous — 13. *P. echinulatum*
cc. Sexual reproduction unknown
d. Conidia and sporangia catenulate — 14. *P. intermedium*
dd. Conidia not catenulate — 15. *P. elongatum*

1. *Pythium gracile* Schenk

Hyphae branched, 1–5μ in diameter (according to Butler forming enlarged places and hyphal clumps in culture). Sporangia filamentous, indistinguishable from the ordinary hyphae, with a long or short tube of discharge. Zoospores three to twenty or more to a vesicle, usually 6–8μ long. Oogonia formed intramatrically, smooth or slightly irregular, spherical or oval, at times with a papilla, usually terminal, 15–26μ in diameter. Oospores not filling the oogonium, smooth, spherical, 12–20μ in diameter, in the mature stage with a heavy wall (usually about 2 μ thick) and a central reserve globule, surrounded by a granular layer of protoplasm in which a small refractive spot is embedded. Antheridia one to two to an oogonium (usually only one), diclinous.

From soil: Germany (20)

United States (90)

2. *Pythium monospermum* Pringsheim

Mycelium forming a cloud around the substratum. Hyphae irregularly branched up to 7μ in diameter, often with numerous bud-like outgrowths laterally. Sporangia single or branched, length very variable, up to 1 mm. Zoospores from a few to forty or more. Oogonia within and outside the substratum, terminal or intercalary, or formed in the lateral buds. Antheridia one or more, club-shaped, arising from the oogonial stalk or from a distinct hypha. Oospores smooth, completely filling the oogonium, the wall of which is often difficult to define, 12–15μ in diameter, germination after a rest, which may be several months in duration, by a hypha which is quickly transformed into a sporangium.

From soil: Ireland (20)

United States: North Carolina (58) (90), Oklahoma (61), Wisconsin (61)

3. *Pythium graminicola* Subramaniam

Mycelium well developed on corn meal agar. Hyphae irregularly branched, 2.5–7.2μ in diameter. Sporangia consisting of irregular, inflated elements, cut off from remainder of mycelium by a cross wall. Tube of discharge slender, often long. Zoospores fifteen to forty or more to a vesicle, reniform, biflagellate, encysting as spherical or ellipsoid bodies, 9.6–12μ in diameter. Conidia spherical 24–36μ in diameter (usually about 30μ). Oospores usually filling oogonium, spherical, rarely elliptic, 21.6–32μ in diameter (usually about 28μ), in mature stage with a heavy wall and a central reserve globule surrounded by a granular layer of protoplasm, in which a small refractive body is embedded. Antheridia one to six to an oogonium, often curved. Antheridial filament often long and irregular, as many as five may arise from one filament, which may arise from the hypha bearing the oogonium or a more distant one.

From soil: United States: Iowa (87)

4. *Pythium papillatum* Matthews

Mycelium dense, extending for a centimeter or more from the substratum, when grown on boiled hemp seed. Hyphae very irregular in size, often with blunt tips; main hyphae up to 24μ in diameter at base, ordinary hyphae about 7.2μ thick, forming numerous bud-like outgrowths in old cultures as in *P. monospermum*. Sporangia filamentous, short or often very long, arising from unbranched hyphae or highly branched ones, usually very slender near the vesicle, often only about 3.6μ thick. Zoospores from a few to twenty or more to a sporangium, usually about $7.2 \times 12\mu$, soon encysting with a diameter of about 12μ. Oogonia 16.8–26.4μ in diameter, terminal on short branches, very often intercalary, often with several in a chain (rarely up to five), spherical to oval, often with a long neck, smooth or with one or two prominent papillae. Oospores 16–24μ in diameter, usually entirely filling the oogonium except for the neck, usually spherical; in mature stage with a heavy wall and a central reserve globule surrounded by a granular layer in which a small refractive body is embedded. Antheridia lacking.

From soil: United States: North Carolina (86)

5. *Pythium proliferum* de Bary

Mycelium in water culture, fine. Hyphae uniform, 4–5μ broad, branching laterally and sparingly in young cultures. The origin of a lateral branch may be somewhat swollen, and rarely, fusiform swellings occur in the course of the hyphae. Sporangia terminal, spherical, rarely oval, vacuolated, very variable in size, 30–58μ in diameter, with a short tube of discharge, rarely equalling one-fourth of the diameter of the sporangium, placed in any position, but usually opposite the stalk. After discharge, growth of the supporting hypha occurs through the emptied sporangium, or immediately below it, laterally; new sporangia being formed within the empty sporangium, or beyond it, in the first case. Zoospores large, three to numerous. Conidia unknown. Oogonia within and outside the substratum, terminal or often intercalary, 19–36μ in diameter. Antheridia one to three or more, usually more than one, from neighboring branches and, less frequently from the oogonial stalk. Oospores spherical, not filling oogonium, 16–27μ in diameter. Germination after a rest of several months, by a hypha which usually branches, the branches being short and clustered and bearing one or rarely two sporangia, sometimes, particularly in small spores, by an unbranched hypha which soon gives rise to a sporangium.

From soil: France (20), India (20)

6. *Pythium rostratum* Butler

Mycelium in water cultures large; hyphae up to 6 or even 8μ in diameter, and tapering gradually at the ends but never prolonged as fine filaments. Branching irregularly racemose. When old, the mycelium is sparingly septate. Sporangia terminal or intercalary, spherical at first, oval later, 28μ in diameter as an average, ranging from 23–34μ. The tube of discharge is very large and broad, usually about equal to the diameter of the sporangium and thickened about half way in its length in a characteristic fashion. It is usually lateral. Conidia rarely as frequent as sporangia and appear usually later. Oogonia usually intercalary or lateral are formed extramatrically. They measure about 21μ in diameter and are slightly longer than wide. They are completely filled by the oospore but the wall of the oogonium can usually be made out. Antheridia usually single, arise from the oogonial hypha. They are often extremely short. Oospores are spherical, smooth, 21μ in diameter on the average, ranging from 12–26μ. Germination not seen.

From soil: Denmark (80), France (20)
United States: Iowa (87)

7. *Pythium vexans* de Bary

The mycelium is slender, finer than that of *P. debaryanum* or *P. rostratum*, but resembling *P. intermedium*, both in thickness of the hyphae and the size of the thallus in water cultures. The hyphae taper at the ends, particularly in the lateral branches, which are given off in a very irregular manner. The branches of the secondary or tertiary order, often extend far beyond the primary hyphae, tapering into very fine filaments at the ends. This character distinguishes it from any other species. Sporangia and conidia are developed on two or three day-old cultures. The former are rare. They occur both terminally and, more rarely, intercalary, and are scarcely ever spherical or oval, but usually irregularly pyriform, ovate, or sub-angular. The sporangial tube of discharge is short. The sporangia and conidia measure 17–24μ in diameter, averaging about 21μ. The conidia are filled with very dense protoplasm, vacuolation being rare. The oogonia are 22–25μ in diameter and formed on the extra-matrical mycelium. They always arise laterally usually on short branches from the main hyphae, or sessile on the latter. The oogonium is inserted on its stalk by a broad base. The antheridia arise from the oogonial stalk and are rarely hypogynous. Usually there is one to each oogonium, rarely two. The antheridial cells are clavate, or rounded, and large in relation to the oogonium. In every case seen it was closely applied to the oogonial wall, so as to fuse with the latter in a large part of its circumference. Oospores free in the oogonium, but larger in relation to it than in *P. debaryanum*, 20–22μ in diameter, smooth, round. Germination often by giving zoospores directly, a thick tube being put out which, after growing to about the length of the diameter of the oospore, blows up at the apex into a bladder in which the contents of the oospore are divided into zoospores. The older spores (five to six months) germinate only by a branched hypha.

From soil: England (20), France (20), Ireland (20)
United States: Iowa (87)

8. *Pythium ultimum* Trow

Mycelium well developed. Hyphae branched, septate in old cultures, 1.7–6.5μ in diameter (average 3.8μ). Conidia usually terminal and spherical, 12–28μ in diameter, intercalary barrel-shaped forms 14–17 $\times$ 22.8–27.8μ also formed in cultures; germination direct. Zoospores never formed. Oogonia smooth, usually terminal, rarely intercalary, spherical, 19.6–22.9μ in diameter. Oospores not filling the oogonium, spherical 14.7–18.3μ in diameter, in mature stage with a heavy wall and a central reserve globule surrounded by a granular layer of protoplasm in which a small refractive body is embedded; germination direct by one or more germ tubes. Antheridia usually one to an oogonium arising from the oogonial stalk immediately below the oogonium.

From soil: United States: Iowa (87)

9. *Pythium debaryanum* Hesse

Mycelium rather coarse, intra- and extramatrical. Hyphae large, branching irregular and free, septate in old cultures. Sporangia spherical or oval, chiefly extramatrical, terminal and intercalary; supporting hypha usually emptied of its contents for a variable distance below the sporangium and separated by septa from the full portion of the hypha and from the sporangium. Tube of discharge lateral, about the diameter of the sporangium in length. Proliferation absent. Conidia usually numerous, intra- and extramatrical, 15–25μ in diameter, round, oval or somewhat irregular in shape and size in old cultures, may germinate at once but more often do so after a short rest. Oogonia usually numerous, intra- and extramatrical, sometimes formed very easily in culture, 20–55μ in diameter, spherical, terminal or intercalary. Antheridia up to three in number, from the same or another hypha as the oogonium, often formed close below the latter and not seldom hypogynous. Oospores, 14–18μ in diameter, not filling the oogonium, spherical, smooth, germinating after a rest of some months by a branching hypha.

From soil: Europe (20)

United States: Iowa (87), Kentucky (61), Mississippi (61), New York (61), North Carolina (58) (61), Oklahoma (61), Wisconsin (60) (61)

10. *Pythium pulchrum* von Minden
Syn. *P. epigynum* Hoehnk

Mycelium vigorous on corn meal agar or hemp seed in distilled water. Hyphae irregularly branched, up to 7.5μ in diameter. Sporangia spherical to pyriform, terminal or intercalary, occasionally two to four in a chain, abundant, usually 36–48μ in diameter, forming zoospores in a vesicle or germinating as conidia with one to many germ tubes. Tube of discharge short or up to 50μ or more in length, often with a swollen place at the base. In old cultures many may form spores inside the sporangium which do not escape but germinate *in situ*. Zoospores from fifteen to twenty-five or more to a vesicle, large, usually 9×14–16μ in the free swimming stage and 12–14μ in diameter when encysted. Oogonia terminal or intercalary (majority intercalary), occasionally two to four in a chain, smooth, spherical to elliptic, 21–38μ in diameter (majority about 30μ, very small oogonia sometimes present); walls fairly thick, may be angular in old stages. Oospores smooth, not filling the oogonium, spherical to elliptic, 16–26.4μ in diameter (majority about 24μ), in the mature stage with a heavy wall, and a central reserve globule surrounded by a layer of granular protoplasm in which a small refractive body is embedded. Antheridia hypogynous, androgynous and diclinous, one or two to an oogonium, often one on each side of an intercalary oogonium.

From soil: United States: Iowa (87)

11. *Pythium irregulare* Buisman

Mycelium well developed on hemp seed or carrot in distilled water and on agar. Hyphae 2–6μ in diameter with many side branches. Sporangia spherical to pyriform, terminal or intercalary, 10–30μ in diameter, not abundant in most cultures, forming spores in a vesicle or germinating directly as conidia. Tube of discharge short or equal to the diameter of the sporangium. Zoospores from a few to fifteen or more to a vesicle, 4–6 × 10–12μ. Oogonia terminal or intercalary, sessile or stalked, spherical to cylindric, varying greatly in shape; the wall smooth or with several projections varying in length, which may or may not be cut off by a cross wall, diameter of oogonium 14–26μ without projections. Oospores smooth, not filling the oogonium, 10–20μ in diameter, in mature stage of typical structure with a heavy wall and a central reserve globule, surrounded by a granular layer in which a refractive body is embedded. Antheridia stalked, one to three to an oogonium (usually only one), androgynous or diclinous, very rarely hypogynous, usually arising from the oogonial stalk.

From soil: United States: Iowa (167)

12. *Pythium mamillatum* Meurs

Mycelium nonseptate, 4–9.3μ in diameter, mostly 5.3–6.7μ, average 5.9μ. Sporangia globose, 14.3–20.7μ in diameter, average 16.3, five to fourteen zoospores produced in a bladder at the mouth of the sporangium. Zoospores reniform, flattened at one side. Oogonia globose, with obtuse, often curved protuberances. Antheridium clavate, arising from the oogonial hypha. After fertilization oogonia with protuberances 20.3–29.3μ in diameter, mostly 22–27μ, average 25.4μ, without protuberances 13–19.3μ in diameter, mostly 15.3–18μ, average 16.4μ. Length of protuberances 2.7–6.0μ, mostly 4–6μ, average 4.4μ. Oospores completely filling oogonium when ripe; oogonium wall 0.8–1.6μ thick, mostly 1–1.33μ.

From soil: United States: North Carolina (86)

13. *Pythium echinulatum* Matthews

Hyphae measuring 2–8μ in diameter. Sporangia spherical to cylindrical or intercalary, often catenulate, three or four in a series, measuring 10–30μ, averaging about 20μ in diameter; zoospores or germ tubes are produced. Oogonia spherical to cylindrical, terminal or intercalary, measuring 14–30μ, average about 24μ, in diameter exclusive of the many spines, 2–8μ in length. Antheridia monoclinous, typically hypogynous, one to four per oogonium, usually one. Oospores not filling the oogonium, 14–24μ, average about 20μ in diameter, one or two per oogonium, possessing a thick wall enclosing a single reserve globule and refringent body.

From soil: United States: Iowa (87), North Carolina (86)

14. *Pythium intermedium* de Bary

Mycelium intra- and extramatrical, forming a regular fine haze around the substratum in water culture. Hyphae very numerous, up to 6μ thick, regular, without intercalary swellings. Branching often at right angles, sometimes dichotomous, more usually lateral. In old cultures septa, with a distinct double contour, are not uncommon. The tips of all free branches usually end in spores. These measure 18–24μ in diameter, and are normally arranged in chains, up to thirteen in a single chain having been observed. When ripe they fall off readily, and can germinate immediately in fresh water. Growth may continue from the hyphae immediately under the spore, which is gradually pushed to one side. The lateral spore may be on a short stalk, or sometimes from a swollen part, immediately under the spore, which often remains, and in this case the lateral spore lies sessile. Sometimes the new hypha arises farther down, leaving the spore or chain of spores supported on a lateral stalk, which may itself give out branches and support new chains of spores. The chains are formed basipetally, the end spore being the oldest. The spores in the chain are usually spherical and divided from each other by short stalks, which may persist as tiny processes, rectangular in outline, on the fallen spores. Sometimes however, they are pyriform, in which case the narrow end of each arises directly from the spore below. The spores of a chain may germinate as sporangia or conidia, both forms occurring in the same chain. In young cultures large numbers of sporangia occur and discharge zoospores on addition of fresh water. In older ones the conidia are the chief organs found. The tube of discharge is always very short, about one-fourth of the diameter of the sporangium, and appears in any position, most frequently laterally. The conidia are often provided with thick walls, showing a distinct double contour. They can preserve their vitality if kept moist, for at least 11 months and can stand freezing. If completely air dried they soon die. Sexual organs have not been observed.

From soil: England (20), France (20), Germany (20), Ireland (20)

15. *Pythium elongatum* Matthews

Mycelium well developed in agar cultures and when grown on hemp seed and carrots in distilled water. Hyphae branched, 2–4μ in diameter. Conidia terminal or intercalary, spherical, pyriform, cylindric or curved, spherical ones 12–50μ in diameter and cylindric ones up to 65μ long, abundant in agar cultures and on hemp seed and boiled carrots in distilled water. Usually germinating by a large number of germ tubes immediately after bits of the mycelium are cut off and placed in pure water. Sporangia similar to conidia, only rarely formed. Tube of discharge very long in all cases observed. Zoospores many to a vesicle, $6 \times 10\text{–}12\mu$, biflagellate. Sexual organs never observed.

From soil: United States: North Carolina (86)

(2) **Pythiomorpha** H. E. Petersen

Mycelium much branched subdichotomously, the purely vegetative part bearing groups of irregular bud-like outgrowths or involved torulose complexes, protoplasm strongly refractive, pallid, bearing occasional conspicuous granules (cellulin?), walls giving a cellulose reaction. Zoosporangium borne on a slender unbranched or occasionally branched sporangiophore, ovate, ellipsoidal, citriform, or somewhat pyriform, with a broad blunt papillate apex, proliferous, the secondarily formed sporangia either sessile and "nested" within the primary sporangiophore or borne on a sporangiophore which extends through the discharge orifice to the outside. Zoospores of the secondary, laterally biflagellate type, emerging fully formed from the sporangium, sometimes surrounded by a quickly evanescent vesicle, capable of repeated emergence. Sexual reproduction oogamous, oospore thick-walled, sometimes developed parthenogenetically, upon germination forming one or more germ tubes which reestablish the thallus.

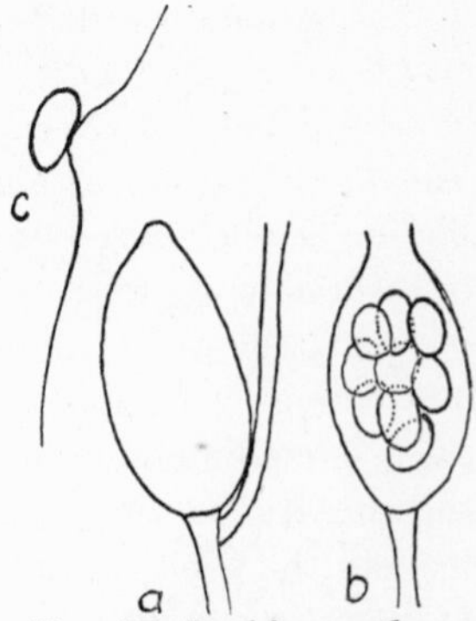

Fig. 37. **Pythiomorpha.** *a* sporangium; *b* sporangium with zoospores; *c* zoospore (after Lund).

A single species treated.

1. *Pythiomorpha undulata* Apinis

Mycelium sparingly branched, the hyphae $3\text{–}8\mu$ in diameter, often undulate, the walls stout, sometimes faintly golden or brownish. Zoosporangia either terminal and borne at the tip of the main hypha or its branches or by cymose branching appearing lateral, narrowly ovoid, $45\text{–}167\mu$ long $\times$ $20\text{–}52\mu$ in diameter, with a prominent terminal papilla, proliferous, the secondary sporangia either sessile and "nested" within the primary one or borne on a strongly undulate sporangiophore which extends through the orifice to the outside. Zoospores formed within the sporangium, $10\text{–}18\mu$ in diameter, with two lateral oppositely directed flagella, capable of repeated emergence. Sexual reproduction not observed.

From soil: Denmark (80)

II. ASCOMYCETES

Mycelium well-developed (except in the Saccharomycetaceae and some Taphrinaceae), thread-like, usually richly branched and septate. Fructification in the typical cases by spores formed in a sac, the ascus, after a nuclear fusion and meiosis. Spores in the majority of cases in definite number in each sac, usually a multiple of two, germinating by a tube, sometimes by yeast-like budding, never by zoospores. Conidia common in various forms.

A. PLECTASCALES

Fungi with well-developed mycelium on which globose fruiting bodies develop either on or in the substrate. These latter possess a sterile, almost always mouthless membrane (peridium). Asci occur as outgrowths or members of irregularly branching hyphae, filling the interior of the fruit-body with their great number and irregular arrangement; as a rule they are globose, two to eight spored; spores one- or many-celled. Conidia are known in many of the species.

a. *GYMNOASCACEAE*

Fruit-body imperfect, formed of many to few regular mycelial knots, whose external sterile parts compose a slightly differentiated, rudimentary, floccose peridium. Asci irregularly distributed, filling the cleistothecium, formed by swelling of lateral hyphae, globose, eight-spored. Conidia present.

KEY TO THE GENERA OF THE GYMNOASCACEAE

a. Peridial threads formed of thin-walled cells
  b. Ascospores hyalin — (1) **Arachniotus**
  bb. Ascospores brown — (2) **Amauroascus**
aa. Peridial threads formed of thick-walled cells
  b. Peridial threads smooth without spines — (3) **Pseudogymnoascus**
  bb. Peridial threads possessing spines — (4) **Gymnoascus**

(1) **Arachniotus** Schroeter

Cleistothecia globose, peridium formed of delicate web-like covering with threads of equal diameter as the vegetative hyphae. Ascospores globose or ellipsoid with walls, hyalin, yellow or red.

A single species treated.

1. *Arachniotus terrestris* Raillo

Mycelium delicate, effuse, white. Cleistothecia, globose, 135–250μ in diameter; peridium formed of white, delicate interlacing hyphae. Asci globose or oval, 13.5μ in diameter, eight-spored. Spores hyalin, oval, 5.4–6 × 3–4μ.

From soil: U. S. S. R. (108)

FIG. 38. **Arachniotus.** *a* cleistothecium; *b* ascus; *c* ascospores

(2) **Amauroascus** Schroeter

Cleistothecia globose; peridium of very delicate, loose web-like hyphae. Ascospores globose or ellipsoid with brown or purple-brown walls.

A single species treated.

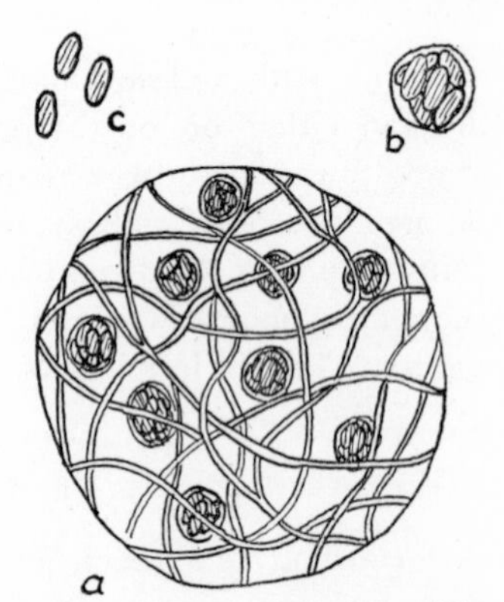

FIG. 39. **Amauroascus.** *a* cleistothecium; *b* ascus; *c* ascospores (after Raillo).

1. *Amauroascus niger* Schroeter

Cleistothecia globose or ellipsoid, white, 0.5–1 mm. in diameter. Asci ellipsoid 11–14 × 10–12μ. Ascospores ellipsoid, 4.5–6 × 3.5–4μ; walls brown, finely echinulate; spore mass black.

From soil: U. S. S. R. (108)

(3) **Pseudogymnoascus** Raillo

Mycelium white, diffuse; cleistothecia globose; peridium formed from thick-walled, smooth, net-like branching hyphae. Asci globose or slightly oval with eight spores. Spores oval or globose. In structure resembles Gymnoascus, differs in the peridium which possesses neither teeth nor spines.

KEY TO THE SPECIES OF THE GENUS PSEUDOGYMNOASCUS

a. Peridium formed from yellow hyphae, of uniform diameter 1. *P. vinaceus*
aa. Peridium formed from red hyphae, thickened at the nodes 2. *P. roseus*

1. *Pseudogymnoascus vinaceus* Raillo

Cleistothecia globose, orange, 135–270μ in diameter; peridium netted, formed of golden-yellow, smooth netted hyphae. Asci globose, 5.4μ in diameter with eight spores. Spores in mass rose-colored, single, hyalin, globose or oval, 2.7–3 $\times$ 2μ.

From soil: U. S. S. R. (108)

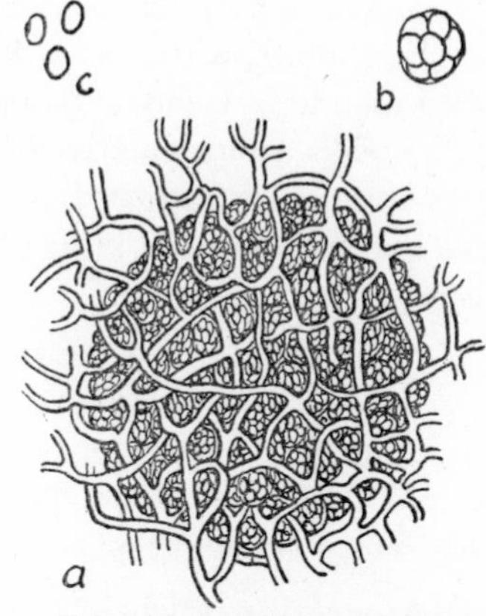

Fig. 40. **Pseudogymnoascus.** *a* cleistothecium; *b* ascus; ascospores (after Raillo).

2. *Pseudogymnoascus roseus* Raillo

Cleistothecia globose, rose-colored, 100–270μ in diameter; peridium netted, formed of red netted hyphae, inflated at the points of juncture. Asci subglobose 5.4μ in diameter, eight-spored. Spores in mass, rose-colored, single, hyalin, globose or oval; 2.7–3 $\times$ 2μ.

From soil: U. S. S. R. (108)

(4) **Gymnoascus** Baranetzky

Cleistothecia globose. Peridium formed from thick-walled, richly branched, lattice-like floccose hyphae; ends of branches pointed or truncate, ending in straight or weakly bent simple teeth or spines. Spores globose, ellipsoid or spindle shaped; walls hyalin or bright colored.

A single species treated.

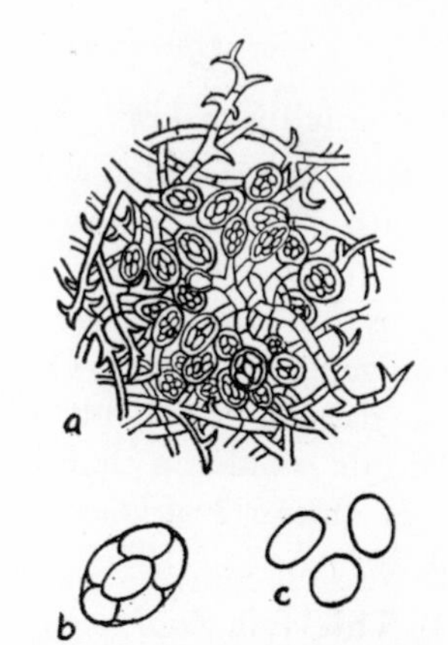

Fig. 41. **Gymnoascus.** *a* cleistothecium; *b* ascus; *c* ascospores (after Brefeld).

1. *Gymnoascus reesii* Baranetzky

Cleistothecia gregarious, globose, straw-colored, yellow-brown or orange, matted, coarse, 0.3–0.5 mm. in diameter. Peridia formed of thick-walled hyphae, richly branching at right angles forming an anastomosing lattice-work, loosely floccose, yellow or red-brown, and possessing short straight or slightly bent spines 10–15μ long. Spores globose or ellipsoid 4–4.5μ long, 3–4μ broad; walls smooth, red-brown.

From soil: Canada (15), U. S. S. R. (108)

b. *ASPERGILLACEAE*

Cleistothecia fully formed, usually very small, globose or knot-like, as a rule sessile, with a thin, dark carbonaceous, or membranous to fleshy, often pseudoparenchymatic peridium, which remains closed at maturity, then breaks up irregularly, seldom opens by a mouth or regular fracture. Asci irregularly crowded in the cleistothecia, globose to pear-shaped, two to eight spored. Spores one- to many-celled. Conidia of very various form.

KEY TO THE GENERA OF THE ASPERGILLACEAE

a. Ascocarps with bright colored walls
  b. Conidiophores with foot-cells and vesicular tips — (1) **Aspergillus**
  bb. Conidiophores with brush-like tips — (2) **Penicillium**
aa. Ascocarps with dark colored walls — (3) **Thielavia**

(1) **Aspergillus** Micheli

Cleistothecia globose, small, without a mouth. Peridium of one or more layers, smooth. Asci when ripe filling the entire interior of the perithecium, ellipsoid or oval, eight-spored. Spores irregularly arranged, one-celled.

See under *Aspergillus* p. 184, Nos. 2, 3, 4, 7, 8, 9

(2) **Penicillium** Link

Cleistothecia, globose, either formed as sclerotia (after a rest-period) or with continuous growth. Peridium pseudoparenchymatous or formed of a felt of loose hyphae. Asci at maturity filling the entire cleistothecium, globose or ellipsoid, four to eight spored. Spores in ascus irregularly arranged, one-celled. Conidiophores many times divided at their tips into parallel or slightly divergent branches which abstrict globose or elongate conidia in chains.

See under *Penicillium* p. 200, Nos. 35, 50, 102.

(3) **Thielavia** Zopf

Cleistothecia globose; walls brown, pseudoparenchymatic, without ostiole, and without appendages. Asci oval, eight-spored. Spores in ascus irregularly arranged, brown, one-celled.

A single species treated.

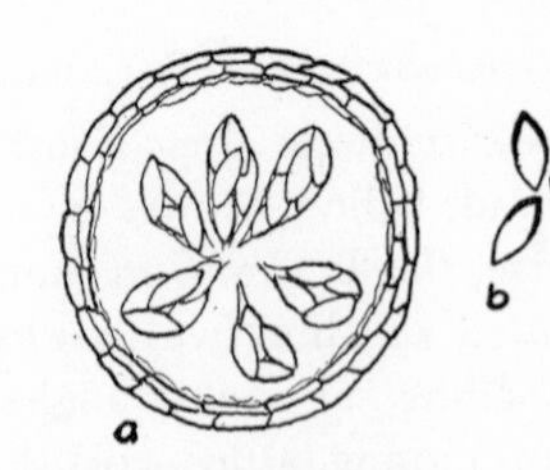

FIG. 42. **Thielavia.** *a* cleistothecium; *b* ascospores.

1. *Thielavia terricola* (Gilman and Abbott) Emmons
Syn. *Coniothyrium terricola* Gilman and Abbott

Colonies on corn meal agar broadly spreading, composed of white, cottony aerial hyphae and submerged hyphae, 1–6μ in diameter, branches constricted at base, homothallic. Ascocarps arising from an ascogonial coil, spherical, without ostiole, 80–125μ in diameter in fresh cultures on corn meal agar, reaching up to 250μ in diameter in old cultures on malt agar, brownish to almost black at maturity, color largely due to masses of dark spores within. Outer wall of cleistothecium composed of two or three layers of uninucleate, rather thick-walled cells, somewhat carbonized; inner wall of cleistothecium composed of thin-walled, flattened cells. Asci oval to pyriform, 16–19 $\times$ 25–35μ, deliquescing within the cleistothecium. Ascospores broadly fusiform or elliptical, slightly apiculate at both ends, dark olivaceous to brown, 7–9 $\times$ 10–16μ, with a wall much thickened at the end opposite the germ pore. No asexual stage known.
From soil: China (81)
United States: Iowa (50), Louisiana (50), Texas (92)

c. *PERISPORIACEAE*

Aerial mycelium superficial, dark, filamentous or lacking, or seldom in a firm stroma. Ascomata in the aerial mycelium or superficial on a stroma, black, more or less globose, seldom elongate, without an ostiole or weathering at the tip or tearing irregularly, without appendages. Peridium usually membranous, seldom carbonous, brittle. Asci occurring in a bush-like arrangement, usually elongate, Spores various. Paraphyses lacking.

A single genus treated.

(1) **Perisporium** Fries

Cleistothecia superficial, loosely gregarious, globose; peridium black, smooth, brittle, carbonous, usually opening irregularly at the tip. Asci club–shaped, stipitate. Spores long cylindrical tapering, four–celled, usually breaking up into part-cells, brownish-black.

A single species treated.

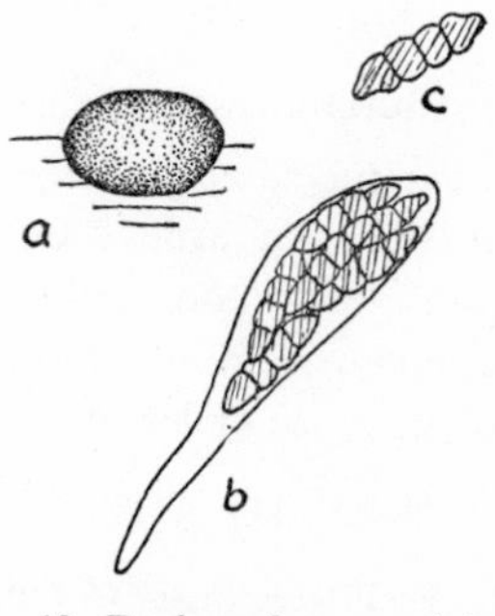

Fig. 43. **Perisporium.** *a* cleistothecium; *b* ascus; *c* ascospores (after Winter).

1. *Perisporium vulgare* Corda

Cleistothecia gregarious, superficial, black, shining. Asci broadly clavate, with a short stipe 35–40 $\times$ 17–19μ, eight-spored. Spores cylindric, four-celled, brown 28 × 5μ, the two middle cells oblong cubical, the end cells almost conical, 6–7 × 5μ.

From soil: U. S. S. R. (108)

B. SPHAERIALES

Mycelium thread-like, superficial or immersed, frequently forming a subiculum or stroma. Ascomata of different form, usually globose, with a more or less elongated mouth, with thin, or leathery, or carbonous, smooth or hairy peridium, free on the substrate, or more or less deeply sunken, or in a subiculum or a stroma. Asci, arising from the base of the perithecium, either single from the ascogenous tissue, or bush-like, opening by an ostiole, sometimes evanescent. Spores of various shape, globose, ovate, elongate, thread-like, one-celled to many-celled by cross walls or muriform, hyalin to yellow, green or brown to black, at times with gelatinous sheath or hyalin appendages. Paraphyses usually present. Asexual spores as conidia on free conidiophores, on acervuli, or in variously formed pycnidia.

a. *CHAETOMIACEAE*

Perithecia entirely superficial, free on a superficial mycelium. Peridium delicate, fragile, with flat, apical mouth (seldom wanting), usually forming a head of characteristically branching hairs. Asci club-shaped, or somewhat cylindrical, eight-spored, evanescent. Spores one-celled, of various shapes, dark colored. Paraphyses absent.

A single genus treated.

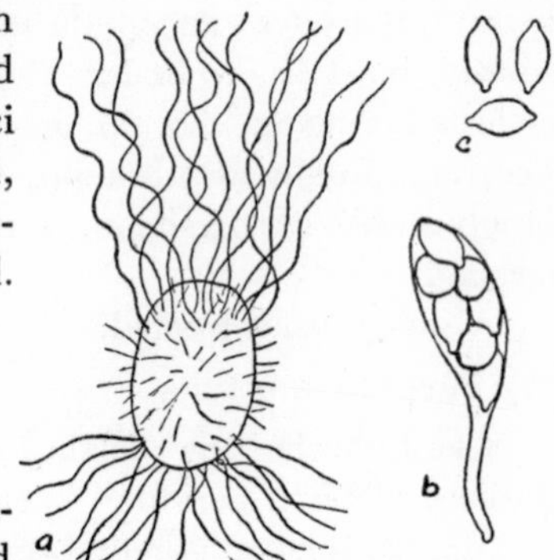

Fig. 44. Chaetomium. *a* perithecium; *b* ascus; *c* ascospores.

(1) **Chaetomium** Kunze and Schmidt

Perithecia superficial, thin-membranous, with an apical tuft of bristles and usually with an ostiole. Asci clavate, evanescent. Spores simple, hyalin to dark brown, more or less compressed.

KEY TO THE SPECIES OF THE GENUS CHAETOMIUM

a. Perithecia ovate to cylindrical, ostiolate
  b. Apical hairs all simple
    c. Hairs straight — 1. *C. subterraneum*
    cc. Hairs not straight; flexuous or contorted
      d. Hairs flexuous, numerous, 500–700μ long, olivaceous — 2. *C. globosum*

dd. Hairs contorted into loops or spirals
  e. Alternate loops in opposite directions — 3. *C. crispatum*
  ee. Spirally coiled
    f. Hairs irregularly and spirally twisted at the tip — 4. *C. cochliodes*
    ff. Hairs more or less regularly coiled at tip
      g. Hairs slender with six to fourteen coils — 5. *C. spirale*
      gg. Hairs short with five to seven coils or two to three loose convolutions — 6. *C. bostrychodes*
bb. Hairs all or partly branched
  c. Hairs up to 500μ in length, deeply incrusted — 7. *C. indicum*
  cc. Hairs up to 375μ in length, smooth or slightly incrusted — 8. *C. funicola*
aa. Perithecia globose, without an ostiole
  b. Hairs short, awl-shaped, incrusted — 9. *C. fimeti*
  bb. Hairs long, straight and spirally coiled — 10. *C. magnum*

1. *Chaetomium subterraneum* Swift and Povah

Forming on Blakeslee's agar circular colonies, at first grayish-white, then slate-blue-green-gray and slightly iridescent, becoming deep olive-gray and at maturity dark olive-gray almost black; reverse deep olive-green. Mycelium at first hyalin, later olivaceous, septate, 2–4μ in diameter, aggregated in rope-like strands from which brown perithecia arise. Perithecia 150–275 $\times$ 70–100μ, spherical when young, becoming ovate or flask-shaped, uniformly covered with mostly simple, straight, attenuate, six- to nine-septate, dark-brown setae, 52–105 $\times$ 3μ. Setae with bulbous base, 4–5μ in diameter, often with elbow-turn just at swelling, sometimes very slightly undulating in upper half; shorter setae 20–30μ long, often surrounding the ostiole. Asci when young clavate, with short hyalin stalk, evanescent, sporogenous portion 21–30 $\times$ 8–14μ, eight-spored, uniseriate. Spores 7–10 $\times$ 5–7μ, citriform, dark olive-green, often containing one or more large oil globules, when young greenish in color and containing droplets of refractive substance.

From soil: United States: Illinois (129)

2. *Chaetomium globosum* Kunze

Syn. *C. olivaceum* Cooke and Ellis
C. setosum Bainier
C. affine Corda

Perithecia scattered or gregarious, broadly ovate or ellipsoid, often pointed at the base, 250–300 $\times$ 200–250μ in fresh condition olivaceous, but in dry specimens dark brown and membranous, thickly and evenly clothed with slender, flexuous hairs. Apical hairs somewhat coarser than the others, simple, sparingly septate, minutely scabrous, 3–4μ thick, often 700μ long, in the fresh condition pale olivaceous, in dry condition light-brown. Asci oblong-clavate, slightly apiculate at both ends, 9–12 × 8–9μ.

From soil: Canada (16), Egypt (115)

United States: New Jersey (146) (147), New York (67)

3. *Chaetomium crispatum* Fuckel

Perithecia more or less gregarious, broadly ovate or subglobose, reaching a height of 400μ, membranous, dark brown, thickly clothed with hairs. Lateral and basal hairs smooth, septate, slender, pale brown. Apical setae forming a dense, black, spherical mass, 700μ in diameter above the perithecium, rigid, densely and minutely incrusted, light brown and distinctly septate at the base, gradually becoming darker toward the tip which is dark brown, 11μ thick, indistinctly septate, and irregularly contorted into six to eight loops, alternate loops being in opposite directions. Asci stipitate, the spore bearing part 82–100 × 10μ; paraphyses simple, elongate-clavate. Spores one-seriate, globose, or globose ovate, apiculate at both ends, 12–13 × 9–11μ, subhyalin or fuscous at maturity.

From soil: England (11), Japan (71)

United States: Louisiana (2)

4. *Chaetomium cochliodes* Palliser

Perithecia scattered or gregarious, broadly ovate to subglobose, 300–400μ in diameter, thin, membranous, dark brown, thickly clothed with hairs. Lateral and basal hairs pale brown, septate, slender, not exceeding 4μ in thickness even at the base, and gradually tapering toward the end. Apical hairs extremely flexuous, almost from the base, at the end becoming irregularly spirally curved, usually smooth but occasionally minutely scabrous at the base, numerous, often forming a densely interwoven mass extending 700μ above the perithecium. Spores subhyalin to pale brown, broadly ovate to subglobose, sometimes scarcely apiculate at the ends, 9–11 × 8–10μ.

From soil: Egypt (115)

United States: California (147), Hawaii (147), New Jersey (146) (147)

5. *Chaetomium spirale* Zopf

Dark brown to black. Perithecia of medium size 150–300μ, globose or ovate with a bluntly pointed base, seated on dark olive-yellow to brown rhizoids. Lateral hairs long, graceful, nearly straight or slightly flexed, very gradually tapering toward the tip, septate throughout, at base 3–5.5μ in thickness, dark olive-brown, sometimes smooth but more frequently roughened by irregular hyalin bodies of varying size and shape, becoming smooth above and fading to a colorless or pale yellow tip. Terminal hairs sparsely septate, dark, rich olive-brown, roughened by minute spines and warts, slightly paler and somewhat less roughened near the tips, straight or only slightly bent below for 300–370μ of their length, 4–6μ in thickness at base, spirally coiled above with six to fourteen turns. Asci clavate, with a short stalk, *pars sporif.* 34–43μ long. Spores citriform, slightly apiculate at either end or irregularly oval or spherical, dark, rich olive-yellow to olive-brown, 9 × 7μ (6–12 × 5.6–9μ), when seen edgewise, 5.5–7μ broad.

From soil: Canada (15), England (11)

6. *Chaetomium bostrychodes* Zopf

Steel-gray. Perithecia of medium size, extremely variable in shape, broadly ovate, globose or nearly cylindrical, generally with a bluntly pointed base, 340–220μ (168–350 × 131–230μ), frequently provided with black, straight, or recurved cirrhi. Lateral hairs not numerous, incrusted, clearly and evenly septate, tapering, at base dark olive-brown and about 3.8μ in thickness, at tips pale yellow or hyalin, frequently collapsed. Terminal hairs incrusted and roughened with spine-like projections throughout, at base straight or very slightly flexed, dark olive-brown to black and about 4μ in thickness, slightly less colored at tips, always more or less spirally coiled but in this respect extremely varied. In the type either regularly coiled with seldom more than five to seven convolutions which diminish almost imperceptibly in diameter toward the extremity, or irregularly coiled with two or three loose, irregular convolutions; in either case irregularly septate, producing along the convolutions one or more branches which in turn are septate and spirally coiled. Asci short, stout, clavate, eight-spored, 50 × 12μ, *pars sporif.* 24μ. Spores when young greenish, hyalin, with granular contents, when mature pale with olive-brown tint, oval to nearly spherical, clearly or obscurely apiculate, or rounded at both ends, frequently with an elliptical, refractive area abreast of each end, a characteristic observed only in this species, 7.4 × 6μ (6.4–8 × 5.6–4μ), when seen edgewise, compressed, 4.8μ broad.

From soil: United States: Iowa (101)

7. *Chaetomium indicum* Corda

Black. Perithecia small, globose to verruciform, 180 × 160μ (105–200 × 101–175μ), firmly attached to the substratum by dark olive-brown to black rhizoids. Lateral hairs comparatively few, rather rigid, septate, tapering to a blunt point or drawn out into a long, hyalin, collapsed tip, at base dark olive-brown to black, and about 5.3μ in thickness. Terminal hairs of two types which can be most clearly distinguished by studying the perithecium at different ages; (a) hairs which first appear from the top of the perithecium and which do not form a dense mass, stout, dichotomously branched with branches reflexed and roughened by spine-like projections, at base dark olive-brown to black and about 7.5μ in thickness, fading only slightly or becoming hyalin at the terminal branches; (b) hairs which appear later, forming at first a tuft about the ostiole, profusely branched at narrow acute angles, branches never reflexed, alternately constricted and inflated, light olive-brown or yellow, finely roughened, terminal branchlets incrusted with clusters of acicular or prismatic crystals. Asci club-shaped, eight-spored, 30 × 9.4μ. Spores hyalin when young, when mature dark, rich olive-brown, ovate to lemon-shaped, slightly apiculate at one or both ends, 5.5 × 4.5μ (5.3–7 × 4.5–5.6μ).

From soil: United States: Louisiana (50)

The following measurements were found for the soil culture: perithecia up to 235μ in diameter; hairs up to 550μ long; spores 5–5.5 × 3.5–4.5μ.

8. *Chaetomium funicola* Cooke

Perithecia more or less scattered, small, broadly ovate, about 150–110μ, dark brown, clothed on all sides with hairs. Lateral hairs simple, comparatively short, smooth, rhizoids slender, pale brown, flexuous. Apical hairs of two kinds, simple and branched; simple hairs lanceolate, extending 375μ above the perithecium, smooth or nearly so, dark brown or almost black at the base, gradually tapering to a point and becoming paler at the tip; branching hairs few in number or forming a mass 180μ above the perithecium, subhyalin or pale brown to dark brown, sometimes incrusted, usually smooth, with numerous ramifications, sometimes regularly dichotomous, more often irregularly branched; branches short, 15–20μ. Spores small, broadly obovate, scarcely apiculate, 4.5–6 × 4.5μ.

From soil: Canada (15)

United States: California (147), Maine (147)

9. *Chaetomium fimeti* Fuckel

Perithecia globose, 0.5–0.75 mm. in diameter, upper half covered with short, awl-shaped, thickened and incrusted brown hairs, and with long and thick cylindrical, wire-like rhizoids arising from the base. Asci club-shaped, very long stipitate, 80–132μ long, 14–18μ thick (*pars sporif*. 40–48μ long). Spores from in front, broad elliptic, 14–16 × 12μ, from the side spindle-shaped, apiculate at both ends, olive-brown.

From soil: U. S. S. R. (108)

10. *Chaetomium magnum* Bainier

Perithecia globose, at first blue-gray, then black, averaging 0.55 mm. in diameter, surrounded by radiating appendages of two types; the primary, essentially straight and slightly enlarged at the base, the others are twisted and remain more or less applied to the perithecial wall, are somewhat spirally coiled toward the base and end in a coil. Asci clavate, long stipitate, evanescent, containing eight spores. Spores subglobose to slightly elliptic with obtusely angled tips, 11.2–14 × 16.8μ.

From soil: U. S. S. R. (108)

b. *FIMETARIACEAE*

Perithecia either superficial, free or more or less deeply sunken in the substrate and erumpent by the neck or ostiole, without stromata or seldom entirely sunken in the stromata, thin-membranous to fragile, dark colored. Ostiole present, circular, without a hairy head. Asci usually cylindric, eight-spored. Spores one- or many-celled, dark colored. Paraphyses present.

KEY TO THE GENERA OF THE FIMETARIACEAE

a. Spores simple
  b. Spores with appendages (1) **Pleurage**
  bb. Spores not appendaged (2) **Fimetaria**
aa. Spores four- to many-celled (3) **Sporormia**

(1) **Pleurage** Fries

Perithecia scattered or aggregate, superficial or sunken; membranous or coriaceous, without stroma. Asci without an apical perforation, stretching at maturity. Paraphyses ventricose or filiform-tubular, usually agglutinate and longer than the asci. Spores ellipsoid; with or without primary appendages, but always having attached to them at maturity two or more hyalin, gelatinous, secondary appendages of variable length.

A single species treated.

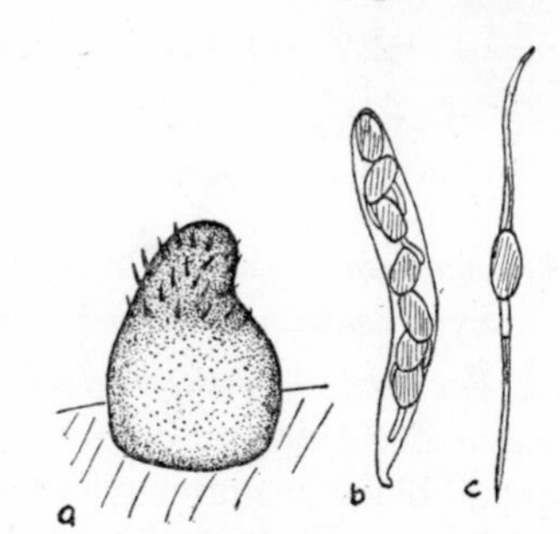

Fig. 45. **Pleurage.** *a* perithecium; *b* ascus; *c* ascospore (after Griffiths).

1. *Pleurage verruculosis* Jensen

Perithecia scattered or aggregated, sunken but becoming partly to entirely exposed at maturity, membranous to carbonous, black, opaque,

350–750 × 225–375μ, pyriform to subglobose, with straight or curved beak. Asci four-spored, cylindrical, broadly rounded at apex and tapering below into a slender stipe, perforate, 90–150 × 11–16μ; paraphyses filiform, slightly tapering upward, longer than the asci, septate to articulate. Spores vertically uniseriate, long ovate when young to subglobose at maturity, obtusely pointed above and broadly rounded to truncate below, germ pore prominent, strongly tuberculate, ranging in color from hyalin when young through brown to black at maturity, 12–14 × 16–18μ, primary appendage 6–8μ and conic shortly after migration of protoplasm from below is completed and the septum is formed; at full maturity it becomes a shrunken hyalin appendage of 3–4μ in length; secondary appendages entirely wanting.

From soil: United States: New York (57)

(2) **Fimetaria** Griffiths and Seaver

Perithecia superficial or sunken, dark and opaque, membranous or coriaceous. Asci with an apical perforation and stretching at maturity. Spores simple, usually dark brown and wholly or partially enclosed in a fugacious envelope.

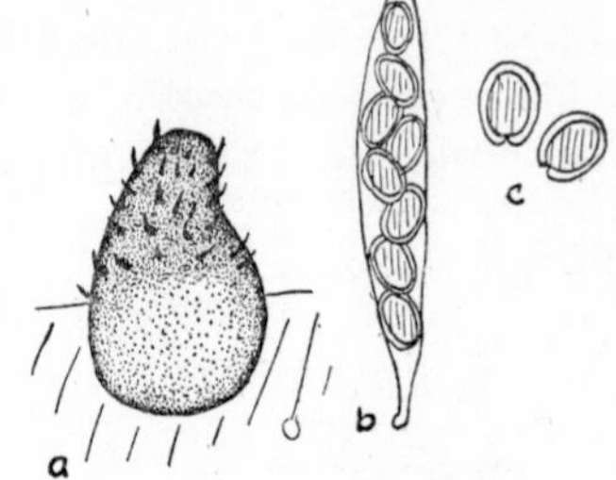

Fig. 46. **Fimetaria.** *a* perithecium; *b* ascus; *c* ascospores.

KEY TO THE SPECIES OF THE GENUS FIMETARIA

a. Paraphyses lacking — 1. *F. sylvatica*
aa. Paraphyses present — 2. *F. fimicola*

1. *Fimetaria sylvatica* (Daszewska) Griffiths and Seaver
Syn. *Sordaria sylvatica* Daszewska

Perithecia black, superficial, 500–800μ in height, 200–300μ at the base. Asci, long cylindric; 160–180 × 12–16μ, eight-spored. Ascospores ovate, black at maturity, 18–24 × 10μ; paraphyses lacking.

From soil: Switzerland (38)

2. *Fimetaria fimicola* (Rob.) Griffiths and Seaver

Perithecia thickly gregarious, partially immersed at the base, later superficial, bulb-like, with a short conical, thick, often slightly curved neck, blackish-brown, about ¼ mm. in diameter, smooth. Asci cylindric, more or less long stipitate, slightly narrowed above, with rounded or slightly truncate tips, and slightly thickened wall, 120–140μ long (*pars sporif.*) 17–19μ thick. Paraphyses thickened. Spores up to eight, in one series or two, elliptic, blackish-brown with gelatinous sheath, 19–22 × 10–12μ.

From soil: Canada (15)

(3) **Sporormia** de Notaris

Perithecia globose or ovate, sunken or less frequently superficial, with papilliform to cylindric beak, membranous to coriaceous and sometimes slightly brittle. Asci, cylindric to clavate with an internal membrane which is usually perforate at the apex. Spores cylindric, three- to many-septate, usually dark brown and opaque and surrounded by a hyalin gelatinous envelope.

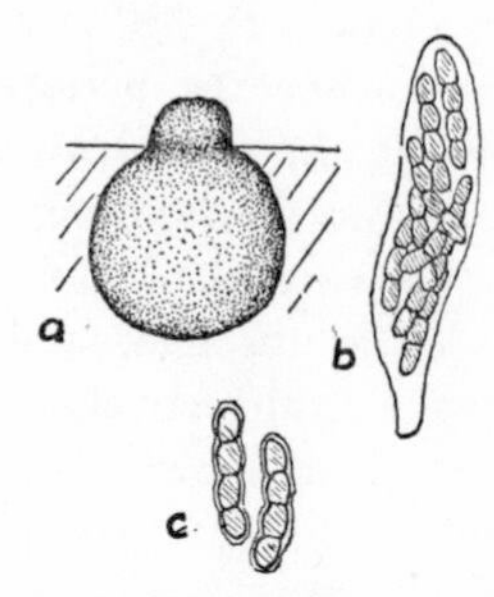

FIG. 47. **Sporormia.** *a* perithecium; *b* ascus; *c* ascospores (after Griffiths).

KEY TO THE SPECIES OF THE GENUS SPORORMIA

a. Base of ascus contracted to form long slender tips — 1. *S. fasciculata*
aa. Base of ascus contracted to form a short, blunt, slightly curved base — 2. *S. intermedia*

1. *Sporormia fasciculata* Jensen

Perithecia scattered or aggregated in small clusters, sunken, with small papilliform beak projecting to the surface, later many become partly exposed, globose, then membranous, inclined to be brittle, black, opaque, 250–525μ in diameter. Asci eight-spored, broad clavate, broadly rounding above and rapidly contracting just below spores to form a long slender stipe 45–60 × 16–30μ; stipe about two-fifths the length of the ascus; paraphyses absent. Ascospores fasciculate, straight or very slightly curved, four-celled, rounded at both ends, deeply constricted and easily separating, 25–30 × 4–7μ, ranging in color from hyalin when young through light brown to dark brown, opaque.

From soil: Canada (15)

United States: New York (67)

2. *Sporormia intermedia* Auerswald

Perithecia scattered, sunken but becoming superficial, pyriform, 385–875 × 205–480μ, dark brown to black and opaque, coriaceous or often slightly brittle, covered even to the tip of the beak with simple, flexuous or bristle-like, septate, smooth, pale-brown hairs, the lower serving as rhizoids and being branched, or often with age the hairs disappear leaving only papillate projections as evidence of their presence. Paraphyses sparingly branched, filiform, numerous, longer than the asci and mixed with them, septate, rather persistent. Asci clavate-cylindric, broadly rounded above and contracted below into a short, blunt, usually curved base, 125–230 × 22–30μ, eight spored, opening by a thimble-like rupture when the perforate membrane becomes plainly visible, rather persistent. Spores in two to three series, overlapping, four-celled, ranging from hyalin when young through pale, olivaceous-yellow, pale brown to dark brown and opaque, cylindric, straight or slightly curved, broadly rounded at the ends and usually deeply constricted, 47.5–65 × 9–15μ, having a hyalin envelope surrounding the entire spore, swelling greatly in water and showing striations continuous with the septa of the spore.

From soil: England (11)

c. *SPHAERIACEAE*

Perithecia single or gregarious, free or surrounded by a thread-like subiculum and then appearing sunken, but without a real stroma. Peridium membranous, leathery, woody, brittle. Ostiole present, papilliform, never long drawn out. Asci with persistent wall. Spores of various shapes, sometimes with appendages.

KEY TO THE GENERA OF THE SPHAERIACEAE

a. Spores one-celled, hyalin, ellipsoid (1) **Trichosphaeria**
aa. Spores three- to many-celled, dark colored (2) **Melanomma**

(1) **Trichosphaeria** Fuckel

Perithecia free, sessile, globose. Walls carbonous or woody, black, covered with hairs. Ostioles level or short wart-like. Asci cylindric, eight-spored. Spores one- or two-celled, hyalin. Paraphyses present.

A single species treated.

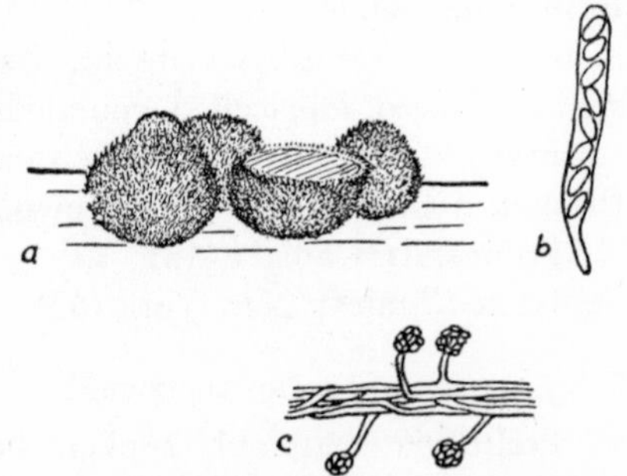

FIG. 48. **Trichosphaeria.** *a* perithecia; *b* ascus; *c* conidiophore and conidia (after Lindau).

1. *Trichosphaeria pilosa* (Persoon) Fuckel

Perithecia small, usually thickly gregarious, forming an extended black cover, seldom scattered, superficial, woody, globose-ovate, black, with a simple ostiole, surrounded thickly by hairs, 0.2 mm. in diameter. Asci cylindric or slightly swollen, tapering below into a stipe, eight-spored, 50–60 × 4–5μ, surrounded by numerous thread-like paraphyses. Spores in a single rank or at times irregularly two-ranked, elliptic, hyalin, one-celled, 5–8 × 3–4μ.

From soil: Egypt (115)

(2) **Melanomma** Fuckel

Perithecium superficial, often erumpent from a turf, globose or ovate. Peridium carbonous, brittle, black, smooth or seldom hairy. Ostiole wart- or keg-shaped. Asci cylindrical to clubbed, eight-spored. Spores long to almost spindle-shaped, with two- to many-septa, brown to black. Paraphyses thread-like.

A single species treated.

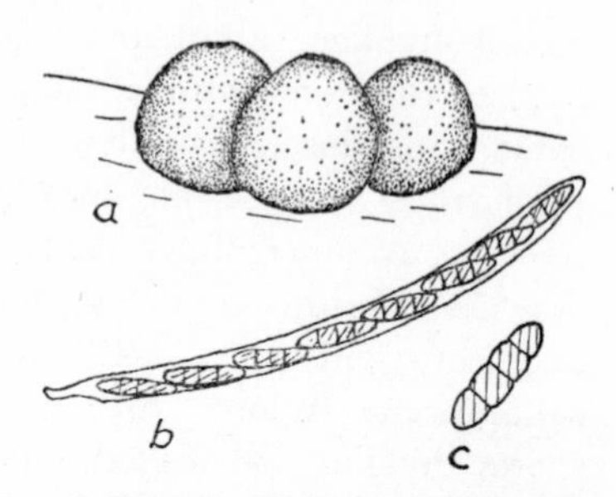

FIG. 49. **Melanomma.** *a* perithecia; *b* ascus; *c* ascospore (after Lindau).

1. *Melanomma sylvanum* Saccardo et Spegazzini

Perithecia gregarious, subsuperficial with bases slightly sunken in the substratum, globose, ¼ mm. in diameter, obtusely papillate ostiolate, then subcupulate, subcarbonous, black. Asci clavate-cylindric, eight-spored, 90–100 × 12μ, short stipitate, with rounded apices. Paraphyses thread-like. Spores in two-ranks, oblong-clavate, with three-septa, brown, 25–30 × 6–7μ.

From soil: U. S. S. R. (108)

d. *CERATOSTOMATACEAE*

In characteristics approaches the Sphaeriaceae. Usually the peridium is not brittle carbonous but more nearly membranous to leathery. The ostiole is always more or less drawn out into a snout, often long, hairy. Spores of different form.

A single genus treated.

(1) **Ceratostomella** Saccardo

Perithecia superficial, free or somewhat sunken in the substrate, globose, with ostiole drawn out into a long beak or hair. Walls, membranous-leathery to carbonous. Asci egg-shaped, eight-spored, very soon evanescent. Spores long, truncate or jointed, one-celled, hyalin.

A single species treated.

1. *Ceratostomella adiposa* (Butler) Sartoris

Syn. *Dematium scabridum* Gilman and Abbott

Mycelium densely woolly, dark, from the brown hyphae, abundantly branched, fertile hyphae unbranched, septate, bearing endoconidia. Endoconidia polymorphic, cylindrical, pyriform or globose, at times hyalin or brown, smooth, at other times dark, rough, 9–25 × 4.5–18μ. Perithecia globose, hairy, black, with an erect rigid neck, 2–6 mm. × 50μ, and a subfimbriate ostiole. Asci evanescent. Ascospores hyalin, one-celled, strongly crescent-shaped, with pointed ends, 6.5 × 3.5μ, immersed in a fatty mucus.

From soil: Louisiana (50)

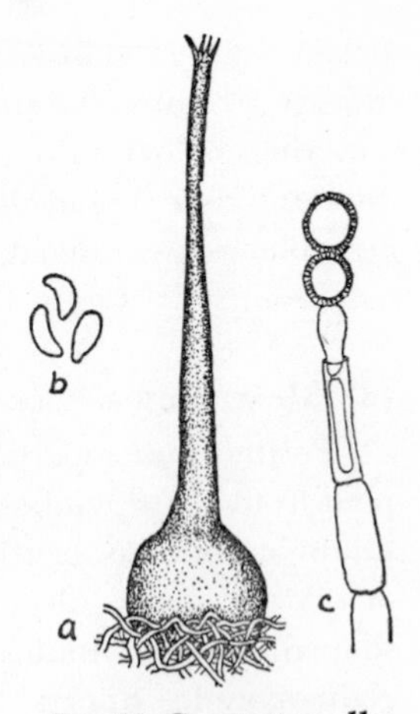

FIG. 50. **Ceratostomella.** *a* perithecium; *b* ascospores; *c* conidiophore and conidia.

C. HYPOCREALES

Mycelium thread-like, septate, superficial or submersed, hyalin or bright colored by its contents, often forming a fleshy stroma of bright, never dark, colors. Perithecia usually spherical, seldom keg- or flask-shaped, wholly free on the substrate or on a subiculum, or on or in a stroma, partly or completely submerged. Peridium bright colored, white, yellow, red, violet, brown, etc., but never black, membranous to fleshy, always delicate, seldom lacking. Ostiole always present, sometimes long, drawn out, but in most cases only papilliform. Asci more or less long, not evanescent, usually eight-spored. Spores of various shapes, frequently thread-like, breaking into many parts in the ascus or budding yeast-like. Paraphyses present or absent.

A single family treated.

a. *NECTRIACEAE*

Perithecia entirely free on the substrate or seated on a fleshy or tubercular stroma, but when the latter is present, perithecia are always superficial, usually in caespitose clusters; stroma often obscured at maturity by the perithecia.

A single genus treated.

(1) **Neonectria** Wollenweber

Perithecia in context and shape close to Nectria, single or gregarious, bright colored. Spores long, thin, ellipsoid, resembling Mycosphaerella, one- to many-septate. Conidia cylindric, referred to Ramularia, or Cylindrocarpon, chlamydospores intercalary.

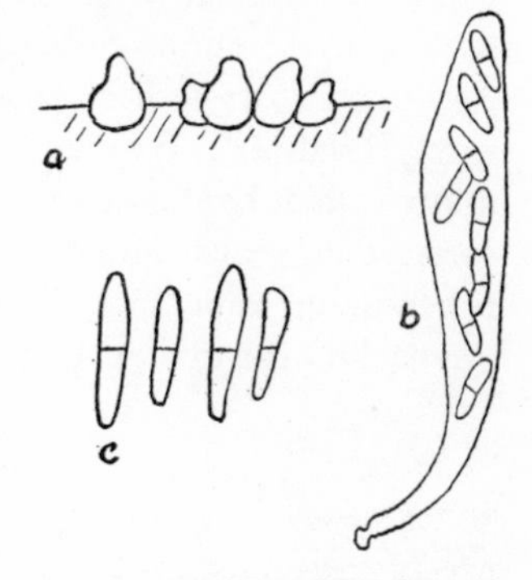

FIG. 51. **Neonectria.** *a* habit, perithecia; *b* ascus; *c* ascospores (after Wollenweber).

1. *Neonectria ramulariae* Wollenweber
   Syn. *Cylindrocarpon magnusiana* (Saccardo) Wollenweber
   Ramularia magnusiana Saccardo

Perithecia solitary or aggregated in acervuli, red, 200–300 × 170–250μ, ovate or globose. Asci eight-spored. Spores in mass pale yellow, singly hyalin, long, typically ellipsoid, one-septate, 12–15 × 3.25–4μ, in state of germination one- to three-septate. Conidia cylindric or slightly dorsiventral, sometimes semiglobose or slightly apiculate at the base, one-septate, 20–27 × 3.5–4.5μ, rarely nonseptate, very rarely three-septate, in tubercular sporodochia or in minute pale yellow columns, erumpent from the host epidermis. Chlamydospores few, intercalary and at times formed within the walls of the conidia.

From soil: United States: Texas (152)

D. PEZIZALES

Ascophores consisting of discoid or cup-shaped apothecia, or more rarely clavate, columnar, or piliate; free or sometimes seated on a subiculum or springing from a sclerotium, ranging in size from a fraction of a millimeter to several centimeters, variously colored. Apothecia concave, plane or convex, and circular or subcircular in form, more rarely elongate or star-shaped. Hymenium either enclosed by the apothecium or the excipulum when young or free from the first. Hypothecium very poorly developed or partially to entirely enclosing the hymenium when young, usually expanding at maturity leaving the hymenium freely exposed or, in a few cases, remaining closed until pierced or ruptured by the maturing asci, the tissue of the hypothecium either composed of loosely interwoven hyphae, prosenchymatous, or giving rise to a parenchyma-like tissue, pseudoparenchyma, or pseudoparenchymatous below and filamentous above. Pileus, when present, bell-shaped, saddle-shaped, but never cup-shaped, even or irregularly convoluted or corrugated, surmounted by the hymenium. Substance fleshy, leathery, cartilaginous or horny. Asci ovate to cylindric, two- to many-spored, operculate or inoperculate, rarely bilabiate. Spores globose, ellipsoid, fusiform or filiform, one- to many-celled, hyalin or variously colored, yellowish, violet, brown or more rarely olivaceous, smooth or variously sculptured, echinulate, verrucose, tuberculate, reticulate, ringed or marked with irregular ridges. Paraphyses filiform to clavate, simple or branched, variously colored.

a. *PEZIZACEAE*

Apothecia cup-shaped to discoid, more rarely convex, sessile or stipitate, variously colored, externally naked or clothed with hairs. Substance fleshy, waxy, leathery, cartilaginous or horny. Hairs varying from a soft tomentum to stiff bristles, hyalin or colored. Asci cylindric to ovate, operculate or more rarely opening by a transverse slit at the apex, giving the open ascus a bilabiate appearance, occasionally marked by a thickened ring or collar near the apex, two- to many-spored. Spores globose, ellipsoid or fusiform, hyalin or colored, the color ranging from yellowish to brown, violet, or more rarely greenish, smooth or variously sculptured. Paraphyses present, filiform or clavate, hyalin or variously colored.

KEY TO THE GENERA OF THE PEZIZACEAE

a. Ascocarps gelatinous (1) **Bulgaria**
aa. Ascocarps fleshy
  b. Spores becoming violet, later brown to blackish (2) **Ascobolus**
  bb. Spores permanently hyalin, or almost pale brown
    c. Vegetative hyphae immersed in the substrate (3) **Humarina**
    cc. Vegetative hyphae superficial (4) **Pyronema**

(1) **Bulgaria** Fries

Cups gregarious with a short thick stem. Externally dark colored, rough, often with short hairs, gelatinous, shrinking when dry. Asci cylindrical, generally eight-spored. Spores elliptical or unequal sided, one-celled, hyalin, then brown. Paraphyses forming a colored epithecium.

A single species treated.

1. *Bulgaria inquinans* Fries

Caespitose, turbinate, firm, gelatinous, externally rough, umber. Hymenium at first concave, becoming plane, black or dark purple. Asci clavate, very long. Ascospores eight, elliptic unequal-sided, often nearly pointed at one end, 10–14 × 5–6μ. Paraphyses filiform slender.

From soil: Egypt (115)

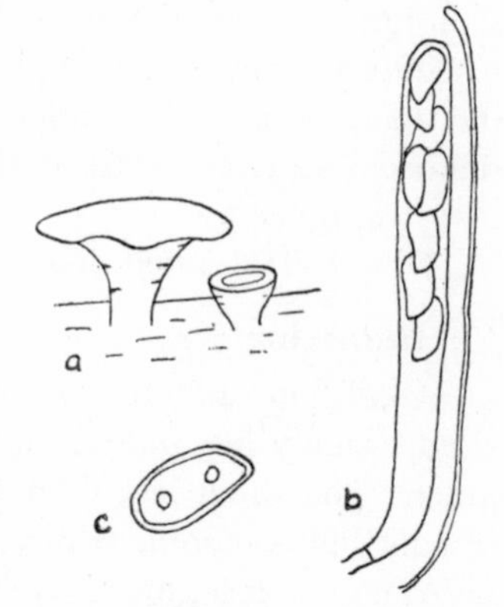

FIG. 52. **Bulgaria.** *a* apothecia; *b* ascus; *c* ascospore.

(2) **Ascobolus** Persoon

Cups sessile or substipitate, superficial or partially immersed in the substrate, externally smooth or pilose. Hymenium concave, plane or convex. Substance soft, fleshy or waxy, usually greenish. Asci cylindric to clavate or subovate, four- to eight-spored. Spores becoming blue or purple, fading to brown or blackish, ellipsoid to subglobose, smooth or becoming sculptured; spore-sculpturing very variable, consisting of warts, ridges, or crevices. Aparaphyses slender and adhering together.

A single species treated.

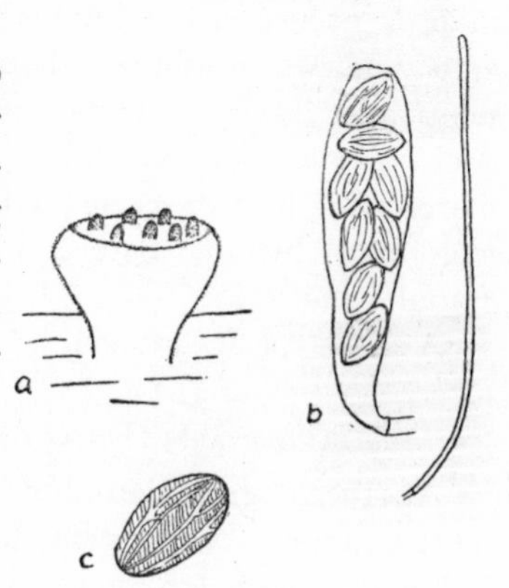

FIG. 53. **Ascobolus.** *a* apothecium; *b* ascus; *c* ascospore.

1. *Ascobolus stercorarius* (Bulliard) Schroeter

Apothecia scattered or thickly gregarious, sessile, often partially buried in the substratum, later becoming superficial or subsuperficial, at first globose or subglobose and closed, opening by a circular aperture and expanding, finally scutellate to discoid, reaching a diameter of 5 mm., externally pale-yellow or greenish (whitish in dried specimens), furfuraceous. Hymenium concave to plane, at first yellowish or greenish, becoming dotted over with the protruding asci which appear black, finally entirely black. Asci clavate, gradually tapering below into a stem-like base, reaching a length of 200–250μ and a diameter of 30μ, eight-spored. Spores partially two-seriate or irregularly disposed, ellipsoid, thick-walled, at first hyalin and more or less granular within, becoming violet, later brown, smooth, becoming sculptured, 11–14 × 20–30μ; spore-sculpturing taking the form of ridges and crevices which have a tendency to be longitudinally disposed, occasionally anastomosing and then giving the spore a reticulate appearance. Paraphyses slender, about 2μ in diameter, scarcely enlarged above, embedded in golden-yellow mucilaginous substance.

From soil: United States: Iowa

(3) **Humarina** Seaver

Apothecia minute to medium-sized, usually less than 1 cm. in diameter and often less than 1 mm.; usually light-colored, white, yellow, orange, or purple, more rarely dark-colored, brownish or blackish, usually discoid or more rarely cup-shaped, occurring on humus often among mosses, or more rarely on the stems and leaves of higher plants. Asci clavate, four- to eight-spored, spores one- to two-seriate, ellipsoid to fusiform, smooth or becoming sculptured, hyalin. Paraphyses slender or stout, usually containing a granular matter.

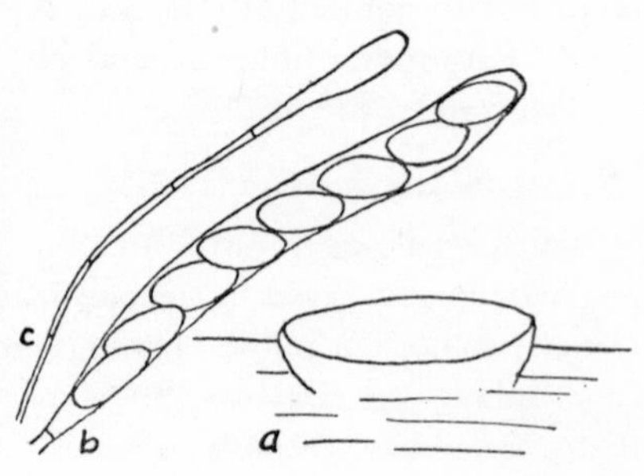

FIG. 54. **Humarina.** *a* apothecium; *b* ascus; *c* paraphysis.

A single species treated.

1. *Humarina convexula* (Persoon) Seaver

Apothecia gregarious but not usually crowded, at first globose or short cylindric, soon opening, pale to bright orange, rarely exceeding 1–2 mm. in diameter and often less. Hymenium becoming convex, similar in color to the outside of the apothecium. Asci cylindric above, tapering rather abruptly below, reaching a length of 200μ and a diameter of $18–20\mu$, eight-spored. Spores obliquely or irregularly disposed, broad ellipsoid, hyalin, smooth, containing either one or two large oil-drops which reach a diameter of about 10μ, entire spore $12–14 \times 22–25\mu$. Paraphyses straight or very slightly curved, rather stout, about 2μ or 3μ thick below, strongly enlarged above when they reach a diameter of $7–8\mu$, filled with rather coarse orange granules.

From soil: Canada (16)

(4) **Pyronema** Carus

Vegetative mycelium superficial, hyalin, thin-walled, septate, branched, the branches proceeding almost at right angles to the main hypha, frequently anastomosing, filled with conspicuous vacuoles, giving rise to paired sex organs. Sex organs occurring in clusters of several pairs each, each pair consisting of a clavate antheridium and an inflated oogonium surmounted by a slender trichogyne, which at a later stage usually fuses with the oogonium, each cluster of sex organs finally giving rise to a compound apothecium. Apothecia open from the first, never enclosed by the excipulum. Asci cylindric, eight-spored. Spores ellipsoid, hyalin, smooth. Paraphyses filiform.

A single species treated.

1. *Pyronema omphalodes* (Bulliard) Fuckel
   Syn. *P. confluens* Persoon

Apothecia small, not usually exceeding 1–2 mm. in diameter, soon becoming confluent and forming congested masses several cm. in diameter. Apothecial masses circular or irregular, often interrupted, pale orange, surrounded by a dense superficial white mycelial growth. Hymenium usually convex, the color varying greatly with conditions, sometimes only slightly yellowish, to bright orange, occasionally with a purplish tinge. Asci cylindric or subcylindric, 150μ long and 10–14μ in diameter. Spores one-seriate, ellipsoid, smooth hyalin, 5–8 $\times$ 10–13μ. Paraphyses rather stout, very slightly enlarged above, reaching a diameter of 6–7μ at their apices, filled with orange granules.

On burnt-over or heated soils.

III. FUNGI IMPERFECTI

Mycelium composed of septate, hyalin or colored hyphae or budding cells. Hyphae distinctly individual or, at times, interwoven into a plectenchyma. Stroma often present. Fructification consisting of singly occurring conidiophores, conidiophores in layers or conidial receptacles (pycnidia), never typical basidia or asci. The Fungi imperfecti are the asexual spore stages of members of other divisions of the fungi, chiefly Ascomycetes.

A. **SPHAEROPSIDALES**

Conidia formed within pycnidia which may remain entirely closed, open by a pore or a long slit or finally become disk-like.

a. *SPHAERIOIDACEAE*

Peridium of the pycnidium, membranous, carbonous or almost leathery, black, never fleshy or bright colored, globose, club-shaped or lens-shaped, remaining closed or opening by a pore. Pycnidia superficial or immersed, with or without a stroma. Seldom as chambers in a stroma without special peridium. Conidiophores of various sorts, often very short, usually simple. Spores of various shapes, hyalin or colored, one- to many-celled.

KEY TO THE GENERA OF THE SPHAERIOIDACEAE

a. Conidia one-celled, hyalin, globose, ovate, or oblong, often curved
  b. Pycnidia separate, smooth
    c. Not beaked; conidia less than 15μ in length — (1) **Phoma**
    cc. Beaked — (2) **Sphaeronaema**
  bb. Pycnidia hairy — (3) **Pyrenochaeta**

aa. Conidia one-celled, dark, globose, ovate or oblong
  b. Pycnidia smooth, membranous, conidia on short conidiophores — (4) **Coniothyrium**
  bb. Pycnidia hairy or setose — (5) **Chaetomella**

(1) **Phoma** (Fries) Desmazieres

Pycnidia globose or slightly lens-shaped with a small papilla at the apex, membranous to leathery or almost carbonous, black. Spores small, egg-shaped, spindle-shaped, cylindric or almost spherical, one-celled, hyalin, usually with two oil drops. Conidiophores thread-like, seldom short, or almost lacking, simple or sometimes forked.

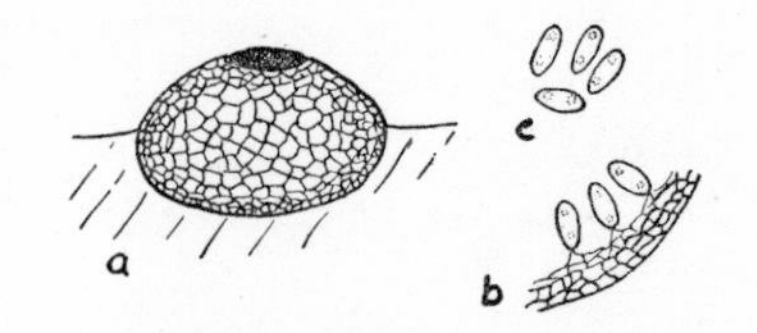

FIG. 55. **Phoma.** *a* pycnidium; *b* pycnidial wall; *c* conidia.

KEY TO THE SPECIES OF THE GENUS PHOMA

a.	Conidia 9.5–12 × 3–4μ	1. *P. humicola*
aa.	Conidia 5–7 × 4μ	2. *P. hibernica*

1. *Phoma humicola* Gilman and Abbott

Colonies dark brown, broadly spreading, largely submerged with little aerial hyphae. Pycnidia dark brown to black, membranous, produced slowly but abundantly, scattered, erumpent through the subicle, subglobose to pyriform, with a short neck, from 150 × 125μ up to 600 × 500μ in size. Conidia oblong or bacillate, with rounded ends, hyalin, 9.5–12.5 × 3–4μ.

From soil: United States: Utah (50)

2. *Phoma hibernica* Grimes, O'Connor and Cummins

Colonies flat, compact with submersed hyphae with pycnidia imbedded on upper surface. Pycnidia with one, two or three ostioles, globular, flask-shaped or lenticular, walls brown, 60–108 × 50–200μ (72–80 × 65–75μ). Spores hyalin, pink in mass, one-celled, oblong, bluntly rounded at both ends, biguttulate, 4 × 5–7μ.

From soil: Canada (16)

(2) **Sphaeronaema** Fries

Pycnidia membranous, leathery or carbonous, immersed or superficial, globose with few or many snout-like or pear-shaped ostioles. Spores ovate or elongate, one-celled, hyalin or almost hyalin, often extruded as a globose mass.

A single species treated.

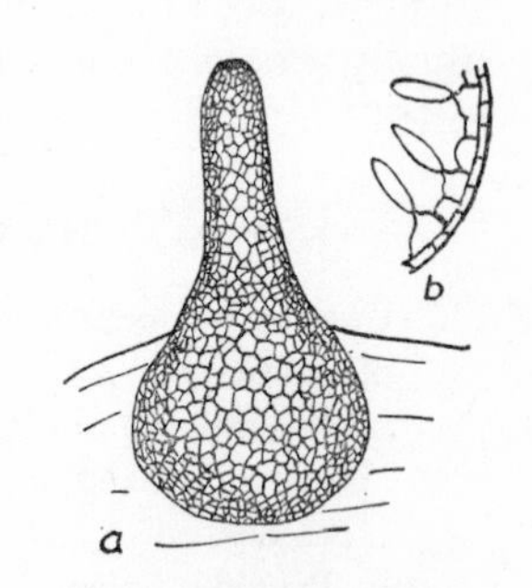

FIG. 56. **Sphaeronaema.** *a* pycnidium; *b* pycnidial wall.

1. *Sphaeronaema spinella* Kalchbrenner

Pycnidia gregarious, pear-shaped, somewhat curved, from an ovate base, black. Spore-mass saffron colored. Spores long-cylindric, rounded at both ends, hyalin.

From soil: Canada (16)

(3) **Pyrenochaeta** de Notaris

Pycnidia spherical, flask-shaped, single and erumpent, membranous or almost carbonous, black, with stiff simple setae chiefly at the top, ostiolate. Spores oval, elongate to cylindrical, almost hyalin, one-celled. Conidiophores, rod-like, branched.

A single species treated.

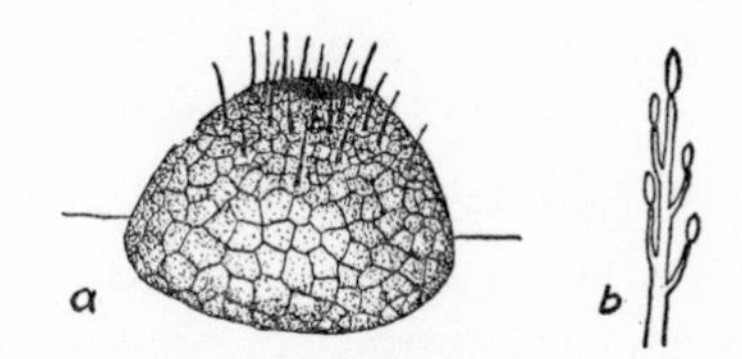

Fig. 57. **Pyrenochaeta.** *a* pycnidium; *b* conidiophore and conidia.

1. *Pyrenochaeta decipiens* Marchal

Pycnidia scattered, superficial, globose, 150–200μ in diameter, thickly parenchymatic, red-brown, tip slightly narrowed, with dark rigid setae, 50–80 × 3.8–4.5μ, little septate, with a minute ostiole, and conidiophores present. Spores abundant, straight, ovate to globose, 3.5–4.5 × 2.0–2.3μ, two-guttulate, hyalin.

From soil: Egypt (115)

(4) **Coniothyrium** Corda

Pycnidia globose or flattened below with papillate mouth, black, membranous to almost carbonous. Spores globose or ellipsoid, small, brown, one-celled. Conidiophores short, simple, or almost lacking.

A single species treated.

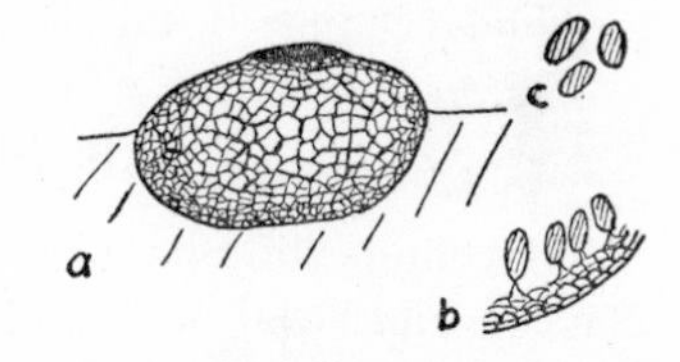

Fig. 58. **Coniothyrium.** *a* pycnidium; *b* pycnidial wall; *c* conidia.

1. *Coniothyrium fuckelii* Saccardo

Colonies at first white, subfloccose, later becoming all black, with white aerial mycelium; reverse at first creamy, later turning black. Pycnidia submerged, scattered, black, 180–200μ in diameter (Waksman (146) gives 240–350μ), depressed, spherical, with scarcely apparent, slightly protruded ostiolar papilla. Spores very numerous, brown, elliptical, apiculate at one end, 2.5–3.5 × 3–5.2μ; sporophores not visible.

From soil: U. S. S. R. (108)

United States: New Jersey (146) (147)

(5) **Chaetomella** Fuckel

Pycnidia superficial, sometimes short stipitate, without a mouth, but covered on the whole surface with long hairs. Spores cylindric or somewhat spindle-shaped, somewhat curved, colored. Sporophores simple or branched.

A single species treated.

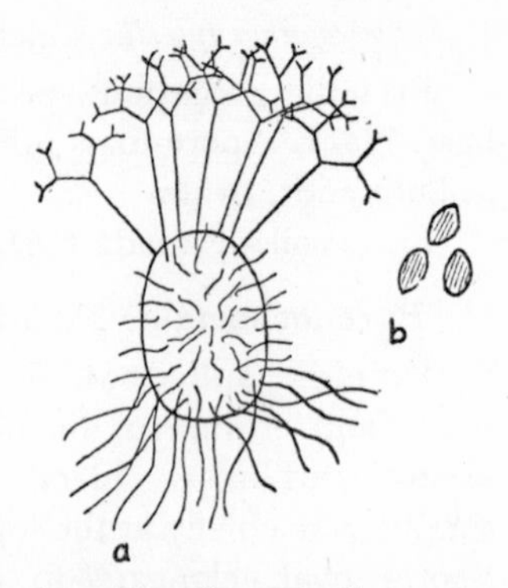

FIG. 59. **Chaetomella.** *a* pycnidium; *b* conidia.

1. *Chaetomella horrida* Oudemans

Mycelium creeping, from white to darkish, branched, septate. Pycnidia 180 × 140μ, superficial, scattered, ovate, without ostiole, brown, in transmitted light dark brown, with setae on all sides. Setae of the old pycnidia rising high, downward black and opaque, upward lighter, dark or dilute olive, septate, when young smooth, when old slightly roughened, once to many times dichotomously branched, ultimate branches awl-shaped. Spores broad elliptical, biconvex, commonly apiculate on both ends, very dilute steel-colored, 5.5–7 × 3.5–4μ. Conidiophores toward the base dark, above hyalin, three times length of spore.

From soil: Holland (100)

United States: Iowa (1) (3)

B. MELANCONIALES

Mycelium diffuse. Conidiophores gathered into a stratum which may be superficial or erumpent, dark or light colored, waxy, horny or even submembranous, accompanied by setae or not. Conidia variable.

A single family.

a. *MELANCONIACEAE*

Characters those of the order.

A single genus treated.

FIG. 60. **Cryptomela.** conidiophore and conidia (after von Beyma).

(1) **Cryptomela** Saccardo

Acervulus submerged or finally erumpent, black, small. Spores spindle-shaped, often curved, black.

A single species treated.

1. *Cryptomela acutispora* v. Beyma

Mycelium white, woolly, with the acervuli in concentric rings. Acervuli black, shining, thick crusts. Sporophores hyalin, forming a loose weft, aggregated towards the periphery, 12–15 × 2–3μ. Spores single, lateral, equidistant from one another or in a head at the end of the conidiophores; spindle-shaped, pointed at both ends, slightly colored, black in mass, 7–7.3 × 2.7–3μ, biguttulate.

From soil: Egypt (115)

C. MONILIALES

Hyphae septate, branched, in or on the substrate, hyalin or dark colored, separate from each other or combined into coremia or sporodochia, seldom remaining sterile. Conidia either as oidia or on slightly differentiated hyphae or in most cases on special erect conidiophores. Conidiophores simple or variously branched. Conidia formed in many ways on the conidiophores or their branches, very different in form and color.

a. *MONILIACEAE*

Hyphae septate usually prostrate, seldom short, hyalin or pale or bright colored, not forming coremia or sporodochia. Conidia seldom as oidia, usually on conidiophores which are not differentiated from the mycelial hyphae or are sharply differentiated. Conidiophores variously formed. Conidia of different shapes, always hyalin or bright colored.

KEY TO THE GENERA OF THE MONILIACEAE

a. Conidia one-celled, globose, ovate, or short cylindric (Hyalosporae)
  b. Conidiophores short, or obsolete, or little different from vegetative hyphae (Micronemeae)
    c. Conidia in chains (Oosporeae)
      d. Conidiophores short, simple or nearly so
        e. Conidia globose to elongate with rounded or truncate, not pointed, ends
          f. Conidiophores decumbent; conidia oval to ellipsoid with rounded ends (1) **Oospora**
          ff. Conidiophores erect, conidia cask-shaped with truncate ends (2) **Geotrichum**
        ee. Conidia with pointed ends (3) **Fusidium**
      dd. Conidiophores longer, distinctly branched (4) **Monilia**
  bb. Conidiophores long and distinct from the vegetative hyphae (Macronemeae)

c. Branching of conidiophores confined to the tip, conidia in heads
  d. Conidia not in chains (Cephalosporieae)
    e. Conidia not enclosed in mucus
      f. Conidiophores simple (5) **Cephalosporium**
      ff. Conidiophores branched at tip (6) **Trichoderma**
    ee. Conidia enclosed in mucus (7) **Hyalopus**
  dd. Conidia in chains (Aspergilleae)
    e. Conidiophores distinctly swollen at the tip, foot cells also prominent (8) **Aspergillus**
    ee. Conidiophores not swollen at the tip, or only slightly; footcells not differentiated; tips verticillately branched
      f. Conidia cask-shaped, not on phialides; conidiophores branching symmetrically (9) **Amblyosporium**
      ff. Conidia globose to elongate on phialides; conidiophores branching symmetrically or asymmetrically at tip
        g. Conidia not enclosed in mucus
          h. Conidia without basal ring and pore (10) **Penicillium**
          hh. Conidia with basal ring and pore (11) **Scopulariopsis**
        gg. Conidia enclosed in mucus, chains of conidia not always distinguishable (12) **Gliocladium**
cc. Branching of conidiophores not confined to the tip
  d. Conidia not formed on special intercalary cells but usually terminal
    e. Branching of the conidiophores various, not in whorls (Botrytideae)
      f. Conidia smooth
        g. Conidiophores decumbent
          h. Lateral branches erect with a single terminal conidium (13) **Acremonium**
          hh. Lateral branches prostrate, conidia terminal and lateral (14) **Sporotrichum**
        gg. Conidiophores erect
          h. Conidia globose or ovate
            i. Conidia borne singly (15) **Monosporium**
            ii. Conidia in heads (16) **Botrytis**
          hh. Conidia cylindric (17) **Cylindrophora**
      ff. Conidia warted (18) **Sepedonium**

ee. Conidiophores branching principally in whorls (Verticillieae)
  f. Apical branches of conidiophores sterile; conidia on lower branches; conidia-bearing branches very short, flask-shaped phialides (19) **Pachybasium**
  ff. All branches of conidiophores bearing spores
    g. Spores not in chains, single or in heads
      h. Spores globose, ellipsoidal or oval
        i. Conidia and branches not enclosed in mucus
          j. Conidia on branchlets occurring singly (20) **Verticillium**
          jj. Conidia-bearing branchlets in pairs at right angles to each other (21) **Verticilliastrum**
        ii. Conidia and branches enclosed in mucus (22) **Acrostalagmus**
      hh. Spores cylindric or long spindle-shaped (23) **Acrocylindrium**
    gg. Spores in chains, on phialides (24) **Spicaria**
dd. Conidia formed on differentiated intercalary cells of the conidiophore (Gonatobotrytideae)
  e. Conidiophores of sterile and fertile cells; sterile cells bone-shaped, swollen at the ends; fertile cells without phialides (25) **Nematogonum**

aa. Conidia more than one-celled, more or less elongate
  b. Conidia two-celled, ovate or short fusoid (Hyalodidymae)
    c. Conidia smooth
      d. Conidia single and terminal on conidiophores sharply differentiated from mycelium (26) **Trichothecium**
    cc. Conidia warty (27) **Mycogone**
  bb. Conidia more than two-celled, oblong, fusoid or elongate (Hyalophragmiae)
    c. Conidiophores distinct from conidia (Macronemeae)
      d. Conidiophores branching in whorls (28) **Dactylium**

(1) **Oospora** (Wallroth) Lindau

Turf spreading or cushiony, thread-like, loose or somewhat close. Hyphae creeping, septate. Fertile hyphae short, mostly simple. Conidia in regular chains, round or ovate, hyalin or bright colored.

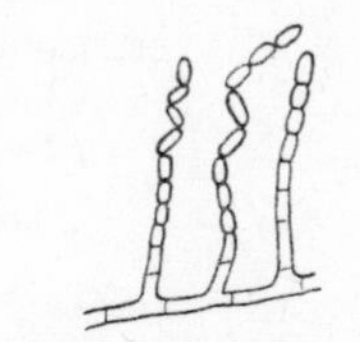

Fig. 61. **Oospora.** conidial habit.

KEY TO THE SPECIES OF THE GENUS OOSPORA

a. Colonies white — 1. *O. variabilis*
aa. Colonies colored
  b. Colonies some shade of yellow
    c. Colonies orange-yellow
      d. Conidia large $31 \times 14\mu$ — 2. *O. lupuli*
      dd. Conidia small $8\text{–}11 \times 2.5\text{–}3\mu$ — 3. *O. roseo-flava*
    cc. Colonies sulphur-yellow — 4. *O. sulphurea*
  bb. Colonies greenish — 5. *O. egyptiaca*

1. *Oospora variabilis* (Lindner) Lindau

Forms a white turf consisting of cells of many different forms. Forming either threads of more or less long cells, which break up easily, or spores with small, round buds, or yeast-like colonies.

From soil: China (81), England (37)

2. *Oospora lupuli* (Matthews and Lott) Lindau

Turf of luxuriant mycelium, finally powdery, orange-yellow. Conidiophores erect, branched, breaking up into oidia-like pieces. Spores cylindric to ovate, rarely globose of very different sizes, up to 31μ long, and 14μ broad, but usually smaller, smooth, thin, reddish, finally yellow.

From soil: England (11)

3. *Oospora roseo-flava* Saccardo

Turf effuse, subpulverulent, rosy-yellow. Sterile hyphae prostrate, fertile branches erect, continuous, $40\text{–}45 \times 4\text{–}4.5\mu$. Conidia oblong-fusoid, biguttulate, in chains, rosy-yellow, $8\text{–}11 \times 2.5\text{–}3\mu$.

From soil: Austria (131)

4. *Oospora sulphurea* (Preuss) Saccardo and Voglino

Turf effuse, sulphur-yellow, cobwebby; hyphae branched, prostrate. Conidiophores nonseptate below, cutting off branching chains of conidia from their tips. Conidia ovate, yellow.

From soil: Austria (131)

5. *Oospora egyptiaca* van Beyma

Colonies of white coremium-like bundles, 1 cm. high, becoming greenish from conidial production. Conidiophores numerous, straight or slightly curved, unbranched, nonseptate, thickly crowded, arising from ascending hyphal bundles, 23–40μ long, 2–2.5μ wide at the base, cutting off long chains of conidia. Conidia numerous, elongate, ovate, rounded at one end, hyalin, greenish in mass, 4.3–6.7 × 1.7–2.3μ.

From soil: Egypt (115)

(2) **Geotrichum** Link

Hyphae prostrate, septate; forming a turf. Conidiophores short, erect or ascending, septate, producing conidia in chains at their apices. Conidia short, cylindrical, truncate at both ends or slightly rounded, hyalin or pale.

A single species treated.

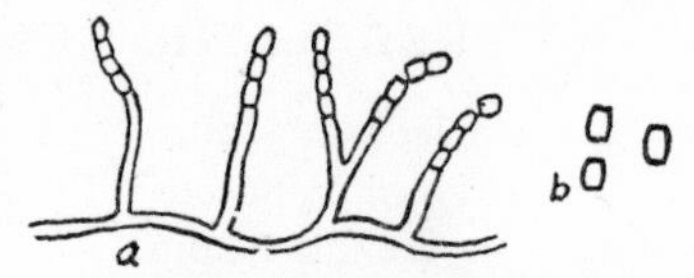

Fig. 62. **Geotrichum.** *a* conidial habit; *b* conidia.

1. *Geotrichum candidum* Link

Syn. *Oospora lactis* (Fresenius) Lindau
Oidium lactis Fresenius

Turf cushion-like, somewhat powdery, white. Hyphae prostrate, with few septa. Conidiophores short, erect. Conidia in chains, short cylindrical, truncate at both ends, 5–10 × 4μ, hyalin.

In soil: Austria (66), Canada (16), Denmark (68), Germany (6)
United States: New York (67)

(3) **Fusidium** Link

Mycelium septate, usually quite indistinct. Conidiophores not differentiated from the mycelium. Conidia usually in chains, spindle-shaped, more or less sharply pointed at both ends, hyalin or bright colored.

A single species treated.

1. *Fusidium viride* Grove

Mycelium white, delicately effuse. Conidia on long chains. Conidia fusiform, straight, with acute tips, pale green, 10 × 3μ.

From soil: England (11)

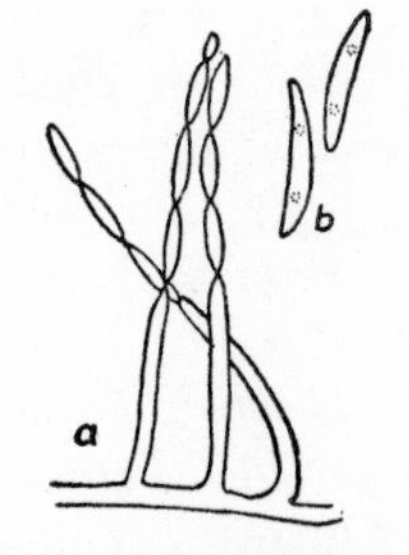

Fig. 63. **Fusidium.** *a* conidiophores; *b* conidia.

(4) **Monilia** (Persoon) Saccardo

Mycelium creeping, septate. Conidiophores ascending or erect with dichotomous, racemose, or irregular branching, which is sparse or abundant. Simple or branched conidial chains borne on the points of the branches or on small, blunt projections near the point. Conidia ovate to elongate, seldom globose, hyalin or light colored, often united by isthmus-like connecting cells.

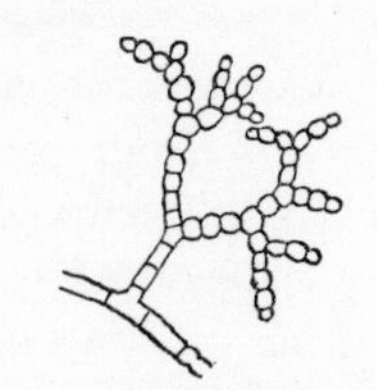

FIG. 64. **Monilia.** conidial habit.

KEY TO THE SPECIES OF THE GENUS MONILIA

a. Colonies pure white
  b. Mycelium floccose
    c. Conidia hyalin
      d. Conidia ovate 12–15 × 8–10μ — 1. *M. acremonium*
      dd. Conidia lens-shaped 2–3 × 1–1.5μ — 2. *M. implicata*
    cc. Conidia orange to red — 3. *M. sitophila*
  bb. Mycelium pruinose — 4. *M. pruinosa*
aa. Colonies colored
  b. Colonies yellow to brown
    c. Colonies ochre-yellow — 5. *M. geophila*
    cc. Colonies buff-brown — 6. *M. brunnea*
  bb. Colonies other colors than yellow to brown
    c. Colonies green — 7. *M. humicola*
    cc. Colonies gray — 8. *M. grisea*

1. *Monilia acremonium* Delacroix

Colonies spreading, somewhat floccose, white. Sterile hyphae creeping, hyalin, sparsely septate, with oil drops present, 4–5μ thick. Conidiophores erect, often united in bundles, with numerous septa, bearing the conidial chains terminally. Conidia ovate-pyriform, somewhat truncate at the base, united by small connecting cells, 12–15 × 8.5–10μ, hyalin.

From soil: Holland (100)

2. *Monilia implicata* Gilman and Abbott

Colonies on Czapek's agar spreading, cottony to floccose, consisting of interwoven, hyalin, aerial hyphae and masses of very long, intertangled conidial chains which spread over the medium; surface pure white, reverse colorless to cream. Conidiophores prostrate, arising laterally from aerial mycelium thickly crowded on the fertile hyphae, tapering gradually toward the apex, hyalin, 20–100μ long. Conidial chains very long. Conidia lense-shaped, apiculate, hyalin, 3–4.3 × 1–1.5μ. Characters on bean agar similar.

From soil: Canada (16), China (81)
United States: Louisiana (50)

3. *Monilia sitophila* (Montagne) Saccardo

Colonies white, floccose, spreading, conidial masses red. Vegetative hyphae hyalin, freely branched, septate; surface mycelium carries short branches from which the conidial chains arise. Conidia ovate to cylindrical, 5.2–13.4μ in diameter.

From soil: Canada (15)
United States: New Jersey (146) (147)

4. *Monilia pruinosa* Cooke and Massee

Effuse, delicate, white pruinose. Conidiophores flexuous, elongate, septate, irregularly branched. Conidia in short chains, subglobose or ovate, hyalin, 14–15 × 12μ.

From soil: England (11)

5. *Monilia geophila* Oudemans

Colonies yellow to ochre-yellow, composed of creeping, hyalin, branched, sparsely septate hyphae loosely floccose. Conidiophores ascending or erect, with numerous septa, toward the apex once or twice forked or irregularly branched, with few short branchlets. Conidial chains borne on the points of the branchlets singly or in twos. Conidia at first globose, then elliptical, yellowish-white, 3–5μ long × 2–3μ.

From soil: Canada (15), Holland (100)

6. *Monilia brunnea* Gilman and Abbott

Colonies on Czapek's agar, round, not spreading, densely floccose; aerial hyphae creeping, densely interwoven, hyalin, 2.5–3μ in diameter; surface pale buff-brown, reverse brown. Conidiophores arise from aerial mycelium, scattered on the hyphae, 8–20μ long, narrowing abruptly near the apex to produce a short sterigmata-like cell which bears the conidial chains. Conidial chains often branched, short. Conidia elliptical, rather sharply pointed, smooth, light-buff, 5.5–7.5 × 3–4μ. Characters on bean agar similar.

From soil: Egypt (115)
United States: Louisiana (50), Texas (93)

7. *Monilia humicola* Oudemans

Colonies orbicular, dense, at first nearly hyalin, later entirely green; sterile hyphae creeping, when young hyalin. Conidiophores ascending or erect, yellow or green, closely septate, branched; branches alternate or nearly opposite, once or twice dichotomously branched. Conidia in short chains, elliptical when mature, both ends apiculate, greenish, 4–10 × 2–5μ.

From soil: Holland (100)

United States: Louisiana (50), New Jersey (146) (147)

8. *Monilia grisea* Daszewska

Mycelium dark brown, forming a gray turf. Hyphae fasciculate, branching, septate, 4μ in diameter. Conidia terminal in chains, globose, hyalin, 6–10μ in diameter.

From soil: Switzerland (38)

Species of doubtful position:

Monilia candida Bonorden

Poorly described. From soil: Germany (6)

Monilia fimicola Costantin and Matruchot

Its isolation from the soil is very uncertain; see (107)

Monilia koningi Oudemans

Listed by Dale (37) as a synonym of *Scopulariopsis rufulus* Bainier.

From soil: England (37), Holland (100), India (49)

United States: Michigan (51)

Monilia lunzinense Szilvinyi

From soil: Austria (131)

Monilia terrestris Daszewska

From soil: Switzerland (38)

(5) **Cephalosporium** Corda

Sterile hyphae creeping. Conidiophores arise as short branches of aerial hyphae, erect, nonseptate, not swollen at the tip. Conidia borne singly at the tips of the conidiophores, being pushed to the side as they are formed successively; conidia usually ovate, hyalin or slightly colored.

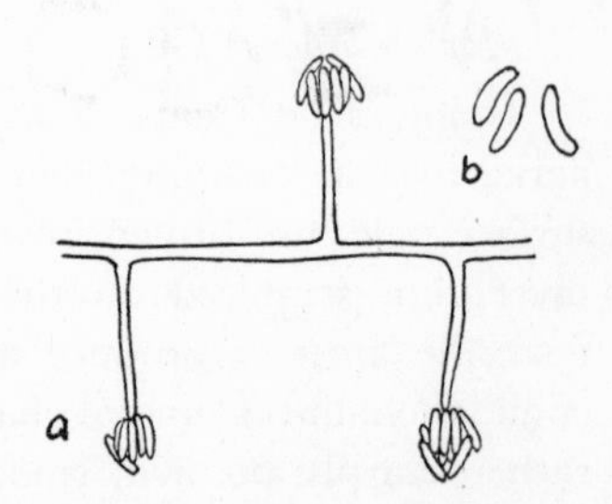

FIG. 65. **Cephalosporium.** *a* conidiophores; *b* conidia (after Corda).

KEY TO THE SPECIES OF THE GENUS CEPHALOSPORIUM

a. Conidia elliptical
  b. Colonies white
    c. Colonies remaining white; conidia 8–10 × 3.5–4.0μ — 1. *C. curtipes*
    cc. Colonies becoming rose-colored; conidia 4 × 1.0–1.5μ — 2. *C. acremonium*
  bb. Colonies gray; conidia 4–6 × 3.2–3.7μ — 3. *C. asperum*
aa. Conidia globose
  b. Colonies white with light rosy center; conidia 2.3–2.5μ — 4. *C. humicola*
  bb. Colonies white becoming gray; conidia 8–8.4μ — 5. *C. coremioides*

1. *Cephalosporium curtipes* Saccardo

Colonies on Czapek's agar spreading, felty to floccose, pure white, consisting of creeping, septate, dichotomously branched hyphae, reverse colorless. Conidiophores arise as branches of aerial hyphae; short, up to 25μ long, conidial heads round. Conidia elongate elliptical, hyalin, 9.0–10.0 × 3.5–4.0μ.

From soil: Canada (15), U. S. S. R. (108)

United States: Colorado (146), Louisiana (50), New Jersey (146) (147), Porto Rico (147), Texas (92)

2. *Cephalosporium acremonium* Corda

Colonies orbicular, dense, floccose, at first white, later very light rose-colored; vegetative hyphae hyalin, sparsely septate, branched. Conidiophores arise as side branches on aerial hyphae, erect, simple, nonseptate, 40–60 × 3μ. Conidia numerous, elliptical or oblong, straight or curved, nearly hyalin, 4.0 × 1.0–1.5μ.

From soil: Austria (131), Canada (15), China (81), England (11) (37), Holland (100), U. S. S. R. (108)

United States: New Jersey (146) (147)

3. *Cephalosporium asperum* Marchal

Colonies gray, floccose, indeterminate; sterile hyphae prostrate, flexuous, branching, 400–600 × 2–4.5μ, sparingly septate. Conidiophores simple or rarely branching, continuous, 15–30μ long, straight, heads often irregular, two- to seven-spored, in a crown. Conidia sessile or with short, delicate stipes, ovate or submoniliform, greenish-hyalin, 4–6 × 3.2–3.7μ, asperulate.

From soil: U. S. S. R. (108)

4. *Cephalosporium humicola* Oudemans

Colonies orbicular, floccose, at first white, later with white margin and dilute rose-colored center. Sterile hyphae septate, branched, hyalin,

3.0–5.0μ thick, intermixed with segments which appear like chlamydospores. Conidiophores erect, 100–200μ long, unbranched, nonseptate; conidial heads 20.0–26.0μ in diameter. Conidia globose, 2.3–2.5μ, almost hyalin.

From soil: Austria (131), Canada (15), England (11), Holland (100)

5. *Cephalosporium coremioides* Raillo

Colonies at first white, becoming grayish-white, characteristically forming coremia. Mycelium hyalin, delicate, 1.2μ in diameter. Conidiophores at times dichotomously branched, slightly thickened at the base, 2.7μ in diameter. Conidia hyalin, cylindrical 4–8.4 × 1.5–2μ.

From soil: U. S. S. R. (108)

Doubtful species: *Cephalosporium koningi* Oudemans (100)

This species, the isolation of which is reported from the soil by Koning in Holland (100) and Waksman from Alaska soil, (147), is believed by Lindau (107) to belong with the Mucoraceae because of the nonseptate mycelium, the presence of chlamydospores, and the loose adherence of the conidia in the head.

(6) **Trichoderma** (Persoon) Harz

Sterile hyphae creeping, septate, forming a flat, firm turf. Conidiophores erect, arising from short, branched side-branches, branching usually opposite, not swollen at the apex and bearing terminally the conidial heads. Conidia small, mostly globose, bright colored or hyalin.

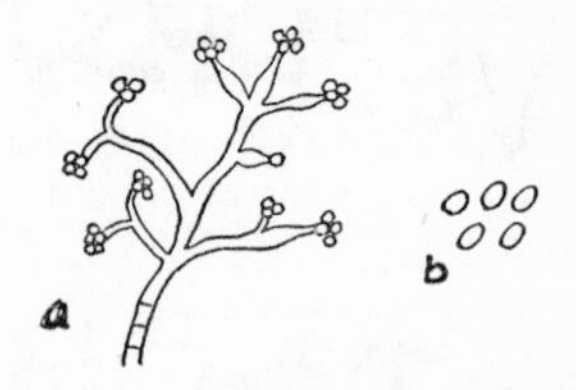

FIG. 66. **Trichoderma.** *a* conidiophores; *b* conidia.

KEY TO THE SPECIES OF THE GENUS TRICHODERMA

a. Colonies with mature fruiting areas white — 1. *T. album*
aa. Colonies with mature fruiting areas green
  b. Colonies floccose, surface light green, conidia elliptical — 2. *T. koningi*
  bb. Colonies tufted, strict, not floccose
    c. Surface deep green, conidia globose — 3. *T. lignorum*
    cc. Surface yellow to yellow-green, conidia ovate to elliptical — 4. *T. glaucum*

1. *Trichoderma album* Preuss

This fungus produces practically no growth on Czapek's agar. A few transparent hyphae spread over the medium, visible only when held up to the light. Scattered tufts of white aerial hyphae develop, with little or no fruiting. On bean agar, colonies, thin, spreading, small, white tufts of aerial mycelium and conidiophores developing in about two weeks. Conidiophores arise from aerial mycelium, branched, up to 25μ or 30μ in length, bearing terminal heads up to 15μ in diameter. Conidia elliptical to oval, hyalin, $2.5–3.2 \times 1.5–2.0\mu$.

From soil: Canada (15), China (81), England (36)

United States: Iowa (50), Louisiana (147), New Jersey (147)

2. *Trichoderma koningi* Oudemans

Colonies on Czapek's agar spreading, floccose, white at first, becoming light green in four to five days; may show various shades of light green, but never becomes deep green; reverse colorless. Vegetative hyphae septate, hyalin. Conidiophores arise as branches of aerial mycelium, alternate or opposite, up to 25μ in height $\times$ 3.0μ in diameter, di- or trichotomously branched. Fruiting heads up to 10μ in diameter; conidia oblong to elliptical, $3.2–4.8\mu$ long $\times$ $1.8–3.0\mu$ wide, smooth, hyalin.

From soil: Austria (66) (131), Canada (15), China (81), Denmark (68), England (11) (36), Holland (100), India (21), Japan (132)

United States: Alaska (147), California (147), Hawaii (147), Illinois (129), Iowa (1) (3) (147), Louisiana (2), Maine (147), Michigan (51), New Jersey (146) (147), New York (67), Oregon (147), Porto Rico (147), Rhode Island (110), Utah (50)

3. *Trichoderma lignorum* (Tode) Harz

Colonies on Czapek's agar broadly spreading, hyalin; fruiting areas appear as tufts, white at first, and becoming various deep green shades with age; reverse colorless. Conidiophores arise as branches of aerial mycelium, septate, up to 70μ in height $\times$ 3.0μ in diameter, di- or trichotomously branched, occasionally forming whorls. Conidial heads to to 10μ in diameter; conidia globose to ovate, smooth, $3.8–3.2\mu$ in diameter.

From soil: Austria (66), Canada (15) (147), China (81), England (11) (36), India (49)

United States: Colorado (74) (76), Iowa (3) (101), Louisiana (2) (147), Maine (147), New Jersey (146) (147), New York (67), Oregon (147), Porto Rico (147), Utah (50)

4. *Trichoderma glaucum* Abbott
Syn. *T. flavus* Abbott

Colonies on Czapek's agar spreading; at first only a thin, sterile, mycelial film covers the surface of the medium. White aerial mycelium develops in five to seven days, followed in ten days by the appearance of the yellow fruiting areas, which change through shades of chartreuse-yellow to citron or lime-green. Vegetative mycelium hyalin, 3.0–6.0μ thick, multiseptate, and freely branched; cells are often short and swollen or barrel shaped. Conidiophores arise as side branches, alternately, oppositely, or irregularly branched; up to 60μ in height × 3.0μ in width. Conidial heads 6.5–10.0μ in diameter. Conidia smooth, hyalin, ovate, 3.8–5.0 × 2.5–3.0μ, mode 4.0 × 3.0μ.

From soil: Canada (15)

United States: Colorado (74), (75), Iowa (3)

Species of uncertain position:

Trichoderma nigrovirens Goddard

From soil: United States: Michigan (52)

(7) **Hyalopus** Corda

Sterile hyphae prostrate, sparse. Conidiophores erect, usually nonseptate, hyalin, not or very slightly swollen at their tips. Conidia hyalin or bright colored, enclosed in mucus to form a head.

A single species treated.

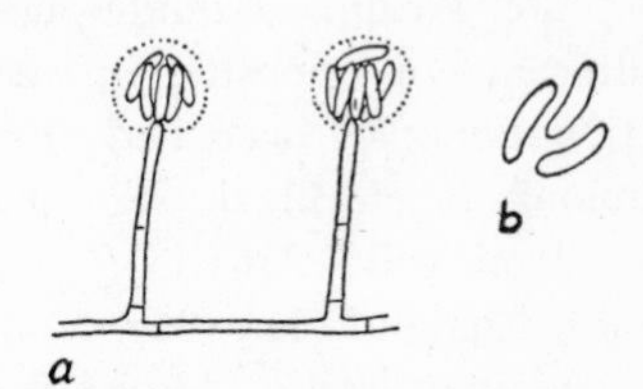

Fig. 67. **Hyalopus.** *a* conidiophores; *b* conidia.

1. *Hyalopus ater* Corda

Turf extended, black, velvety. Hyphae delicate, yellow. Conidiophores erect, in dense clusters, thread-like, septate, dark below, bright above. Conidia single to many embedded in a mucus to form a terminal white head at first, later yellow, long ellipsoid, truncate, greenish white, transparent, 6μ long.

From soil: Canada (16)

(8) **Aspergillus** (Micheli) Corda

Vegetative mycelium consisting of septate branching hyphae, colorless. Conidial apparatus developed as stalks and heads from specialized, enlarged, thick-walled hyphal cells (foot-cells) producing conidiophores as branches approximately perpendicular to the long axis of the foot-cells. Conidiophores nonseptate or septate, usually enlarging upward and broadening into elliptical, hemispherical, or

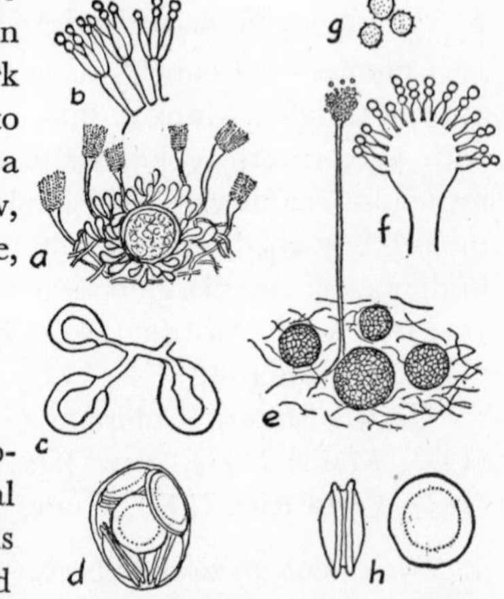

Fig. 68. **Aspergillus.** *a* habit; *b* phialides; *c* "hullen-cellen"; *d* ascus; *e* habit; *f* conidial head; *g* conidia; *h* ascospores.

globose fertile vesicles bearing phialides, either parallel and clustered in terminal groups or radiating from the entire surface. Phialides in one series, or as a primary series, each bearing a cluster of two to several secondary phialides at the apex. Conidia varying greatly in color, size, shape, and markings, successively cut off from the tips of the phialides by cross-walls, and forming unbranched chains arranged into radiate heads or packed into columnar masses. Cleistothecia found in certain species only. Sclerotia regularly found in some strains, occasionally found in others.

KEY TO THE SPECIES OF ASPERGILLUS

Key	Species
a. Conidial heads some shade of green	
b. Conidiophores with smooth walls	
c. Vesicle cylindrical clavate, conidiophores coarse	1. *A. clavatus*
cc. Vesicle flask-shaped or globose, not cylindrical clavate	
d. Phialides in one series	
e. Conidia mostly elliptical and more than 4μ long	
f. Yellow cleistothecia commonly present (*A. glaucus* group)	
g. Ascospores without equatorial ridges	2. *A. repens*
gg. Ascospores with equatorial ridges	
h. Ridges low and rounded	3. *A. ruber*
hh. Ridges thin and flexuous	4. *A. chevalieri*
ff. Cleistothecia not found; heads enclosed in slime when ripe	5. *A. conicus*
ee. Conidia mostly globose, 4μ or less in diameter; conidial heads in narrow solid columns	6. *A. fumigatus*
dd. Phialides in two series	
e. Conidial chains in columns	
f. Cleistothecia present; surrounded by hülle cells	
g. Ascospores with smooth walls	
h. Equatorial ridges two in number	7. *A. nidulans*
hh. Equatorial ridges four in number	8. *A. quadrilineatus*
gg. Ascospores with rough walls	9. *A. rugulosus*
ff. Cleistothecia absent; replaced by sterile, thick-walled hyphae	10. *A. unguis*
ee. Conidial chains in radiate heads	
f. Heads blue-green	11. *A. sydowi*
ff. Heads glaucus-green or yellow-green	12. *A. versicolor*
fff. Heads gray-green to brown-green	13. *A. minutus*

bb. Conidiophores pitted, often appearing rough; conidial heads yellow-green — 14. *A. flavus*
aa. Conidial heads never green
b. Conidiophore walls smooth
c. Conidial heads brown
d. Conidial heads avellaneous, or fuliginous
e. Conidial heads columnar
f. Conidial heads avellaneous — 15. *A. terreus*
ff. Conidial heads pale buff, stalks yellow, commonly with sulphur-yellow sclerotia — 16. *A. flavipes*
ee. Conidial heads radiate
f. Conidia with smooth walls — 17. *A. humicola*
ff. Conidia with rough or spinulose walls
g. Conidia 3–4μ in diameter — 18. *A. ustus*
gg. Conidia up to 5μ in diameter — 19. *A. insuetus*
dd. Conidial heads dark brown to black
e. Heads chocolate-brown — 20. *A. luchuensis*
ee. Heads black
f. Primary phialides about 20–30μ long — 21. *A. niger*
ff. Primary phialides about 40–60μ long — 22. *A. phoenicis*
cc. Conidial heads not brown
d. Conidial heads yellow, orange to umber
e. Conidial heads orange to umber, colonies usually floccose — 23. *A. wentii*
ee. Conidial heads sulphur-yellow
f. Sclerotia not reported — 24. *A. sulphureus*
ff. Sclerotia present
g. Sclerotia becoming cinnamon — 23. *A. sachari*
gg. Sclerotia becoming black — 26. *A. alliaceus*
dd. Heads white or pink, radiate
e. Heads white
f. Colony surface powdery, phialides in two series
g. Colonies white to cream in age — 27. *A. candidus*
gg. Colonies white to sulphur-yellow in age — 28. *A. okasakii*
ff. Colony surface floccose, phialides in a single series — 29. *A. koningi*
ee. Heads pink (*A. candidus* group)
bb. Conidiophore walls pitted or apparently rough
c. Heads yellow to ochre, radiate
d. Heads in ochraceous shades
e. Phialides in two series — 30. *A. ochraceous*
ee. Phialides in one series — 31. *A. terricola*
dd. Heads orange to umber or brown — 32. *A. tamarii*

1. *Aspergillus clavatus* Desmazieres

Colonies on Czapek's agar gray-green to dark green, densely matted, heavy, rapidly growing and spreading; reverse and agar more or less brownish. Conidiophores with walls smooth, colorless, up to one to several millimeters in length, commonly 15–20μ in diameter, gradually enlarged at the apex into a clavate vesicle which is fertile over an area up to 150μ long × 20–25μ in longest diameter. Phialides 7–10 × 2–3μ, in a single series, densely covering the fertile area, bearing long chains which frequently adhere more or less into masses. Conidia elliptical, green, 2.5–3 × 3.4–4.5μ, smooth. Cleistothecia not found.

From soil: United States: Colorado (76), Iowa (1) (3), New Jersey (146) (147), Texas (147)

2. *Aspergillus repens* (Corda) de Bary

Colonies on Czapek's agar (20 per cent sucrose) spreading broadly, plane or slightly wrinkled, orange-yellow, with broad zones of dull green conidial heads. Surface growth consisting of loosely woven hyphae studded with orange granules enmeshing abundant yellow perithecia among projecting conidial heads; reverse yellow-orange to deep maroon.

Cleistothecia very abundant, borne in loose network of yellow to orange-red hyphae, yellow, spherical to subspherical, 75–100μ (125μ) in diameter. Asci 10–12μ. Ascospores lenticular, 4.8–5.6 × 3.8–4.4μ, smooth walled, with equatorial area rounded or somewhat flattened, occasionally slightly furrowed but without crests or ridges.

Conidial heads abundant, 125–175μ in diameter in different strains, consisting of divergent chains of conidia radiating from a hemispherical vesicle, 25–40μ in diameter. Stalks smooth, colorless, 500–1,000μ in length. Phialides in one series, 7–10 × 3.4–4.5μ. Conidia elliptical to subglobose, spinulose 5–6.5μ.

From soil: Denmark (68), England (36)
United States: New Jersey (146)

3. *Aspergillus ruber* (Spieckermann and Brenner) Thom and Church

Colonies on Czapek's agar (20 per cent glucose) spreading and regular or uneven on margin, plane, predominantly red, ranging from ferruginous to morocco-red. Cleistothecia abundant in a dense layer at the agar surface and largely concealed within a felt of red-incrusted hyphae. Conidial heads above the felt pale gray to green to deep olive-gray, more or less abundant, and generally crowded near the center of the colony; reverse in shades of dark red-brown.

Cleistothecia, yellow to orange-red, spherical to subspherical 80–120μ (140μ) in diameter. Asci 12–15μ. Ascospores lenticular, 5.2–6.0 × 4.4–

4.8μ, with furrow as a broad depression around equator, ridges low, walls smooth except along ridges.

Conidial heads numerous in localized areas or scattered thinly, pale blue-green, radiate, 150–250μ in diameter. Stalk smooth, colorless to orange-brown, 500–750μ in length, terminating in a vesicle, 25–35μ in diameter. Phialides in a single series 7–9 $\times$ 4–5μ. Conidia elliptical to subglobose, closely spinulose, 5–6μ in long axis.

From soil: U. S. S. R. (108)

4. *Aspergillus chevalieri* (Mangin) Thom and Church

Syn. *Eurotium chevalieri* Mangin

Colonies on Czapek's agar (3 per cent sucrose) restricted, plane, closely felted, bluish-gray in center with typical heads and cleistothecia largely confined to marginal area; reverse maroon in center to orange at margin. Cleistothecia abundant and closely enmeshed in a felt of orange-red incrusted hyphae, mostly 100–140μ, occasionally up to 150μ in diameter, globose to subglobose, yellow or orange. Asci 9–10μ. Ascospores lenticular, 4.6–5 $\times$ 3.4–3.8μ, with walls smooth, with equatorial crests prominent, thin and often recurved, and with furrow consisting more of a trough between parallel crests than an equatorial depression in the spore body. Conidial heads abundant, pale blue-green, appearing radiate from divergent conidial chains, mostly 125–175μ in diameter, occasionally larger. Stalks mostly 700–850μ long, enlarging to a vesicular apex, somewhat globose, 25–35μ in diameter. Phialides in a single series, closely packed, 5–7 $\times$ 3–3.5μ. Conidia subglobose, spinulose, 4.5–5.5μ in diameter.

From soil: India (49)

5. *Aspergillus conicus* Blochwitz

Colonies slow growing, forming a slimy, convoluted or buckled mass with submerged irregular margin, colorless, then green to dark green or almost black. Reverse dark green to black. Stalks and heads sometimes erect and free from slime at first, but later enveloped in a mass of greasy slime. Conidiophores short, 100–200μ long; vesicles up to 20μ in diameter, fertile mostly at the apex only. Phialides of the *A. glaucus* type, 5–10 $\times$ 2–4μ; in some strains developing into secondary stalks bearing little heads. Conidia elliptical, 4–6 $\times$ 3–3.5μ, in some strains smooth, in others up to 8μ long, becoming thick walled and rough. Cleistothecia not found.

From soil: England (36), Europe (139)

United States: Iowa (101)

6. *Aspergillus fumigatus* Fresenius
Syn. Probably *A. calyptratus* Oudemans

Colonies on Czapek's agar in some strains strictly velvety, in others with varying amounts of tufted aerial mycelium up to felted floccose forms, green to dark green, becoming almost black in age, spreading. Reverse and substratum, colorless to yellow. Conidiophores short, usually densely crowded, up to 300μ (occasionally 500μ), $\times$ 2–8μ in diameter, arising directly from submerged hyphae or as branches from aerial hyphae, septate or nonseptate, gradually enlarged, upward, with apical flask-shaped vesicles up to 20–30μ in diameter, fertile usually only on the upper half, bearing phialides in one series, usually 6–8 $\times$ 2–3μ, crowded, closely packed, with axis roughly parallel to axis of the stalk. Chains of conidia form solid columns up to 400 $\times$ 50μ. Conidia dark green in mass, globose, 2–3.5μ, mostly 2.5–3μ.

From soil: Austria (66) (131), Canada (15), China (81), Denmark (68), England (11), Greenland (96), India (21) (49), Japan (132)

United States: Colorado (76), Illinois (129), Iowa (1) (3) (101) (147), Louisiana (2), North Dakota (147), New York (67), New Jersey (146) (147), Oregon (146), Texas (92) (146) (147) (152), Utah (50)

7. *Aspergillus nidulans* (Eidam) Winter

Colonies on Czapek's agar plane, spreading broadly, dark cress-green (Ridgway Pl. XXXI) from abundant conidial heads during the first two weeks; cleistothecia developing from the center of the colony outward, separately produced, often abundant; sectoring occasionally. Reverse of colony in shades of purplish-red, becoming dark in age.

Cleistothecia developed separately within or upon the conidial layer, globose, 100–175μ in diameter, with outer layer a yellowish to cinnamon colored envelope of scattered hyphae bearing "hülle" cells up to 25μ in diameter; wall a single layer of cells, dark reddish-purple; in ripening becoming a mass of eight-spored asci which break down leaving the ascospores free. Ascospores purple-red, lenticular, smooth walled with two equatorial crests, 3.8–4.5 $\times$ 3.5–4μ, crests plaited with sinuous margins 0.5–1μ in width.

Conidial heads short columnar, 40–80 $\times$ 25–40μ. Stalks sinuous, with smooth walls, in shades of cinnamon-brown, ranging from 60–130μ, about 2.5–3μ in diameter near foot, to 3.5–5μ below the terminal vesicle; vesicle 8–10μ in diameter. Phialides in two series, primary 5–6 $\times$ 2–3μ and secondary 5–6 $\times$ 2–2.5μ. Conidia globose, rugulose, 3–3.5μ in diameter, green in mass.

From soil: Canada (15), China (81), Egypt (115), India (21) (49), Japan (132)

United States: Colorado (146) (147), Iowa (1) (3) (101) (147), Michigan (51) (52), New Jersey (146) (147)

8. *Aspergillus quadrilineatus* Thom and Raper

Colonies on Czapek's agar spreading, plane or slightly wrinkled, with a tendency toward floccosity, central area gray with a definite purplish tinge, and olive-green conidial areas toward the margin, occasionally as sectors. Reverse purplish-red.

Cleistothecia developing separately but abundantly throughout the colony; enveloped by hülle cells, light brownish in color, spherical, partially embedded in the mycelial felt, 125–150μ in diameter including the hülle cell layer. Cleistothecial wall one-cell layer in thickness, breaking down with ripe asci to leave ascospores free. Ascospores purple-red, lenticular, with smooth wall, 4–4.8 $\times$ 3.4–3.8μ (spore body) and with two plaited equatorial crests about 0.5μ in width paralleled by a secondary narrower pair which are sometimes indistinct.

Conidial heads short columnar, green, 60–70 $\times$ 30–35μ. Stalks sinuate, smooth walled, dull brownish, 50–75μ in length $\times$ 3.5–4.5μ wide, broadening to 7.5–9μ at the apical hemispherical vesicles. Phialides in two series, primary phialides 5–6 $\times$ 2–3μ, secondary 5–7 $\times$ 2–2.5μ. Conidia globose, pale yellow-green, rugulose, 3–4μ in diameter.

From soil: United States: Colorado, Louisiana, Maryland, New Jersey, Texas (140)

9. *Aspergillus rugulosus* Thom and Raper

Colonies on Czapek's agar slowly and restrictedly growing, buckled or wrinkled, 2–3 mm. deep, enveloping abundant cleistothecia at different depths, purple-gray to purple-brown in age, with green conidial heads sparsely produced, occasionally as small groups or marginal extensions into drying media; reverse in shades of deep purple-red.

Cleistothecia very abundant, often in two or three layers and each surrounded by hyphae and dark brown hülle cells, globose, 225–350μ in diameter including mycelial coverings, with dark reddish-purple walls, one-cell thick, breaking down to leave ascospores free. Asci 10–11μ in long axis. Ascospores purplish-red, lenticular, walls conspicuously rugulose, 4–4.4 $\times$ 3.6–3.8μ (spore body) with two plaited equatorial crests with sinuate and entire margins 0.5–0.6μ in width.

Conidial heads short columnar, 75–100 $\times$ 30–40μ. Stalks sinuous, smooth-walled, pale brownish, 50–80μ long, slender, up to 5μ in width then enlarging into apical vesicles 8–10μ in diameter. Phialides in two series; primary phialides 7–8 × 3–3.5μ, secondary 6–7 × 2.5–3μ. Conidia globose, green, rugulose 3–4μ.

From soil: United States: Nebraska, New Jersey, Texas (140)

10. *Aspergillus unguis* (Emile Weil and Gauden) Thom and Raper

Colonies on Czapek's agar restricted, plane, spreading at the margin as irregular lobes, yellowish-green, green to dark green becoming brown in age; without cleistothecia or hülle cells. Mycelial preparations show striking sterile, thick-walled hyphae with walls in brown shades, irregularly roughened, tapering to blunt point, arising from foot-cells or mycelial cells, often 1,000μ or more in length, slanting upward but rising slightly above conidial area.

Conidial heads columnar, 75–150 × 40–50μ. Stalks smooth-walled, dull brown in color, 45–65μ in length × 3–5μ in diameter; enlarging to apical vesicles 9–12μ in diameter. Phialides in two series; primary 5–6 × 2.5–3μ, secondary 5–6 × 2.0–2.5μ. Conidia globose, rugulose, dull green, 2.5–3.5μ in diameter.

From soil: United States (140)

11. *Aspergillus sydowi* (Bainier and Sartory) Thom and Church

Colonies on Czapek's agar blue-green, with the bluish effect prominent, velvety with some aerial interlacing and trailing hyphae. Reverse and substratum, shades of orange to red, becoming almost black. Conidiophores mostly arise from submerged hyphae, up to 500 × 4–8μ, colorless, smooth, thick-walled. Heads radiate or globose; vesicles 12–20μ in diameter. Phialides radiate in two series primary up to 7 × 2–3μ, secondary 7–10 × 2μ. Conidia globose 2.5–3.5μ, spinulose. No sclerotia or cleistothecia found.

From soil: Austria (131), Canada (16), China (81), India (21) (49) United States: Iowa (101), Louisiana (50), Utah (50)

12. *Aspergillus versicolor* (Vuillemin) Tiraboschi

Syn. *A. diversicolor* Waksman

A. globosus Jensen

A. tiraboschi Corbone

Colonies on Czapek's agar white, passing through shades of yellow, orange-yellow, buff to pea-green or sage-green, with green color occasionally entirely suppressed to produce orange-buff to almost flesh colored strains. Reverse and agar from yellow through orange to rose or red. Surface growth in some strains consisting almost entirely of conidiophores (velvety), in others showing marked development of floccose, sterile hyphae. Conidiophores, when arising separately from the substratum, up to 500–700μ long × 5–10μ in diameter, walls smooth, 1–1.5μ thick. Heads becoming 100–125μ in diameter, subglobose to globose or more or less calyptrate; vesicles 12–20μ in diameter, occasionally globose, usually flask-shaped, fertile on the upper two-thirds, with radiating phialides in two series; primary phialides 3–5 × 3–10μ; secondary phialides 1.5–2 × 5–10μ. Conidia globose, usually delicately roughened, 2.5–3μ or 4μ, usually in loosely radiating chains.

From soils: Austria (131), Canada (15), China (81), Egypt (115), England (11) (36), India (21)

United States: Alaska (147), California (147), Iowa (3) (101), Louisiana (2), New Jersey (146) (147), New York (67), Utah (50)

13. *Aspergillus minutus* Abbott

Colonies on Czapek's agar white at first, becoming neutral gray with the appearance of fruiting areas, and changing through shades of green-gray or brown-gray, finally becoming dark olive to brown or almost black in some strains. Surface velvety to more or less cottony or closely floccose. Reverse various shades of yellow to orange, often deepening to brown with age. Conidiophores septate, arising as short side branches of aerial mycelium 30–60μ long × 3μ in diameter, rarely attaining a height of 125μ; also arise directly from substratum, up to 250μ. Heads round and radiate in young cultures, later tending toward calyptriform; vesicles small, 8–18μ in diameter, globose; phialides in two series, primary 4.8–6.5 × 3.5–3.8μ, secondary 4.8–3.2μ. Conidia globose verrucose, light green-brown in mass, 3.2–4.5μ in diameter, mode 3.5μ. Sclerotia or cleistothecia not found.

From soil: China (81), Egypt (115)

United States: Colorado (75) (76), Illinois (50, Iowa (3), Louisiana (2)

14. *Aspergillus flavus* Link

Syn. *A. humus*

Colonies on Czapek's agar widely spreading, with floccosity limited to scanty growth of a few aerial hyphae in older areas. Conidial areas ranging in color from sea-foam yellow through chartreuse-yellow, citron-green or lime-green to mignonette-green. Reverse and agar uncolored or yellow, or buff. Conidiophores arise separately from the substratum, 400–700μ or 1,000μ long × 5–15μ in diameter, broadening upward, walls so pitted as to appear rough or spiny with low magnification, occasionally granular, gradually enlarging upward to form a vesicle 10–30μ

or 40μ in diameter. Heads in every colony vary from small with a few chains of conidia to large columnar masses or both mixed in the same area; small heads with small dome-like vesicles and single series of a few phialides up to 10–15 $\times$ 3–5μ; larger heads partly with simple phialides, partly with branched or double series, or with both in the same head; primary phialides 7–10 $\times$ 3–4μ; secondary 7–10 $\times$ 2.5–3.5μ. Conidia pyriform to almost globose, colorless to yellow-green, sometimes almost smooth, usually rough, varying from 2 $\times$ 3, 3 $\times$ 4, 4 $\times$ 5μ, or 5 $\times$ 6μ in diameter, or even larger. Sclerotia at first white, then brown, hard, parenchymatous. Cleistothecia not found.

From soil: Canada (15), China (81), Egypt (115), England (11), India (49)

United States: California (146) (147), Colorado (74) (75), Hawaii (147), Iowa (1) (3), Louisiana (2) (146), New Jersey (146), Porto Rico (147), Texas (92) (147), Utah (50)

15. *Aspergillus terreus* Thom
Syn. *A. venetus* Massee

Colonies on Czapek's agar from tints of pinkish-cinnamon through cinnamon to deeper brown shades in age, spreading, velvety, or in some strains developing definite floccosity of anastomosing ropes of aerial hyphae. Reverse and agar from pale or bright yellow to fairly deep browns. Conidiophores to 150μ or even 250μ long $\times$ 5–8μ, more or less flexuous, with walls smooth, septate or nonseptate, with apex enlarged to form a vesicle commonly 12–18μ, occasionally up to 25μ in diameter, bearing phialides usually in two series upon its dome-like upper surface; primary phialides 7–9 $\times$ 2–2.5μ; secondary 5–7 $\times$ 2–2.5μ; closely packed. Heads becoming solid columnar masses up to 500μ long $\times$ 50μ in diameter. Conidia elliptical to globose, 2.2–2.5μ or even 3μ in diameter, smooth, in long, parallel, adherent chains. Cleistothecia not found.

From soil: Austria (131), China (81), Egypt (115), India (21) (49), Sumatra (130)

United States: California (137) (139), Connecticut (137), Iowa (3) (101), Louisiana (2), New Jersey (137), Texas (137) (139) (152), Utah (50), Virginia (137)

16. *Aspergillus flavipes* Bainier and Sartory

Colonies on Czapek's agar white at first, becoming yellowish, in some strains forming more or less abundant, closely woven, yellow masses containing many helicoid to horse-shoe shaped, thick-walled cells ("hülle" cells). Reverse yellow to orange or brown. Heads mostly columnar or calyptriform masses, commonly persistently white, but with some strains in pale avellaneous shades to deep avellaneous. Conidiophores 300–500 × 4–5μ, or up to 2–3 mm. in length and 8–10μ in diameter, smooth; vesicles subglobose or elliptical up to 20 × 30μ; phialides in two series, primary 4–7μ or 8 × 2μ or 3μ, secondary 5–8 × 1.5–2μ. Conidia 2–3μ, smooth, subglobose, colorless or nearly so.

The soil strains of this species are usually characterized by the abundant production of sulphur yellow sclerotial masses of cells.

From soil: Canada (15), China (81), Egypt (115), India (21) (49) United States: Iowa (3), Louisiana (2)

17. *Aspergillus humicola* Chaudhuri

Colonies at first white, passing through shades of grayish-olive, gray, (Ridgway Pl. XLVI, 21 O-Yy); reverse and substratum persistently some shade of yellow, colony velvety at margins and floccose towards the center where conidiophores are borne as short branches of the hyphae. The height of these conidiophores average 70μ; the conidiophores which arise separately from the substratum are up to 30μ in length and 2–5.4μ in diameter, with walls smooth and almost colorless; vesicles 9–15μ in diameter and flask-shaped, with phialides borne in two series all over the larger vesicles and on upper one-third portion only in smaller ones. Primary phialides 3.6–5.4 × 1.8–2μ, secondary phialides 3.5–1.8μ, mostly; primary and secondary phialides are almost equal in length. Conidia globose, smooth, 2–3μ in diameter, forming closely radiating chains. Heads are radiate.

From soil: India (21)

18. *Aspergillus ustus* (Bainier) Thom and Church

Colonies more or less felted, floccose, with fine hyphae, from white through shades of gray, olive-gray, yellow, yellow-brown toward fuscous, with often a greenish cast, but no true green color, in old cultures purplish, vinaceous at times. Reverse through shades of yellow, orange and brown. Stalks when rising from submerged hyphae up to 1,000μ, most branches of aerial hyphae up to 500 × 5–10μ, few septate, sinuous, with walls rather thin, smooth, usually partly colored some shade of brown; vesicles 10–20μ in diameter; heads hemispherical to almost columnar; phialides colorless, semi-radiate, loosely arranged into two series, primary 5–8 × 3μ, secondary 7–9 × 2–2.5μ. Conidia globose about 3.6μ (3.5–4μ) spinulose, or with fine faint bars of rosy, reddish-yellow or vinaceous color, with chains forming fairly compact columns in old cultures. Some strains show sterile clusters of thick-walled helicoid cells, comparable to the "hülle" cells of *A. nidulans*, but cleistothecia have not been found.

From soil: Canada (15), China (81), Egypt (115)

19. *Aspergillus insuetus* (Bainer) Thom and Church

Colonies entirely fuliginous, with more or less abundant branching aerial mycelium bearing the conidiophores as branches usually short; vesicles 11–16μ in diameter, phialides either regularly produced as primary phialides 5–8μ, with very short secondary phialides or with proliferation of primary phialides to form a second or more series before the conidia-bearing series is reached. Conidia globose, rough, up to 5μ in diameter, fuliginous, with the color mostly aggregated into echinulations of the cell wall, and even forming bars and tubercles at times.

From soil: Egypt (115)

20. *Aspergillus luchuensis* Inui

This form differs from *A. niger* in showing a single series of phialides 7–9 $\times$ 5μ, with conidia 4–4.5μ and finely roughened. Conidiophores up to 2.5 mm. $\times$ 10–15μ, smooth; vesicles 30–40μ in diameter, showing pores or marking where phialides fall off; phialides in one series, 6–3μ

From soil: Austria (131), China (81)

United States: Illinois (129), Louisiana (2), Texas (92)

21. *Aspergillus niger* van Tieghem

Syn. *A. fuscus*, Schumann

Colonies on Czapek's agar rapidly growing with abundant submerged mycelium, in some strains with more or less yellow color in the hyphae aerial hyphae usually scantily produced. Reverse usually without color. Conidiophores mostly arise directly from the substratum, smooth, septate or nonseptate, varying greatly in length and diameter, 200–400 $\times$ 7–10μ, or several millimeters long and 20μ in diameter. Conidial heads fuscous, blackish-brown, purple-brown, in every shade to carbonous black, varying from small, almost columnar masses of a few conidial chains to the more common globose or radiate heads, up to 300, 500, or 1,000μ long; vesicles globose, commonly 20–50μ, up to 100μ in diameter; phialides typically in two series, thickly covering the vesicle, primary varying greatly in length, secondary 6–10 $\times$ 2–3μ. Conidia globose, at first smooth, but later spinulose with coloring substance, mostly 2.5–4μ, less frequently 5μ. Globose, superficial sclerotia produced in some strains, but not common.

From soil: Austria (66) (131), Canada (15), China (81), Denmark

(68), Egypt (115), England (36) (37), India (21) (49), Japan (132), U. S. S. R. (108)

United States: California (146) (147), Colorado (74) (76), Illinois (129), Iowa (1) (3) (101), Louisiana (2) (146), New Jersey (146) (147), Porto Rico (147), Rhode Island (110), Texas (92) (147), Utah (50)

22. *Aspergillus phoenicis* (Corda) Thom and Church

Characters similar to those of *A. niger* except that the primary phialides are about twice as long, 20–30μ, as the primary phialides of that species.

From soil: Austria (131)

23. *Aspergillus wentii* Wehmer

Colonies on Czapek's agar deeply floccose, spreading, with sterile hyphae, white or yellowish, and with heads white at first, changing through cream, cream-buff, honey-yellow, old gold, to light brownish-olive, medal-bronze, or in old cultures sometimes snuff-brown (Ridgway LV, XVI, XXX, and XXIX); in some strains producing large masses of aerial mycelium which in tubes may fill the lumen 3 cm. above the substratum. Reverse yellowish at first, becoming reddish-brown when old; agar frequently colored yellow. Conidiophores 2 or 3 or up to 5 mm. long, 10–12 or 25μ in diameter, one- to two-septate, walls thick, smooth, enlarged at tips to vesicles varying up to 80μ in diameter; heads large, yellow to brown, radiate; phialides usually in two series, primary varying greatly, 6–8μ, occasionally to 15 $\times$ 3–5μ; secondary 6–8 $\times$ 3μ. Conidia pyriform to globose, usually 4–5μ, less commonly up to 5 or 6μ; walls often pitted or furrowed, frequently appearing smooth or nearly so. Cleistothecia have not been found; sclerotia have been occasionally but not uniformly produced.

From soil: China (81), Egypt (115), U. S. S. R. (108)

United States: Colorado (75) (76), Iowa (3), Louisiana (2), Oregon (138)

24. *Aspergillus sulphureus* (Fresenius) Thom and Church

Colonies on Czapek's agar powdery, sulphur-yellow in color, reverse brown. Conidiophores arise from aerial hyphae, up to 200μ in length (up to 1,000μ on bean agar), stalk walls smooth. Heads loose columns of conidial chains, rarely radiate; phialides in two series. Conidia globose, thick-walled, smooth, 3.0–3.5μ.

From soil: United States: Iowa (50), Louisiana (50)

25. *Aspergillus sachari* Chaudhuri

Colony color naphthalene yellow (Ridgway Pl. XVI; 23 yellow f.) reverse colorless in the beginning, later on some shade of yellow; surface somewhat floccose, sclerotia begin to form in a week's time; they are white hard bodies to begin with, then assume cinnamon color through shades of yellow. Conidial heads are abundant and sclerotia are scattered among them as globose or elliptical bodies up to 2 mm. in long axis with a depression in the center. Conidiophores with smooth and colorless walls about 1.5μ thick, 700–900μ long and about 8μ in diameter, nonseptate; vesicles globose, about 29μ in diameter with phialides in two series scattered all over the surface, primary phialides 5.4 $\times$ 2–2.5μ and secondary phialides 7.2 $\times$ 1.8μ. Conidia colorless, smooth, globose, 2–2.7μ in diameter. Heads radiate, 50–85μ in diameter but becoming somewhat columnar with age, about 160–100μ.

From soil: India (21)

26. *Aspergillus alliaceus* Thom and Church

Colonies on Czapek's agar with white floccose mycelium spreading rapidly over the surface of the substratum, and quickly producing abundant sclerotia, at first white, later becoming black without yellow or orange colors, ovate to elliptical up to 500–700μ in horizontal diameter, up to 1,000μ or more in vertical axis, with a depression or pore(?) at the apex. Ascigerous forms not found. Producing few and scattered stalks with heads up to 200μ in diameter, yellow or becoming ochre to brown in age; stalks up to 1,500μ by up to 15μ, with walls colorless, smooth, 1.5μ in thickness, breaking with rough or ragged edges; vesicles up to 40–50μ in diameter with wall 1.8–2μ in thickness and showing prominent pores at bases of phialides; phialides, primary 7–12 $\times$ 2–4μ, secondary 7–8 $\times$ 2μ, colorless. Conidia faintly yellowish, elliptical to globose, 2.5 $\times$ 3μ in diameter.

From soil: United States: Texas (92)

27. *Aspergillus candidus* Link

Colonies on Czapek's agar white, or becoming cream or yellowish-cream in age; surface growth usually stalks and heads with scanty sterile mycelium or anastomosing ropes of hyphae bearing short fertile stalks. Conidiophores vary with the strain, less than 500μ long up to 1,000μ or longer $\times$ 5 or 10 or 20μ in diameter, walls thick, smooth. Heads white, globose, radiate, varying from large globose masses 200–300μ in diameter, to small heads less than 100μ in diameter; vesicles typically globose, up to 50μ in very large heads, fertile over the whole surface; phialides typically in two series, primary 5–10μ or even 15–20μ long, secondary 5–8 $\times$

2–2.5 or 3μ. Conidia colorless, globose, smooth, 2.5–3.5 or 4μ. Sclerotia occasionally produced.

A fungus having the general morphology of the *A. candidus* group but producing bright pink conidial heads has been found commonly in Louisiana soils. Because of the similar morphology it was classified with the *A. candidus* group. Some strains produce yellow sclerotia which are similar to those of *A. candidus*.

Production of bright yellow sclerotial masses have been observed in certain white strains of *A. candidus* which are found commonly in the soil.

From soil: Canada (15), China (81), Egypt (115), England (36), India (21), U. S. S. R. (108)

United States: Colorado (75), Iowa (3), Louisiana (2), Texas (152)

28. *Aspergillus okazakii* Okazaki

Colonies white to sulphur-yellow. Stalks hyalin, straight or sinuate, smooth or asperulate, 200–500 × 8–12μ figured as undulate, especially toward the base, with walls 2–3μ thick; heads 80–100μ in diameter; vesicles 12–40μ in diameter; phialides primary 15–20 × 6–8μ, secondary 8–14 × 2.5–4μ. Conidia globose, hyalin, 2.5–5.4μ smooth, with connective.

From soil: Canada (15)

29. *Aspergillus koningi* Oudemans

Colonies on Czapek's agar spreading, closely floccose, (never powdery as in *A. candidus*) vegetative hyphae creeping, septate, hyalin; surface creamy white, reverse colorless. Aerial mycelium abundant, conidiophores comparatively sparsely produced. Conidiophores arise from aerial mycelium, nonseptate, hyalin, straight or flexuous, smooth; vesicles 16–20μ in diameter; heads radiate; phialides in one series, 8–10 × 2.5μ. Conidia globose, cream colored, smooth, 3μ in diameter.

From soil: China (81)

United States: Iowa (3), New York (67)

30. *Aspergillus ochraceous* Wilhelm

Colonies on Czapek's agar ochraceous shades, consisting of conidiophores and conidial heads with little aerial mycelium. Conidiophores variable in length, commonly several millimeters, rough or pitted, yellow, bearing large, radiate conidial heads. Vesicles globose to 60–75μ in diameter; phialides in two series, primary commonly 15–30μ long, although sometimes longer, secondary 7–10 × 1.5–2μ. Conidia globose to elliptical, smooth or delicately spinulose, yellow, 3.5–5μ or 3.5–4 or 4.5μ. Orange to vinaceous or purple sclerotia commonly present.

From soil: Egypt (115), India (49)
United States: Colorado (75), Iowa (50), Louisiana (2), Texas (152)

31. *Aspergillus terricola* Marchal

Colonies on Czapek's agar with colorless submerged mycelium; conidial areas at first yellow, then golden, and finally fulvous; stalks 1–2 mm. high up to 20–25μ in diameter, septate; heads up to 500μ in diameter; vesicles 30–50μ in diameter, nearly globose, and fertile over nearly the entire surface; phialides in one series, 8–12 $\times$ 3–4μ, with long, loosely radiating conidial chains. Conidia yellow or golden, then brown, lemon-shaped, 5–9 $\times$ 5–6μ, rough from irregularly branching ridges of yellow to brown coloring matter between the inner and outer wall. Sclerotia are occasionally found.

From soil: China (81)

31a. *Aspergillus terricola* var. *americana* Marchal

Colonies on Czapek's agar from shades near yellow-ochre to brown or umber; aerial growth consisting of crowded conidiophores 300–600 $\times$ 6–8μ, walls pitted. Heads radiate; vesicles up to 20μ in diameter; phialides in one series, 7–10 $\times$ 2–4μ. Conidia tuberculate from the presence of color bars, ovate from 3 $\times$ 5μ up to 5–7μ or nearly globose, usually about 5.5μ, occasionally 5–8μ.

From soil: United States: Georgia (118)

32. *Aspergillus tamarii* Kita

Colonies on Czapek's agar spreading broadly, with vegetative hyphae mostly submerged, with fruiting areas at first colorless, then passing through orange-yellow shades to brown in old colonies, light brownish-olive, buffy-citrine, medal-bronze, or raw umber, (Ridgway XXX, XVI, LV, and III) not showing true green. Reverse uncolored or occasionally pinkish. Stalks arising from submerged hyphae, up to 1–2 mm. in length, 10–20μ in diameter, increasing in diameter toward the apex and passing rather abruptly into vesicles; vesicles 25–50μ in diameter; heads vary greatly in size in the same fruiting area, from more or less columnar to nearly but not completely globose and up to 350μ in diameter, with radiating chains and columns of conidia; phialides in one series in small heads, in two series in large heads, primary commonly 7–10 $\times$ 3–4μ, secondary 7–10 $\times$ 3μ. Conidia more or less pyriform to globose, 5, 6, or up to 8μ in diameter, rough from masses of coloring matter. Sclerotia occasionally produced, usually purple or reddish-purple, globose to pyriform with apex white.

From soil: China (81), India (49)
United States: Texas (92)

Species of uncertain position:

Aspergillus minimus Wehmer

From soil: Austria (131)

Aspergillus fumigatoides Saccardo

From soil: England (11)

(9) **Amblyosporium** Fresenius

Sterile hyphae creeping, septate, branched. Conidiophores erect, septate, not swollen at the apex, but terminating in a number of irregular branches, on which the conidial chains are borne. Conidia long, barrel-shaped, in chains, bright colored.

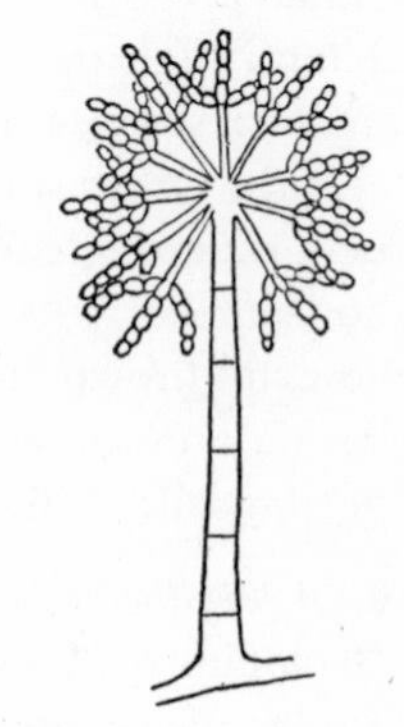

FIG. 69. **Amblyosporium.** conidiophore and conidia (after Lindau).

1. *Amblyosporium echinulatum* Oudemans

Colonies orbicular, gray-green; vegetative hyphae hyalin, articulate, branched. Fertile hyphae swollen at tip, up to 200μ high, hyalin toward the base, toward the apex with dilute gray-green branches; branches basidia-like, closely and repeatedly verticillate or spirally arranged, lageniform, continuous, 25μ high. Conidia catenulate, at first hyalin and globose, afterward dilute gray-green and ovate or broadly elliptical, truncate at ends, apiculate, very minutely spiny, 8–12 $\times$ 6–9μ.

From soil: China (81), Holland (100)

(10) **Penicillium** Link

Vegetative hyphae creeping, septate, branched. Conidiophores erect, usually unbranched, septate, at the apex with a verticil of erect primary branches, each with a verticil of secondary (metulae) and sometimes tertiary branchlets or with a verticil of conidia-bearing cells (phialides) borne directly on the slightly inflated apex of the conidiophores, sometimes with secondary conidiophores borne on the apex of the main conidiophore. Conidia borne in chains which typically form a brush-like head, not enclosed in slime; well-differentiated foot-cells not present. Conidia globose, ovate, or elliptical, smooth or rough.

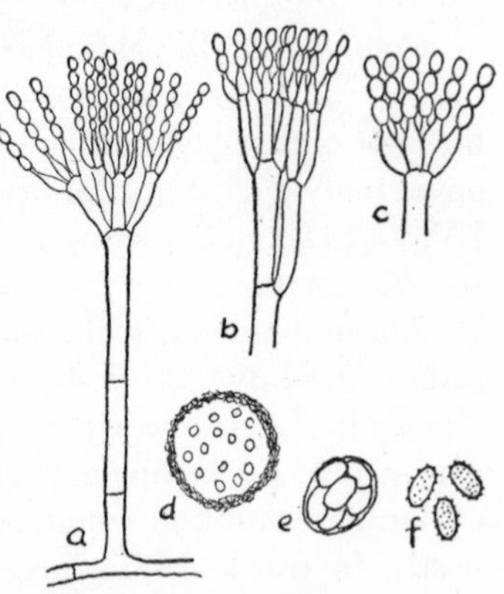

FIG. 70. **Penicillium.** *a* conidiophore; *b*, *c* conidial heads; *d* cleistothecium; *e* ascus; *f* ascospores.

The treatment of this complex genus and its relatives follows closely that of Thom (136) whose works should be consulted for many details that could not be incorporated into this volume.

KEY TO THE SECTIONS OF THE GENUS PENICILLIUM

a. Penicilli typically in single verticils of phialides borne on branches which maintain the identity of each verticil — I. Monoverticillata
aa. Penicilli of more than one series of elements
  b. Penicilli consisting of two or more series of elements including phialides and metulae with or without branches of one or more series, branching asymmetrically — II. Asymmetrica
  bb. Penicilli typically consisting of one symmetrical verticil of metulae, bearing symmetrical verticils of phialides — III. Biverticillata-symmetrica
  bbb. Penicilli consisting of three or more series of verticils forming a fairly symmetrical mass — IV. Polyverticillata-symmetrica

SECTION I. MONOVERTICILLATA

KEY TO THE SPECIES

a. Penicilli typically in single verticils of phialides borne on branches which maintain the identity of each verticil
  b. Conidiophores simple or occasionally branched
    c. Sclerotia present
      d. Conidia elliptical
        e. Conidia 3.5 × 2μ; sclerotia pink — 1. *P. thomii*
        ee. Conidia 3–3.5 × 2–2.8μ; sclerotia yellow — 25. *P. turbatum*
      dd. Conidia globose, 2.5–3μ, sclerotia doubtful — 16. *P. griseum*
    cc. Sclerotia not present
      d. Colonies floccose, extensive simple aerial hyphae, not ropy or velvety
        e. Colonies neutral gray to black, never green
          f. Reverse uncolored — 2. *P. restrictum*
          ff. Reverse orange — 24. *P. nigricans*
        ee. Colonies with green or blue-green fruiting areas
          f. Conidia larger than 4.5μ in length

g. Reverse or substrate red — 3. *P. purpurrescens*
gg. Reverse of substrate not red — 4. *P. fuscum*
ff. Conidia less than 4.5μ in length
g. Conidia elliptical — 5. *P. cyaneum*
gg. Conidia globose
h. Reverse colorless
i. Conidia spinulose
j. Colonies dull green with white areas of mycelium exposed — 6. *P. spinulosum*
jj. Colonies dark gray-green — 7. *P. pfefferianum*
ii. Conidia smooth — 8. *P. trzebinskii*
hh. Reverse red
i. Conidial areas pale green — 9. *P. sanguifluum*
ii. Conidial areas blue-green — 10. *P. cesiae*
dd. Colonies not floccose but with part of aerial hyphae as ropes or funiculose masses or velvety
e. Colonies with part of aerial hyphae as ropes or funiculose masses
f. Ropiness well marked
g. Conidia elliptical
h. Colonies scarcely green, reverse yellow — 11. *P. carmino-violaceum*
hh. Colonies greenish-white, reverse sordid orange to red, almost black — 12. *P. chermesinum*
gg. Conidia globose
h. Colonies spreading, without white margin
i. Colonies zonate — 13. *P. adametzi*
ii. Colonies not zonate — 14. *P. niklewskii*
hh. Colonies forming felts
i. Felts up to 3–4 mm. deep, with white margin; reverse vinaceous — 15. *P. vinaceum*
ii. Felts thin, tough; reverse deep orange — 16. *P. griseum*
ff. Ropiness reduced to trailing or more or less fascicled hyphae in colonies appearing velvety
g. Conidia subglobose to globose, smooth
h. Colonies appearing velvety
i. Colonies white to drab — 20. *P. decumbens*

ii. Colonies bluish to yellowish-green to drab; reverse yellowish to greenish to deep blackish-green — 17. *P. fellutanum*
iii. Colonies pale bluish-green to green to brown; reverse to orange — 18. *P. citreo-viride*
hh. Colonies subfloccose, gray with greenish marginal zones — 19. *P. cinerascens*
gg. Conidia subglobose to globose; rough
h. Colonies bluish-green becoming yellowish-gray in age — 21. *P. paczoskii*
hh. Colonies colorless to pale flesh color, with green conidial area in center — 22. *P. terlikowskii*
ee. Colonies velvety
f. Conidia elliptical
g. Conidia rough
h. Colonies deep blue-green, reverse yellow to tan — 23. *P. lividum*
hh. Colonies dark gray to fuliginous — 24. *P. nigricans*
gg. Conidia smooth — 25. *P. turbatum*
ff. Conidia globose to subglobose
g. Colonies restrictedly growing
h. Reverse uncolored, colonies pale gray to olive-green — 6. *P. spinulosum*
gg. Colonies spreading, velvety conidial chains forming columns
h. Colonies blue-green to green to brown, reverse yellow to orange to reddish-orange
i. Colonies not zonate
j. Conidia smooth
k. Conidiophores short, less than 100μ long
l. Reverse brown — 26. *P. glabrum*
ll. Reverse orange brown — 27. *P. aurantiobrunneum*
lll. Reverse pale yellow to brick-red — 28. *P. sublateritium*
kk. Conidiophores longer, up to 200μ long

l. Colonies dark green to gray, reverse orange. Conidiophores 50–150μ long — 29. *P. oledzkii*

ll. Colonies green to orange-brown, reverse orange-yellow, conidiophores 100–300μ long — 30. *P. szulczewskii*

lll. Colonies blue-green to olive to dark brown, reverse yellow, conidiophores 100–125μ long — 31. *P. frequentans*

jj. Conidia rough

k. Colonies blue-green to brown — 32. *P. flavi-dorsum*

kk. Colonies blue-green to gray-green — 33. *P. baiiolum*

ii. Colonies zonate

j. Phialides long, 30μ — 34. *P. geophilum*

jj. Phialides short, 6–12μ — 35. *P. egyptiacum*

bb. Conidiophores, prostrate or ascending, mostly branching, each branch bearing a terminal penicillus

c. Conidia elliptical — 5. *P. cyaneum*

cc. Conidia globose

d. Conidia 2μ — 36. *P. affine*

dd. Conidia 2.5–3μ

e. Colonies dull gray-green, conidia 2.5–3μ — 37. *P. waksmani*

ee. Colonies gray or greenish-gray, deeply floccose, conidia 2–2.5μ — 38. *P. siemaszki*

1. *Penicillium thomii* Maire

Colonies in bean agar with cane sugar, sordid (amoene) gray-green, quickly and broadly spreading in the substratum with growing margin broad, white. Aerial growth composed of crowded conidiophores. Reverse of colony white, then salmon. Conidiophores 40–300 $\times$ 1.5–2.5μ unbranched, dilated to 4μ at the apex and bearing a verticil of four to eight phialides, 7–9 $\times$ 1.5–2μ, subfusiform. Conidia in chains loosely columnar in arrangement, 3–3.5 $\times$ 2–2.5μ, very pale greenish, thin walled, smooth. Sclerotia salmon, subglobose to ellipsoid, 100–250 $\times$ 11–200μ, frequently confluent. Gelatin not liquified; cultures acid to litmus.

From soil: Austria (131), Canada (15) (16)
United States: Texas (92)

2. *Penicillium restrictum* Gilman and Abbott

Colonies on Czapek's agar small, round, restricted, raised, velvety; surface deep to dark grayish-olive; reverse colorless. Conidiophores arise from aerial hyphae, hyalin, smooth, unbranched, 10–50 × 2–2.5μ, rarely reaching a height of 75μ. The apex of the conidiophore is usually only very slightly inflated, from 0.5–1μ larger than the diameter of the conidiophore, and bears a verticil of crowded, flask-shaped phialides, 5–7 × 2–3.2μ. The heads are loose columns of five to ten chains up to 85μ long. Conidia globose, delicately echinulate under oil, 2–2.5μ in diameter.

From soil: Canada (15), China (81)
United States: Louisiana (50), Texas (92)

3. *Penicillium purpurrescens* Sopp

Colonies on meat-peptone-sugar-gelatin, with thick mass of mycelium somewhat folded or wrinkled, woolly or floccose, white, studded with dark red drops, and producing dark olive-green conidial areas as spots or marginal areas. Reverse at first somewhat reddish, then purple, finally black, with color production reduced or delayed in some cultures. Hyphae commonly fine; conidiophores moderately coarse, septate, mostly producing one verticil of phialides, occasionally with one to several branches at the uppermost node; phialides flask-shaped, in great numbers in the verticil, occasionally bearing secondary phialides. Conidia globose, echinulate, about 6μ in diameter, produced in great abundance and from the figure apparently forming columns. Cleistothecia and sclerotia not found.

From soil: Canada (15)
United States: Texas (92)

4. *Penicillium fuscum* (Sopp) Thom
Syn. *Citromyces fuscus* Sopp

Colonies upon meat-peptone-sugar-gelatin with mycelium spreading rapidly over the substratum, forming a thin close-felted but tough wrinkled mass, quickly becoming dark olive-green with the development of the conidial area which becomes dark or fuscous in age. Reverse at first reddish-yellow (chamois) later with greenish shades ultimately almost black; gelatin colored reddish-brown; odor little or none. Conidiophores arising from prostrate coarse hyphae, at the surface of the mycelium septate, fairly slender, and moderately long, with small vesicular enlargement at the apex bearing one to fifteen phialides. Phialides borne all over the vesicular area as in Aspergillus, and with their chains of conidia divergent, occasionally branched and bearing secondary phialides. Conidia showing a definite connective, echinulate, globose, brown, in age smoother than when young, 5–6μ in diameter. Cleistothecia not found.

From soil: United States: Texas (92)

5. *Penicillium cyaneum* (Bainier and Sartory) Biourge

Colonies upon licorice sticks forming small colonies with mycelium spreading slowly, floccose or hirsute with rather long erect or ascending, branching hyphae, becoming conspicuously bright blue in color (C.d.C. Nos. 392, 397, 398) with conidium formation, with a sterile margin canary-yellow. Reverse yellow-orange to red. Conidiophores mostly arising as diverging branches from erect or ascending hyphae each with its branches terminated by an obconic vesicular apex with a verticil of eight to twelve phialides, about 11.2μ in length producing chains of conidia packed into a columnar mass. Conidia elliptical, about 4 × 2μ.

From soil: Austria (131)

6. *Penicillium spinulosum* Thom

Colonies upon gelatin, bean agar or Czapek's agar deep green, spreading broadly in the substratum with broad sterile margin when young; aerial portion consisting of conidiophores and sparsely floccose aerial hyphae. Reverse of colony not discolored or at times showing a pinkish tinge. Conidiophores 150–300μ or longer by 3–3.5μ, with apex enlarged to 5μ in diameter, bearing a single verticil of phialides 9–11 × 2–3μ; penicillus a loose column of conidial chains up to 300μ or even 500μ in length × 15–30μ. Conidia pyriform to globose, 3.2–3.5 × 3.6–4μ; very thin walled, smooth at first then delicately spinulose or verrucose, yellowish-green, then almost smoky; liquifying gelatin slowly, with strongly acid reaction.

From soil: Austria (66) (131), Canada (15)

var. *ramigena* Szilvinyi

From soil: Austria (131)

7. *Penicillium pfefferianum* Wehmer

Colonies loose cottony, spreading; surface green to gray-green, with growth of superficial hyphae. Sterile hyphae hyalin, septate, ascending, branched. Conidiophores simple or branched, up to 70μ long × 3μ in width, and inflated at the apex to form a swelling 4–8μ in diameter.

Conidial chains long, borne on a verticil of phialides which cover the swelling, pointed at the apex, 9–14 × 2–4μ. Conidia globose, smooth, hyalin, 2.3–3.8μ in diameter, light green to gray or brownish in mass.

From soil: Denmark (68)

United States: Iowa (146) (147), New Jersey (146) (147), North Dakota (147), Oregon (146) (147)

8. *Penicillium trzebinskii* Zaleski

Colonies on neutral Raulin with 10 per cent gelatin in petri dishes, fairly rapidly growing becoming 35–40 mm. in diameter in twelve days, liquefying the gelatin slowly but completely, velvety or slightly subfloccose, showing traces of zonation only in the outer areas, within thrown into wrinkles cerebriform in the center becoming radiate, broad, distant, and progressively shallower outwardly; in color conidial areas near the margin blue-green C.d.C. Nos. 378B, 428A, 428B, becoming green 346, 347, 343, 318, with the green fading out in age and the colonies in dark grays tinged with orange-yellow such as 194, 198. Reverse in pale orange or orange-yellow shades such as 153C, 146, 128D, 138C, 121, 122; odor none. Conidiophores 100, 200, to 400 or up to 600μ, by 2.5–3μ, with apices commonly inflated 4.5–6μ or 8μ, more or less upright, unbranched, sometimes slightly enlarged upward; phialides about 10–11 ×-2.5–3μ, in verticils of 6, 10 to 20, or 25. Conidia about 2.5–3.5μ, smooth, globose or subglobose, showing connectives in the chains.

From soil: Austria (131), Poland (167), Sumatra (130)

9. *Penicillium sanguifluum* Sopp

Colonies on meat-peptone-sugar-gelatin, with tough leathery mycelium irregularly wrinkled and folded, at first more or less yellow, later changing to reddish then red, producing loosely velvety, pale greenish conidial areas, and exuding abundant blood-red drops of fluid. Reverse and gelatin at first yellowish-red, then progressively deeper red until almost black with coloring matter difficult to remove from cloth. Hyphae comparatively fine; conidiophores slender, uniform, much branched with branches short, diverging and bearing single verticils of phialides, usually curved, and few in the verticil. Conidia globose 1.5μ (in diagnostic description, given as 3μ in Sopp's "Key") in divergent chains. Cleistothecia and sclerotia not found.

From soil: Austria (131)

var. *lunzinense* Szilvinyi

From soil: Austria (131)

10. *Penicillium cesiae* Bainier and Sartory

Colonies upon licorice sticks floccose, in a comparative thick mass spreading rapidly with conidial areas in blue or blue-green to dark grayish-blue, quickly producing drops of transpired fluid, rose or red-orange or red on potato cultures. Reverse and substratum such as agar or potato, orange-rose, shades of reddish or red. Conidiophores produced as branches or as terminal segments of creeping or ascending (not erect) hyphae, about 2.8μ in diameter, and uneven in length, only slightly enlarged at the apex, and bearing verticils of six to twelve phialides about 8μ in length. Conidia globose, about 2μ, in loosely parallel to divergent chains.

From soil: Austria (131), Sumatra (130)

11. *Penicillium carmino-violaceum* Dierckx

Colonies in wort gelatin restricted in growth, commonly velvety in appearance but with a felt of creeping hyphae and ropes of hyphae, bluish-gray-green, to brown, with secondary rosy spots or areas from overgrowth of hyphae from the submerged mycelium; coremia none. Reverse at first yellowish, then orange, red-orange, or violet-red; odor none. Conidiophores 2–3.5μ in diameter arising from ropes of hyphae; penicillus 10–15 or 30–50μ, with all walls smooth, figured as simple verticils and branches (?metulae) when present as diverging and independent rather than parts of a penicillus since main axis and branches are septate and without marks suggestive of a single fruiting structure; metulae 13–18 $\times$ 2.5μ, single or few; phialides 7–10 $\times$ 2–3μ in verticils of three to many. Conidia ovate 3 $\times$ 2μ, soon 4 $\times$ 2.8μ, finally 5.5 $\times$ 4μ.

From soil: Canada (15)

12. *Penicillium chermesinum* Biourge

Colonies on wort gelatin restricted in growth, with margin white, crenulate, 0.5–1 mm. broad, bright green, then gray-olive, and finally brown; coremia none. Reverse orange-yellow to purplish-red shades to dark reddish-brown; odor none. Conidiophores 1.5–2.5μ in diameter, arising from creeping hyphae; penicillus figured as simple crowded verticils of phialides on unbranched stalks, or duplicated by divergent branches of varying length at varying distances from the verticil on the main axis; metulae (or branches C.T.) very diverse 8–25–40–65μ long, in pairs or single; phialides 8–12 $\times$ 2–3.5μ, frequently incurved, in groups of three to ten. Conidia elliptical 2.5–4 $\times$ 1.5–2.5μ, with ultimate and penultimate ones 5 $\times$ 3μ.

From soil: Austria (131)

13. *Penicillium adametzi* Zaleski

Colonies in neutral Raulin with 10 per cent gelatin in petri dishes, slowly growing, becoming 25–27 mm. in diameter in twelve days, liquefying gelatin slowly but completely, zonate, with thallus in radiate wrinkles, and surface growth consisting of anastomosing hyphae and ropes of hyphae, which become coremia in acid media; conidial areas in blue-green shades such as C.d.C. Nos. 378B, 371, 372, 373, 367, to green 346, fading to shades of gray; margin white 2–3 mm. wide in the growing colony; crystals abundantly produced in media initially acid in reaction; drops small, yellow, abundant in the furrows toward the margin. Reverse pale orange-yellow shades such as C.d.C. 153c, 171, 196; odor none. Conidiophores 30–40, 80μ, or up to 100 $\times$ 2–2.3μ, more or less inflated at the apex, straight, or slightly flexuous, rarely branched, arising from trailing hyphae or ropes of hyphae; phialides about 7–8 $\times$ 2–2.3μ in compact verticils of 4, 8 to 12, or up to 16 with short tubes. Conidia about 2–2.3μ, smooth, globose, fairly uniform in size and shape, showing connectives.

From soil: Poland (168)

14. *Penicillium niklewskii* Zaleski

Colonies upon neutral Raulin with 10 per cent gelatin in petri dishes, slowly growing, becoming 25–28 mm. in diameter in twelve days, liquefying the gelatin only slightly and tardily, thin, with superficial growth of trailing and anastomosing hyphae and ropes of hyphae, zonation more or less evident, with central area more or less radiately wrinkled; marginal zone 1.5–2 mm. wide white; in color conidial areas at the margin blue-green shades such as C.d.C. 378B and C, then 372, and green 347, with the green fading in age giving dark orange-gray shades such as 139, 143. Reverse in light yellow shades such as 203B, 171, 166; odor none. Conidiophores 10–150μ or up to 300 $\times$ 2–2.5μ, unbranched, straight or flexuous, erect or ascending, commonly enlarging upwards to a more or less definite vesicle, commonly arising from ropes of hyphae; phialides about 9–10 $\times$ 2.3–2.8μ, in verticils of four, eight to twenty, more or less diverging, described as with marginal phialides incurved. Conidia 2–2.5μ, smooth, globose, showing connectives in the growing chains.

From soil: Poland (168)

15. *Penicillium vinaceum* Gilman and Abbott

Colonies on Czapek's agar spreading, floccose, consisting of densely interwoven, and sometimes roped hyphae. Surface white at first, becoming vinaceous to lavender in some strains, and in others vinaceous-gray or green as fruiting areas develop. Reverse white at first, becoming vinaceous or deep wine-red in some strains. Conidiophores arise from aerial mycelium or directly from the substratum, 20–110μ long, mode 50–75μ. Conidial heads a few chains in loose columns up to 60μ long when young, but becoming up to 200μ long when mature. Chains of conidia borne on a single verticil of phialides, 6.5–11.5 × 2–2.5μ. Conidia ovate to elliptical, smooth, echinulate, 2.5–3.5μ in diameter.

From soil: China (81)
United States: Utah (50)

16. *Penicillium griseum* (Sopp) Thom
Syn. *Citromyces griseus* Sopp

Colonies in meat-peptone-sugar-gelatin producing close felted but thin, tough, wrinkled, paper-like mycelium, gray-green to bluish-green or mouse-gray, in age brown. Reverse chamois-red (deep orange); odor weak or doubtful. Conidiophores erect, short, slender, septate, often with one or more diverging branches, without vesicular apex, figured as if partly arising from trailing hyphae(?); phialides small, short and figured as few in the verticil. Conidia (from Sopp's tabulation, not given in description) 2.5–3μ, globose, smooth. Cleistothecia not found.

From soil: Austria (66)

17. *Penicillium fellutanum* Biourge

Colonies on wort gelatin restricted in growth, at first transiently bluish-green, then gray-olive; coremia none. Reverse at first pale yellow, then sordid rosy; odor weak. Conidiophores rising from creeping hyphae 10–35 × 2–3.5μ; penicillus 20–35μ long, with all walls smooth, figured variously from monoverticillate (Citromyces-like) to a dense group of metulae, always on short stalks from creeping hyphae; metulae 9–13μ or even to 20 × 2–3μ, in groups of two to five; phialides 6–11 × 1.5–4μ in verticils of two to eight. Conidia oblong 2–3 × 1.8–2.5μ.

From soil: United States: Texas (92)

18. *Penicillium citreo-viride* Biourge

Colonies in wort gelatin, undulate, wrinkled, forming a thin fibrous felt, pale bluish-green, quickly passing to sordid dark violaceous shades, at length subfuscous; coremia none. Reverse at first white, then pale yellow to greenish-yellow, or orange; odor none. Conidiophores 25–80μ or longer × 1.5–2μ, arising from creeping hyphae; penicillus very short as a simple verticil of phialides, or a main axis with phialides in a terminal verticil or in sessile or stalked groups from lower nodes; phialides 6–9 ×

1.5–3μ, occasionally much larger, in apical verticils of two to ten, at lower nodes single or clustered or on true metulae occasionally. Conidia globose 2–3μ.

From soil: Austria (66)

19. *Penicillium cinerascens* Biourge

Colonies restricted in growth, with trailing or ascending hyphae, at first slightly bluish-green then gray-green to gray, at length reddish-brown; coremia none. Reverse pale yellow; odor none. Conidiophores smooth, about 35μ long and 1.5–2.8μ in diameter, only slightly swollen at the apex, arising from aerial hyphae; phialides commonly attenuated at both ends, 6–11.5 × 1.5–2.8μ, in verticils of four to ten, smooth, rather divergent, conidia oblong to globose, 2–3.8 × 2.2–3μ, rough echinulate when ripe, with the end cell much larger than the others in the chain.

From soil: Austria (66), China (81)

United States: Utah (50)

20. *Penicillium decumbens* Thom

Colonies on potato agar white to gray, gray-green ultimately yellowish-brown, green in cultures with cane sugar. Reverse colorless. Surface growth consisting of trailing stolon-like hyphae sparsely developed and so close to the substratum as to appear only as fertile hyphae, bearing the conidiophores as short branches 20–100μ in length; in old cultures with dense tufts of sterile secondary mycelium scattered over the surface. Conidial fructifications consisting of single verticils of phialides 7–9 × 2–3μ, bearing conidial chains first in loose columns up to 100μ in length, but soon enveloped and broken up in the drops of fluid secreted abundantly from the mycelium. Conidia globose 2.5–3μ, smooth, pale green, then brownish in mass.

From soil: Egypt (115), England (11)

United States: Colorado (147), Iowa (1), New Jersey (146) (147), Oregon (53)

21. *Penicillium paczoskii* Zaleski

Colonies upon neutral Raulin with 10 per cent gelatin in petri dishes, fairly rapidly growing, becoming 38–42 mm. in diameter in twelve days, liquefying the gelatin slowly but completely, thin, plane, velvety-figured as with trailing hyphae and ropes of hyphae, zonate, more or less radiate wrinkled within and more or less overgrown with secondary mycelium in center; with marginal band 6–8 mm. wide, white or with faint bluish color, central areas bluish-green C.d.C. Nos. 372, 422, 423, 375, 396, losing the blue-green in age and becoming yellowish-gray shades such as 164, 160; drops numerous in the intermediate zones, uncolored. Reverse in pale orange shades such as C.d.C. Nos. 146, 121, 97, 128A; odor none. Conidiophores 50, 100–200μ or up to 300 × 2.2, 2.5μ, or 3μ, commonly inflated at apex to 4–5μ, arising largely from coarse ropes of hyphae, mostly unbranched, straight or flexuous; phialides about 9–10 × 2–2.5μ, in crowded verticils of 8, 10–15 or 20 with short tubes. Conidia 2.5–3μ, or up to 3.5μ, smooth, subglobose to ovate, showing connectives in the chains.

From soil: Poland (168)

22. *Penicillium terlikowskii* Zaleski

Colonies in neutral Raulin with 10 per cent gelatin in petri dishes, fairly rapidly growing becoming 32–36 mm. in diameter in 12 days, liquefying gelatin, velvety, usually but not always zonate, with the central area in cerebriform wrinkles changing to radiate in the intermediate area; marginal zone rosy; abundant crystals produced in acid media; in color conidial areas near the margin in blue-green shades such as C.d.C. Nos. 378A, 378B, becoming 367 or green 342, 318, 139, within and in age dark orange shades such as 134, 143. Reverse orange-yellow and orange shades such as 196, 157, 166, 171, 109, 110; odor none. Conidiophores 100–150μ or up to 250 × 2.5–3μ, with or without inflation of the apices to 4–5μ or 6μ; phialides about 9–11 × 2.2–2.5μ, in compact verticils of 3, 8–12, or up to 15, with short tubes. Conidia 2.2–2.5μ or 3μ, smooth, more or less globose, showing connectives distinctly.

From soil: Canada (16), Poland (168)

23. *Penicillium lividum* Westling

Colonies show abundant development of floccose mycelium, with the colony surrounded by a rather broad margin; surface blue-green, becoming darker, and finally dark brown in age. Reverse white to yellowish. Conidiophores arise from submerged mycelium, up to 450μ long, smooth, usually unbranched, although sometimes with one side branch; apex inflated up to 6.5μ in diameter; phialides numerous, 9–12 × 2–2.4μ. Heads 45–150μ long. Conidia oval to ovate, 2.5–4 × 2.2–2.6μ or up to 3.6μ.

From soil: Canada (146), England (37), U. S. S. R. (108)
United States: New Jersey (146) (147), Texas (92)

var. *lunzinense* Szilvinyi
From soil: Austria (131)

24. *Penicillium nigricans* Bainier
Syn. *P. echinatum* Dale

Colonies forming a close textured fairly deep felt of fine or delicate trailing hyphae and occasionally but not regularly, bundles or ropes of hyphae, plane or increasingly wrinkled at higher temperatures toward 30° C., azonate at first, then more or less zonate at margin; conidial areas in various shades of gray, steel-gray, dark olive-gray, hathi-gray, without or with only traces of greenish, to mouse-gray in age. Reverse yellow to deep orange to deep ferruginous shades; odor, strong, drops abundant, colorless or slightly yellowish. Conidia-bearing hyphae variously short branches of aerial hyphae, or whole trailing hyphae showing thickened walls and bearing short branches with penicilli; or as separate conidiophores arising directly from submerged hyphae in marginal area; penicilli terminal on trailing hyphae or on short branches, about 50μ long, consisting of variously diverging branchlets bearing phialides, few to many, and chains of conidia, parallel, to more or less divergent or tangled in age. Conidia 3–3.5μ in diameter, globose, spiny.

In soil: Austria (66), China (81), England (36)
United States: Colorado (36), Idaho (106), Iowa (101)

25. *Penicillium turbatum* Westling

Colonies in prune gelatin closely velvety in appearance but with a basal network of aërial branches forming thin mycelium, with conidial areas at first green (C.d.C. 347) then gray-green (347–372), becoming clear brown only after a month or more, with white margin very narrow. Reverse uncolored; gelatin slowly liquefied, beginning in five to six days, with a neutral or weakly alkaline reaction; odor wanting. Conidiophores arising from creeping hyphae usually very short but not over 120μ long $\times$ 3–4.5μ in diameter; penicillus 20–85μ long, either a single verticil of phialides or a verticil of metulae each bearing a monoverticillate conidial mass; metulae when present two to three in the verticil and 12–20 $\times$ 2.8–4μ; phialides 8–10.5 $\times$ 2–2.6μ, figured as diverging at the tips, numerous in the verticil. Conidia elliptical, smooth 3–3.5 $\times$ 2.2–2.8μ in germination 4.5–6 $\times$ 6–7.5μ. Cleistothecia (=sclerotia?) colorless to yellowish, ovate to globose, 55--105μ in diameter appearing after seven to eight days, becoming numerous and giving a granular look to the colony.

From soil: Austria (66)

26. *Penicillium glabrum* Wehmer
Syn. *Citromyces glaber* Wehmer

Colonies smooth, velvety, spreading, with a narrow sterile margin;

surface at first bright gray-green, becoming darker, and deepening to brown-green with age; reverse brown. Conidiophores inflated at the apex to form a swelling up to 15μ in diameter; phialides 7.5–11.6μ or 12 $\times$ 2–2.8μ. Conidia globose, smooth, 2–3μ in diameter.

Neither Lindau nor Westling give the length of the conidiophores. The species is similar to *P. pfefferianum*, however, and the conidiophores are probably less than 75μ long.

From soil: Austria (66) (131), England (11) (36), U. S. S. R. (108) United States: New Jersey (146) (147)

27. *Penicillium aurantio-brunneum* Dierckx

Colonies on wort gelatin rapidly spreading, somewhat wrinkled; variously gray-green, blue-green, to dark olive-green, or olive-fuscous; coremia none. Reverse yellow verging toward reddish, or orange-brown; odor none or tardily somewhat ammoniacal. Conidiophores unbranched about 50 $\times$ 2–3μ, with all walls smooth; phialides 9–16 $\times$ 2μ. Conidia described as globose (rotundis) but given as 3.8 (to 5.5) $\times$ 2.5μ, subglobose 3μ (Thom), smooth.

From soil: Canada (15), Egypt (115)

28. *Penicillium sublateritium* Biourge

Colonies on wort gelatin restricted in growth, more or less wrinkled, velvety bright bluish-green, sometimes truly blue-green, then dark blue-green, finally fuscous or fuscous-gray; coremia none. Reverse pale yellowish to pale orange, at length almost brick-red; odor weak or none. Conidiophores about 70 $\times$ 1.8–3.2μ, with cell walls smooth, increasing from 1.8μ at base to the uninflated apex at 3.2μ, arising from decumbent hyphae, figured mostly as unbranched but with an occasional branch at the topmost node bearing a secondary fruit; phialides 10.5–15 $\times$ 2.8–3.3μ in verticils of 3–10. Conidia partly ovate, partly globose, 2.5–3.5 $\times$ 1.8–2.4μ or 2.8–3.2μ.

From soil: Austria (66)

29. *Penicillium oledzkii* Zaleski

Colonies on neutral Raulin with 10 per cent gelatin fairly quickly growing becoming 40–45 mm. in diameter in twelve days, liquefying the gelatin quickly and completely, velvety, at times, zonate in the outer area, central area somewhat convex and more or less wrinkled mostly in a radiate manner; white marginal band 1–1.5 mm. wide; in color conidial areas near the margin bluish-green then green, such shades as C.d.C. 342, 338, 343, becoming dark gray shades of orange-yellow such as 168, 173 in old cultures. Reverse in yellow to orange shades such as 211, 191, 181; odor none. Conidiophores 50–150μ or 200 $\times$ 2.5–3.5μ, commonly inflated at the apex, figured as arising mostly as branches from trailing aerial hyphae, straight or flexuous, commonly enlarging more or less from base toward apices; phialides about 11–12 × 2.5–3μ, mostly in verticils of 5, 8, to 12 or 15, in compact verticils. Conidia about 2.5–3.5μ, smooth, globose or subglobose, showing rather long connectives in the newly formed chains.

From soil: Poland (168)

30. *Penicillium szulczewskii* Zaleski

Colonies in neutral Raulin with 10 per cent gelatin in petri dishes, fairly rapidly growing becoming 33–35 mm. in diameter in twelve days, liquefying the gelatin quickly and completely, velvety, plane and zonate in the outer area, within showing cerebriform wrinkles, marginal zone white or rosy, 2–3 mm. wide in the growing colony; in color, green shades such as C.d.C. 347, 342, 338, 339, 318, 314, with the green fading out in age to dark orange-gray shades such as 199, 195. Reverse in orange-yellow and orange shades such as 153C, 146, 142, 133, 128, 109; odor none. Conidiophores 100–300μ or up to 400 × 2–2.5μ with apices inflated to 6–7μ; penicilli 12–15μ in length; phialides about 9–11 × 2.5–3μ, in compact verticils of as few as six, more commonly sixteen to twenty-four, with short broad tubes. Conidia about 2.5–3μ, smooth, globose, showing a short but definite connective when young.

From soil: Poland (168)

31. *Penicillium frequentans* Westling

Colonies blue-green to dark blue-green or olive-green, deepening with age to dark brown-green; velvety or powdery, not floccose or felty. Reverse yellow to reddish-yellow; colony surrounded by a sterile white margin. Conidiophores arise from aerial hyphae, up to 500μ long; often short, 60–225μ; enlarging at the apex to form a swelling 3–4.5μ in diameter. Heads 45–115μ long. The conidial chains are borne on phialides only, 8–11.5 × 2.2–3.2μ. Conidia globose, smooth or slightly verrucose, 2.6–4μ in diameter.

From soil: Austria (66) (131), Canada (15), U. S. S. R. (108)
United States: New Jersey (147)

32. *Penicillium flavi-dorsum* Biourge

Colonies in wort gelatin wrinkled, velvety with central overgrowth spreading with broad white margin (4 mm.); blue-green toward violet, at length dark fuscous (C.d.C. given as very dark orange-yellow shades 164, 165); coremia none. Reverse from sordid yellow to fulvous; odor none. Conidiophores 2.5–3μ in diameter with walls squamulose; penicillus figured as a simple verticil of phialides packed parallel with apparently chains of conidia parallel (or adherent into a column?): metulae rare, long, 30–45 × 2.2–2.8μ; phialides 9–14μ long, with walls echinulate-squamulose, in verticils of four to ten. Conidia subglobose 2.5–4.4μ.

From soil: Austria (66), Canada (16)

33. *Penicillium baiiolum* Biourge

Colonies in wort gelatin, velvety to sublanose, wrinkled, pale blue-green, then grayish-blue-green, gradually darkening to a reddish-fuscous; coremia none. Reverse sordid yellow, to yellow-brown; odor none. Conidiophores 2–3μ in diameter, increasing gradually from base to apex, with walls smooth, penicillus a simple verticil of phialides, occasionally with extra phialides lower on the stalk than the vesicular apex; phialides 9–12 × 3.5μ in verticils of four to ten. Conidia elliptical, then nearly globose, 2.8–4.8 × 2.5–4.8μ, in germinating 6 × 5.2μ, smooth at first, rugose echinulate when ripe.

From soil: Austria (131)

34. *Penicillium geophilum* Oudemans

Colonies circular, with alternate zones white and gray-green; vegetative hyphae 4–8μ in diameter. Conidiophores 360 × 6μ, hyalin, sparsely septate; penicillus a single verticil of phialides; phialides about 30μ long, commonly nine in number, flask-shaped. Conidia globose, 3–4μ in diameter, almost hyalin, produced in long chains packed into columnar masses.

From soil: Holland (100)

35. *Penicillium egyptiacum* van Beyma

Colonies slightly raised in the center, with a white woolly mycelium covered with cleistothecia, yellowish-green, with a white margin, zonate; odor slightly moldy. Reverse yellowish-green. Conidiophores occurring as short lateral branches 30 × 4μ, smooth, enlarging slightly toward the apex. Metulae usually absent, phialides flask-shaped with a short neck, 6–12 × 2–3μ, conidia in long chains without disjunctors, globose, smooth 2.3–2.7 (—4)μ. Cleistothecia numerous, giving the colony a granular appearance, globose to elongated ¼ mm. in diameter. Asci more or less globose, usually with six ascospores, 8.7 × 7.3μ. Ascospores rounded elongated, bivalvate, with two rather indistinct ridges projecting at the poles, barrel-shaped, somewhat flattened on the side of the ridges, 3.2 × 2.7μ.

From soil: Egypt (115)

36. *Penicillium affine* Bainier and Sartory

Colonies on licorice sticks spreading broadly, forming a thick floccose, heavy growth, at first white then green, finally sordid dark green from the great quantities of conidia produced. Conidiophores about 2μ in diameter, smooth, produced as branches and systems of branches from aerial hyphae, figured as sinuate, each commonly tipped with a verticil of phialides then the terminal segment equalled or surpassed by a secondary, diverging branch produced at the next node below, bearing second conidial mass, and the process repeated often two to three times. Conidia about 2μ in diameter.

From soil: Austria (131)

37. *Penicillium waksmani* Zaleski

Colonies on neutral Raulin with 10 per cent gelatin in petri dishes, slowly growing, becoming 2.8–3.3 cm. in twelve days, with the gelatin slowly but completely liquefied, in surface velvety, occasionally showing zonation in the outer areas, with regularly radiate wrinkles 2–4 mm. wide and 1–2 mm. in height, and central area elevated and more or less overgrown with secondary mycelium; margin white 1 mm. in width during the growing period; in color conidial areas blue-green C.d.C. in such shades as 378B, 396, 347, 367, 366, becoming dark orange-brown colors such as 139, 140 in old colonies. Reverse and liquid showing more or less definite zones, pale yellow to orange in such shades as 146, 171, 131, 136, 137; odor faint, agreeable. Conidiophores varying in length 100μ, more commonly $200–300 \times 2.5–3.5\mu$, straight or flexuous, simple or sparingly branched, with apices vesicle-like, with walls smooth; penicillus $20–28\mu$ in length; metulae varying in length in the verticil $10–16\mu$ or even $18 \times 3–4\mu$ in groups of about three to six, often increasing in diameter from base to apex and dilated at the apex; phialides about $8.5–10 \times 2–2.5\mu$, in verticils of about six to twelve, straight, fusiform, showing mostly a slender tube. Conidia $2.2–2.5\mu$ (up to 2.8μ), smooth, globose, showing connectives.

From soil: Poland (168)

38. *Penicillium siemaszki* Zaleski

Colonies in neutral Raulin with 10 per cent gelatin in petri dishes, slowly growing becoming 25–30 mm. in diameter in one to two days, liquefying the gelatin quickly and completely, velvety or slightly floccose, plane and zonate only in the outer area, central area more or less depressed and thrown into broad wrinkles; white margin about 2 mm. wide in the growing colony; in color pale blue-green such as C.d.C. 378B when young, later 372, 368, to green 347, 348, and in age orange-brown shades such as 173 and 148. Reverse and liquefied gelatin in orange-yellow shades such as 171, 156, 161, 128; odor strong, suggesting potatoes rotted by fungi. Conidiophores 100–300μ, or even to 500 × 2.5–3μ, usually flexuous, simple or sparingly branched, with apices inflated and with all walls smooth; penicilli 10–12μ long when simple, 20–28μ long when branched; metulae 10, 12–16μ or even to 18 × 2.5–3μ, commonly two or three in the group and unequal in length; phialides about 9–10 × 2.3–2.8μ, commonly in verticils of 5–12 or 15, with short tubes. Conidia 2.3–2.5μ or even to 3μ, smooth, globose, showing a definite connective.

From soil: Poland (168)

SECTION II. ASYMMETRICA

KEY TO THE SUBSECTIONS OF THE SECTION ASYMMETRICA

a. Aerial hyphae simple or branched but not combined into ropes or fascicles
  b. Colonies velvety — i. *Velutina*
  bb. Colonies floccose or lanose
    c. Penicilli with short compact base and diverging phialides — ii. *Brevi-compacta*
    cc. Penicilli of complex branching, aggregated to form a typical brush
      d. Brush not biverticillate — iii. *Lanata-typica*
      dd. Brush asymmetrically biverticillate — iv. *Lanata-divaricata*
aa. Aerial hyphae combined, in part at least, into ropes, fascicles or coremia
  b. Conidiophores borne separately as branches from trailing ropes of hyphae — v. *Funiculosa*
  bb. Conidiophores combined into fascicles or erect coremia — vi. *Fasciculata*

SUBSECTION i. *VELUTINA*

KEY TO THE SPECIES

a. Penicilli with metulae divaricate
  b. Conidia more than 4.5μ in long axis
    c. Colonies green or olive
      d. Colonies olive-green — 39. *P. digitatum*
      dd. Colonies dark green — 40. *P. oxalicum*
  bb. Conidia smaller
    c. Conidia elliptical
      d. Colonies dull dark green, reverse becoming purple-brown to almost black — 41. *P. atramentosum*

cc. Conidia globose to subglobose
d. Conidial chains in compact columns
e. Reverse in yellow shades
f. Conidia 2–2.5μ — 42. *P. steckii*
ff. Conidia 2.4–3.5μ — 43. *P. citrinum*
ee. Reverse uncolored, greenish to bluish-black — 44. *P. umbonatum*
dd. Conidial chains divergent and tangled in age
e. Colonies spreading broadly — 45. *P. westlingi*
ee. Colonies restrictedly growing — 17. *P. fellutanum*
aa. Penicilli more compact in characteristic brush
b. Colonies with ascending rather than erect conidiophores
c. Drops usually colorless
d. Conidia small 2.6–3.2μ — 46. *P. notatum*
dd. Conidia larger 3.2–4.5μ
e. Conidia elliptical, 3.3–4μ
f. Reverse yellow or yellowish — 47. *P. chrysogenum*
ff. Reverse colorless to reddish — 48. *P. meleagrinum*
ee. Conidia globose, 3μ, occasionally larger — 49. *P. griseo-roseum*
dd. Drops usually yellow — 50. *P. baculatum*
bb. Colonies with conidiophores erect
c. Colonies with an arachnoid (cobwebby) margin — 51. *P. roqueforti*
cc. Colonies without an arachnoid margin
d. Conidiophores less than 100μ long — 52. *P. melinii*
dd. Conidiophores 100–200μ long — 53. *P. puberulum*

39. *Penicillium digitatum* Saccardo

Colonies on bean agar grayish-olive, aerial portion consisting only of very short conidiophores and conidia; surface grayish-olive. Reverse commonly brown to black. Conidiophores arise directly from the substratum 30–100 × 4–5μ, usually very short. Conidial fructifications a few tangled conidial chains up to 160μ in length, in two stages; phialides 13–16 × 3–4μ. Conidia cylindrical to almost globose, 4–7 × 6–8μ, often uneven in size and shape in the same chain.

From soil: United States: Colorado (75) (76), New Jersey (146) (147)

40. *Penicillium oxalicum* Thom

Colonies on Czapek's agar, ivy-green, velvety, spreading widely with surface growth of conidiophores only; reverse pale yellow; agar uncolored or only slightly colored. Conidiophores up to 200 × 3.3–5.4μ, enlarged to 5μ at the apex. Conidial fructifications consist of a single verticil of two to three branches (metulae) 15–20 × 3.5μ, appressed, bearing verticils of phialides 10–14 × 2.5–3.5μ, in parallel whorls which bear conidial chains in close columns. Conidia at first cylindrical, then elliptical, from 2×3, 2×4, $2 \times 5\mu$, up to $3.5 \times 5\mu$.

From soil: China (81)

United States: Iowa (50), New Jersey (146) (147)

41. *Penicillium atramentosum* Thom

Colonies on bean agar bright green, becoming brown when old. Reverse uncolored or yellow; color in medium, none. Conidiophores 240–400μ, averaging about 300μ in length. Conidial fructifications up to 200μ in length, usually 100μ or less; metulae and phialides in divergent verticils; the conidial chains from each verticil form a dense column, which diverges more or less from the other columns when old; phialides 8–10μ long, closely parallel. Conidia elliptical, $3.5\text{–}4 \times 2.5\text{–}3\mu$ smooth, light yellowish-green.

From soil: United States: Iowa (101), New Jersey (146) (147)

42. *Penicillium steckii* Zaleski

Colonies in neutral Raulin with 10 per cent gelatin in petri dishes, slowly growing becoming 28–30 mm. in diameter in twelve days, quickly and completely liquefying the gelatin; surface uneven, velvety or slightly floccose, azonate, thrown into a few radiate wrinkles with a raised central area; margin white, velvety, 2–3 mm. or even 4 mm. wide when young; in color blue-green such as C.d.C. 367 to green 348, to orange-brown shades such as 169 in age. Reverse in pale yellow shades 171, 166, 161; drops few, uncolored, in the center or scattered over the whole surface; odor none. Conidiophores varying 100, 200, 300μ, or even 400μ long $\times$ 2.2–3μ, with walls smooth, simple or sparingly branched; penicilli about 18–24μ long, as figured monoverticillate or divaricately and asymmetrically biverticillate; metulae when present about 10–14μ or even $18 \times 2.5\text{–}3.5\mu$, usually in asymmetrical verticils of four to six, shown in figures as divaricate, and with vesicle-like apices; phialides about $9\text{–}10 \times 2.3\text{–}2.8\mu$, with a definite more or less coarse tube in verticils of five to ten, or occurring singly on the main axis or in the verticil with metulae. Conidia 2.2–2.5μ (even to 2.8μ), smooth, subglobose to globose.

From soil: Poland (168)

43. *Penicillium citrinum* Thom

Syn. *P. aurifluum* Biourge

Colonies on bean agar bluish-green to clear green, becoming olive to brownish-olive when old, usually with sterile white margin; reverse yellow. Aerial part of colony consists of densely standing conidiophores except in the center, where tufts of aerial hyphae arise. Conidiophores arise separately from submerged hyphae or from mycelium on the surface, usually up to 150μ in length (rarely 300μ). Conidial fructifications up to 150μ in length, in two stages; metulae 16–30 $\times$ 3μ, enlarged at the apex to 5μ, each producing a compact verticil of phialides, 6–7 $\times$ 2–3μ. Conidial chains in columns, a separate column arising from each verticil of cells, so that the fructification may appear double, triple, or more complex. Conidia globose, 2.4–3μ or 3.5μ, green, slightly granular.

From soil: Austria (66) (131), China (81), Egypt (115), U. S. S. R. (108)

United States: Colorado (75) (76), Iowa (3), Louisiana (2)

44. *Penicillium umbonatum* Sopp

Colonies on meat-peptone-sugar-gelatin, gray-green, with a marked projection or umbo in center, due to buckling of the mycelium, and producing small disk-like colonies with broad sterile marginal zones; reverse colorless or with a tinge of Naples yellow, especially in center, but colors potato black; hyphae comparatively coarse; odor strong, moldy, peculiar. Conidiophores small, slender, with branches producing either monoverticillate conidial apparatus or verticils of short or long metulae from a vesicular apex, each producing a monoverticillate conidial mass. Conidia 3–3.5μ. Cleistothecia not found.

From soil: U. S. S. R. (108)

45. *Penicillium westlingi* Zaleski

Colonies slow growing, becoming 24–26 mm. in diameter in twelve days, with liquifaction of gelatin beginning about the twelfth day, commonly velvety, azonate, with distant radiate wrinkles in the outer part of the colony and central area more or less elevated; margin about 1 mm. wide in the older colony; color blue-green becoming dark shades of orange-brown in age. Reverse and liquified gelatin pale yellow; drops small, uncolored, very numerous in the growing conidial areas; odor none. Conidiophores 200, 300–500 $\times$ 2.5–3.5μ; apices enlarged capitate, all walls smooth, sparingly branched; penicilli 20–25μ long, (branched 40–5μ) with phialides and metulae often mixed in the verticil; branches rare, 10–40μ long and one or two at a node; metulae varying in length and asymmetrically arranged, 8–16 $\times$ 2.5–3.5μ, with apices enlarged to 4.5μ, about three to six in a group; phialides 9–10 $\times$ 2.3–2.8μ. Conidia 2.3 $\times$ 2.8μ, smooth, subglobose to globose with connective evident.

From soil: Austria (131), Poland (168)

var. *lunzinense* Szilvinyi

From soil: Austria (131)

46. *Penicillium notatum* Westling

Colonies on gelatin spreading, floccose, surface bright blue-green, later becoming darker; reverse yellow. Conidiophores usually arise from submerged mycelium but also from aerial hyphae, sometimes branched, up to 750μ long $\times$ 2.8–4.6μ broad. Heads 45–135μ long; fructifications in three stages (see Biourge (14); Westling's drawing indicates two); metulae 10.5–14 $\times$ 3–4.6μ; phialides 7–8 $\times$ 2.2–3μ. Conidiophores and elements of fructifications smooth. Conidia globose to oval, 2.6–3.2μ in diameter.

From soil: Austria (66) (131), Egypt (115), U. S. S. R. (108)

United States: California (147), New Jersey (146) (147)

47. *Penicillium chrysogenum* Thom

Colonies on bean or Czapek's agar gray-green or mixed green and gray, becoming brownish when old, cottony to subfloccose, broadly spreading, with broad sterile margin when young. Reverse commonly yellow; medium uncolored. Conidiophores mostly arising separately, up to 300μ long; some as short branches of aerial hyphae. Conidial fructifications 100–200μ long, with one or two alternate divergent branches; usually in two stages, but may also have three; phialides 8 $\times$ 2.5μ. Conidia elliptical, becoming globose, 3–4μ, pale green.

From soil: Austria (131), Canada (15), China (81), U. S. S. R. (108)

United States: California (146), Colorado (74) (75) (76) (147), Iowa (1) (3) (101), Louisiana (2) (147), New Jersey (147), New York (67), North Dakota (147), Porto Rico (147), Texas (92), Utah (50)

48. *Penicillium meleagrinum* Biourge

Colonies on wort gelatin semifloccose to tomentose-lanose, wrinkled and buckled, conidial areas blue, such as C.d.C. 422, 397, 392, then blue-green, 362, with marginal white zone 1 mm. broad, zonation indicated by placing but not described; coremia none. Reverse colorless or pale yellow; odor none or indefinite. Conidiophores about 3μ in diameter, with wall smooth; penicillus about 40μ long (20–25μ when branches are wanting), figured as main stalk and one branch, unequal, more or less divergent, bearing compact verticils of metulae and rather coarse phialides but in Biourge's sketch several diverging branches at various levels are suggested; branches two or three or none; metulae 7–14μ, commonly 11–12 $\times$ 2–3μ, enlarging toward the apex, in verticils of two to four; phialides 7–9 $\times$ 3–3.5μ, in groups of two to four. Conidia 4.5 $\times$ 3.2–3.6μ, scarcely deciduous.

From soil: Austria (131)

49. *Penicillium griseo-roseum* Dierckx

Colonies on wort gelatin, velvety, with broad white margin and conidial areas blue-green, such as C.d.C. 361, 362, in age overgrown with rosy mycelium; becoming wrinkled and buckled, with the liquefaction of the gelatin; reverse at first uncolored, then yellow, in age; (in Czapek's agar, colonies velvety up to 500μ deep with radiating creeping partly submerged partly aerial hyphae, and in age overgrowth of mycelium in the center, with white margin of the growing colony about 3 mm. broad with superficial creeping hyphae to the very edge). Conidial areas blue-green to green with abundant but loose mealy masses (not crusts) of conidia; reverse and agar pale yellow; drops abundant, large, colorless. Conidiophores about 4μ in diameter, with all walls smooth (in Czapek's agar up to 350–500μ in length); penicillus about 60μ long, figured, as a main axis with a terminal verticil of metulae with paired or single branches either monoverticillate or biverticillate, short or long, commonly more or less diverging at one to several nodes of the main axis; branches paired 20–30μ long; metulae mostly 10–12μ or even 16 $\times$ 3μ, occasionally very much longer; phialides 7–10 $\times$ 3μ in verticils of three to five (or seven). Conidia globose, 3μ or elliptical 3.5 $\times$ 4μ.

From soil: Canada (15)

50. *Penicillium baculatum* Westling

Colonies in prune gelatin, floccose, white then bluish-green, such as C.d.C. 366–388, 389, 393, with a white floccose margin; gelatin rather slowly liquified and alkaline; sessile hyphae colorless, 2–5μ in diameter, and among them in the substratum needles or columnar crystals of calcium oxalate, mostly in bundles or packets; odor almost wanting. Conidiophores with walls smooth, 50–800μ long $\times$ 3.4–5μ; penicillus figured as a terminal verticil of metulae; metulae 10–14 $\times$ 4.2–6μ; phialides 6.5–9 $\times$ 3–3.4μ. Conidia smooth, oval to elliptical, 3.8–4.6 $\times$ 3–3.6μ, in germination up to 5.3μ or even 7.5μ. Cleistothecia (rarely produced) yellow, thin-walled. Asci hyalin, globose to ovate. Ascospores 4.2–4.8 $\times$ 5.2–6μ lenticular (two-valved) with edge subcanaliculate, as described and as the ascospores were figured these cleistothecia belong in Aspergillus.

From soil: U. S. S. R. (108)

51. *Penicillium roqueforti* Thom

Colonies on bean agar broadly spreading, velvety, gray-green, to clear green, becoming brownish when old. Reverse colorless or cream to yellowish, color in medium none. Conidiophores arise from submerged hyphae, 200–300μ long. Conidial fructifications 90–160 × 30–60μ, usually appearing double by the divergence of the lowest branch, usually in three stages; metulae irregularly verticillate, bearing crowded verticils of appressed phialides, 9–11 × 2.5μ, with long, divergent chains of conidia. Conidia bluish-green, cylindrical to globose, smooth, 4–5μ in diameter.

From soil: U. S. S. R. (108)
United States: Iowa (1) (3), Maine (146)

52. *Penicillium melinii* Thom

Colonies in Czapek's agar, rather slowly growing 25 mm. in diameter in eight days, velvety, plane or slightly raised in the central area with centers more or less raised or sometimes umbonate, greenish-gray or with a bluish effect (Ridgway's Hathi gray LII); with margin (fimbria) white 0.5–1 mm. wide. Reverse and agar citrine at first becoming a shade of orange-brown or russet (Ridgway XV); drops yellow, small, abundant over the whole area, later running together and becoming a shade of orange-yellow or tawny or russet, (Ridgway XV). Conidiophores with walls granular tuberculate, usually less than 100μ long × 2–3μ; penicilli terminal and monoverticillate or a terminal group of diverging and unequal branches or metulae 10–20 × 2–2.5μ, or with branches irregularly developed at lower nodes, at the tip of the main stem and prolonging it by each successive branch outgrowing the older one in length; phialides 6–9 × 2μ, with beak-like tubes, with chains of conidia divergent or tangled. Conidia rough tuberculate 3–3.5μ in diameter, with connectives evident.

From soil: United States (136)

53. *Penicillium puberulum* Bainier

Colonies upon licorice sticks, vigorous, blue-green, becoming dark blue-green in age. Conidiophores (length not given but noted only as "plus ou moins allonge") about 5.6μ in diameter, figured as somewhat sinuous, not always smooth, at times with walls showing "fine granulations" difficultly visible, commonly also many vacuoles, occasionally bearing a long branch toward the base; penicillus figured and described as usually sparingly branched with rather long more or less diverging primary branches, bearing metulae or secondary branches or both in the same verticil; phialides not described. Conidia globose, more or less unequal in diameter but averaging 4.2μ, swelling slightly before germinating and emitting one or more germ tubes.

From soil: Austria (131)

SUBSECTION ii. *BREVI-COMPACTA*

KEY TO THE SPECIES

a. Penicilli short with elements compact at base; with tips and conidial chains divergent
  b. Conidia globose
    c. Colonies forming deep mycelial masses with long conidiophores — 54. *P. stoloniferum*
    cc. Colonies velvety, conidiophores short
      d. Colonies narrowly restricted — 55. *P. biourgeianum*
      dd. Colonies moderately broad
        e. Dull olive-green — 56. *P. hagemi*
        ee. Yellowish-green — 57. *P. szaferi*
  bb. Conidia elliptical
    c. Colonies yellowish-green to purple-brown in age — 58. *P. patris-mei*
    cc. Colonies gray-green to bluish-green to brown in age — 59. *P. bialowiezense*

54. *Penicillium stoloniferum* Thom

Colonies on bean agar green or yellowish-green, becoming gray-green or gray when old, floccose, spreading by aerial stolons. Conidiophores arise as short branches of aerial hyphae up to 100μ, or arising separately 300μ or more in length. Conidial fructifications 40–80μ or rarely up to 170μ long; usually in three stages, phialides $10 \times 3\mu$. Conidia slightly elliptical or globose, 2.8–3.4μ, smooth, yellowish-green in mass.

From soil: England (36)

United States: Colorado (75) (76), Illinois (129)

55. *Penicillium biourgeianum* Zaleski

Colonies on neutral Raulin with 10 per cent gelatin, about 30–35 mm. in diameter after twelve days, forming a symmetrical mass elevated 3–5 mm. above the substratum, with gelatin unchanged during the first twelve days but later entirely and quickly liquefied, surface growth velvety, azonate, with a broad marginal area white, then a pale blue-green ring, such as C.d.C. 428B, about a central area partly depressed, partly elevated, and in blue-green shades such as 367, 372, or later green 342, 343. Reverse in pale yellow to orange colors. Conidiophores about 500–700 $\times$ 3–4μ, straight or slightly flexuous, simple or rarely branched, with apices inflated to 4–5μ, phialides about 9–10 $\times$ 2.5–3μ, commonly eight to twelve in the verticil, straight or incurved with long tubes. Conidia 2–2.5μ or 3μ, smooth, globose with connectives distinctly seen in the chains.

From soil: Poland (168)

56. *Penicillium hagemi* Zaleski

Colonies on neutral Raulin with 10 per cent gelatin in petri dishes, thin and flexuous, slowly growing, becoming 2.5–3 cm. diameter in twelve days, at about which time the gelatin quickly liquefies; surface velvety, for the most part superficially zonate, becoming convex with the marginal areas slightly or considerably raised, radiately wrinkled, with center also elevated, with the whole surface sprinkled or covered with white areas suggesting the residues of evanescent drops, margin loose, 0.5–1 mm. or even 2 mm. wide in the growing colony; in color at first pale green (at fruiting margin) such as C.d.C. 341, 342, 337, 338, becoming 289 to 314 or similar shades in age. Reverse more or less zonate pale yellow such as 221, 171, 196; drops numerous uncolored, arising in the margin of the young colony; odor none. Conidiophores about 300μ or 400 $\times$ 3.5–4μ, simple or rarely branched, with walls smooth or rough; penicilli long, usually asymmetrically arranged and 40–50μ long; branches about 16–26 $\times$ 3–4μ, with walls occasionally roughened, in groups of two or rarely three; metulae about 12–15 $\times$ 2.5–3.5μ in groups of three to five; phialides about 9–10 $\times$ 2–2.5μ, in verticils of five to ten. Conidia 2.5–3.5 $\times$ 2.5–3μ, smooth, variously ovate to subglobose, long adhering in masses.

From soil: Poland (168)

57. *Penicillium szaferi* Zaleski

Colonies on neutral Raulin's 10 per cent gelatin, slowly growing, becoming 2.5–3 cm. in diameter in twelve days, with surface wrinkled or undulate and gelatin strongly liquified, appearance velutinous or velvety, with crowded but indistinct zones, radiately wrinkled, whole surface hirsute with secondary mycelium, with central area elevated, margin (fimbria) 1–2 mm. wide when young, indefinite in age; in color pale green at first, such as C.d.C. 342, 343, to very dark yellowish-green, and olive shades such as 269, 274, 148, 140. Reverse and liquefied gelatin in orange, yellow tints, 151, 157. Conidiophores commonly 300–500 $\times$ 4–4.5μ, straight or slightly flexuous with walls slightly asperulate; penicilli 40–50μ, occasionally 60μ long, with walls mostly smooth; branches (rami) 18–30 $\times$ 3.5–4μ, asymmetrically arranged in groups of two or three, straight or slightly incurved with apices inflated or vesicle-like up to 4.5μ in diameter; with walls somewhat asperulate; metulae about 12–14 $\times$ 3.5–4μ, commonly somewhat enlarged and rounded at the apex and four to six in the group; phialides 9–10 $\times$ 2.2–2.5μ, straight, cylindrical, crowded in verticils of eight to twelve. Conidia subglobose to globose, 2.5–3μ or even 3.5μ, smooth, long adherent in masses and showing connectives between members of the chains.

From soil: Poland (168)

var. *lunzinense* Szilvinyi
From soil: Austria (131)

58. *Penicillium patris-mei* Zaleski

Colonies in neutral Raulin with 10 per cent gelatin in petri dishes, slowly growing, becoming 26–30μ in diameter in twelve days, liquefying the gelatin only slowly and partially, velvety or slightly floccose, azonate or with a trace of zonation near the margin, whole central area depressed, thrown into cerebriform wrinkles, irregular in outline with a marginal zone and slowly showing indistinct green shades such as C.d.C. 322 in areas within. Reverse and liquefied gelatin in orange-yellow and orange shades 171, 166, 104, 132; odor none. Conidiophores about 50, 100–300μ, or up to 600 × 2.0–2.5μ, commonly somewhat enlarged at apex, unbranched, erect or ascending, flexuous; penicilli 8–10μ long when simple, 15–25μ when branched, with all walls smooth; metulae occasionally present, about 8, 10–16μ, or 18 × 2–2.5μ, commonly somewhat enlarged at the apex, unequal in length and asymmetrically arranged; phialides about 7–8 × 2–2.3μ, in verticils of 3, 6–15, or 20, or occasionally occurring singly, with tubes usually small and short. Conidia 2–2.5μ, smooth, subglobose to globose, with connectives evident.

From soil: Poland (168)

59. *Penicillium bialowiezense* Zaleski

Colonies on neutral Raulin with 10 per cent gelatin in petri dishes, slowly growing, becoming 2.5–3 cm. in diameter in twelve days, with liquefication of the gelatin beginning about the tenth day and progressing rapidly, thin, plane, velvety, distinctly zonate only in the outer areas; central areas becoming convex showing the marks and dried residues of drops, also showing a central elevation or umbilicus (?) and some overgrowth of sordid white aerial mycelium; margin 2 mm. wide at first then 1 mm. during the remainder of the growing period; in color pale blue-green at first, such as C.d.C. 371, 372, then green, 323. Reverse and liquefied gelatin in pale yellow 166, 156, 171, and related tints; drops few, uncolored arising near the marginal area; odor none. Conidiophores more or less straight, 400–600 × 4–5μ, the next series 12–20 × 3–4μ, with walls delicately roughened; metulae about 10–12 × 3–4μ in groups of about four to six, enlarging upward (clavate); phialides about 10–11 × 2.5–3μ, in verticils of five to eight. Conidia smooth, mostly ovate, some subglobose 2.5–3.5 × 2.5–3μ, with connectives evident between members of the chains.

From soil: Poland (168)

SUBSECTION iii. *LANATA-TYPICA*

KEY TO THE SPECIES

a. Colonies with zonation well marked
  b. Drops uncolored — 60. *P. commune*
  bb. Drops yellow — 61. *P. roseo-citreum*
aa. Colonies azonate
  b. Ripe conidia globose or subglobose
    c. Reverse pale yellow or orange-yellow — 62. *P. raciborskii*
    cc. Reverse uncolored or slightly yellow — 63. *P. lanosum*
  bb. Ripe conidia elliptical — 64. *P. biforme*

60. *Penicillium commune* Thom

Colonies on bean agar dull green, becoming brown when old, broadly spreading, with broad white growing margin composed only of conidiophores, in the older parts becoming floccose masses of interwoven hyphae; reverse not colored. Conidiophores commonly 300μ or less in length, sometimes up to 700μ. Conidial fructifications commonly 100–200μ in length, in three stages, compact at the base and broadening above, variously branched, with branches appressed; phialides 8–9 $\times$ 3μ. Conidia elliptical to globose, 3–4μ, smooth, green.

From soil: Austria (131), China (81), England (11), Germany (104)
United States: Colorado (75), Iowa (1) (3), New Jersey (146) (147)

var. *lunzinense* Szilvinyi
From soil: Austria (131)

61. *Penicillium roseo-citreum* Biourge

Colonies in wort gelatin, floccose felted, at first gray-greenish, then dull green and finally brown with frequent overgrowth rosy. Reverse at first yellow (aureus) soon red, or dark red; gelatin liquefied and reddened; drops golden yellow; odor none. Conidiophore 3–3.5μ in diameter, penicillus 20μ without branching, 40–50μ with branching, with all walls smooth, figured as a main axis with terminal verticil of metulae more or less diverging, and with or without a branch from the next node bearing a secondary fruit; branches none or in verticils 25μ long, metulae 8–12μ or even 15 $\times$ 2.5–3μ; phialides 8–9 $\times$ 2.5–3.5μ. Conidia globose or subglobose 3–4μ.

From soil: Austria (131)

62. *Penicillium raciborskii* Zaleski

Colonies on neutral Raulin with 10 per cent gelatin, slowly growing becoming 3–3.5 cm. in diameter in twelve days, with gelatin liquefied, fairly coarse, lanose, not zonate, strongly and irregularly undulate; margin (fimbria) consisting of loose hyphae, 1–1.5 mm. wide during the growing period; in color becoming blue-green, such as C.d.C. 368 in center at about six days with a broad white margin 3–5 mm. wide, with conidial areas becoming progressively 372, 347, 343, and finally dark orange-brown shades in age such as 139, 140, 143. Reverse and liquefied gelatin in pale yellows 271, 266, or orange-yellows such as 161, 157; drops few uncolored, seen mostly in the marginal areas; odor none. Conidiophores about 40–600 × 3–3.5μ, straight, simple or somewhat branched, with all walls smooth; branches about 12–20 × 3–4μ, two or rarely three in the group, with vesicle-like apices; metulae about 10–14 × 2.5–3μ, in groups of five to seven, somewhat claviform with enlarged or vesicle-like apices; phialides about 8–9 × 2–2.5μ, in verticils of six to twelve, more or less straight. Conidia 2.3–2.8μ or even 3μ, smooth, globose, with connectives evident, not persisting in chains in fluid mounts.

From soil: Poland (168)

var. *lunzinense* Szilvinyi

From soil: Austria (131)

63. *Penicillium lanosum* Westling

Colonies in prune gelatin, lanose, at first white then with the center becoming slowly gray-green, such as C.d.C. 347, 318, 323, 343, 299, with a broad white margin, the whole often overgrown with white mycelium, becoming darker shades in age and finally dark brown. Reverse uncolored or slightly yellow; gelatin slowly and partly liquefied in fourteen days, with an acid reaction when litmus is used; odor weak and scarcely definite; abundant drops appeared on the surface of the colonies. Conidiophores smooth, up to 1 mm. × 3.4–4.6μ, metulae 12–14 × 3–4.6μ; phialides 7–9 × 2–2.7μ. Conidia globose, uniform, smooth or slightly roughened 2.2–3μ in diameter, swelling in germination to 5–6μ, in chains separating easily.

From soil: Austria (131)

var. *lunzinense* Szilvinyi

From soil: Austria (131)

64. *Penicillium biforme* Thom

Colonies on Czapek's agar broadly spreading, gray-green, becoming brownish to olive, restricted in growth, densely floccose. Reverse cream; color in medium, none. Conidiophores arise from aerial mycelium, 60–150μ long. Conidial fructifications usually in three stages, 60–240μ

long; phialides 8–13 × 3 μ; conidia elliptical to globose, 3.2–3.5 × 4–4.3 μ or 5 μ.

From soil: England (36), Greenland (96), U. S. S. R. (108)
United States: Iowa (101)

var. *lunzinense* Szilvinyi
From soil: Austria (131)

SUBSECTION iv. *LANATA-DIVARICATA*

KEY TO THE SPECIES

a. Colonies in various shades of lilac
  b. Colonies lilacinus — 65. *P. lilacinum*
  bb. Colonies light lobelia-violet — (*Spicaria violacea*)
aa. Colonies mostly showing green or greenish conidial areas, penicilli upon diverging branchlets
  b. Conidia globose, smooth or granulated under oil immersion
    c. Reverse of colonies uncolored
      d. Colonies broadly spreading — 66. *P. simplicissimum*
      dd. Colonies felted, restricted — 67. *P. chrzaszczi*
    cc. Reverse of colonies in yellow to orange shades
      d. Chains of conidia divergent
        e. Colonies spreading, shallow
          f. Conidia 2.4–3 μ — 68. *P. janthinellum*
          ff. Conidia 2–2.5 μ — 69. *P. rivolii*
        ee. Colonies floccose
          f. Colonies with drops
            g. With yellow to amber drops — 70. *P. guttulosum*
            gg. With larger colorless drops — 71. *P. gilmanii*
          ff. Colonies without drops
            g. Conidiophores with vesicular cells filled with red granules — 72. *P. glauco-roseum*
            gg. Conidiophores showing only traces of granulation — 73. *P. soppi*
      dd. Chains of conidia adherent into columns
        e. Colonies with blue tints
          f. Conidia 3–3.5 μ — 74. *P. glauco-griseum*
          ff. Conidia 2.5–3 μ — 75. *P. canescens*
        ee. Colonies not bluish
          f. Reverse colorless to slowly rosy — 76. *P. jenseni*
          ff. Reverse yellow to orange-brown — 77. *P. matris-meae*
  bb. Conidia globose, rough or spinulose
    c. Conidia rough

d. Colonies forming shallow felts green to olive to brown — 78. *P. albidum*
dd. Colonies tending to funiculose, gray or olive-gray with scarcely a trace of green — 24. *P. nigricans*
ddd. Colonies almost velvety, gray-green
  e. Conidia 3–3.5μ, rough — 79. *P. swiecickii*
  ee. Conidia 3μ with roughening, granules more or less in bands — 80. *P. janczewskii*
cc. Conidia delicately spinulose — 81. *P. kapuscinskii*

65. *Penicillium lilacinum* Thom

Colonies on bean agar white to pale lilac, more or less loosely floccose with hyphae branched, septate, ascending, 3μ in diameter, producing conidial masses upon very short branches irregularly distributed. Reverse not discolored. Heads up to 100μ in length, consisting of solitary, sessile phialides or verticils of phialides, or short branches bearing one, two or three verticils of phialides with long, tangled chains of conidia. Phialides flask-shaped, divergent at the apices, acuminate, 7–10μ long. Conidia elliptical, smooth, pale lilac, 2.5–3 $\times$ 2μ.

From soil: Austria (66) (131), Canada (15), England (16), U. S. S. R. (108)

United States: Colorado (74), Iowa (3), Louisiana (2), New Jersey (146) (147), Texas (92)

66. *Penicillium simplicissimum* (Oudemans) Thom

Colonies orbicular with alternating zones of cream-yellow and dirty gray, occasionally showing a tinge of violet; vegetative hyphae creeping, very thin, septate, hyalin, dichtomously branched. Conidiophores erect, 40μ high, septate, hyalin, usually unbranched, at the most with a small side branch, on the tip constantly producing three branchlets; branchlets nonseptate 8–12μ long, verticillate; conidia 2–3μ in short chains, globose.

Colonies upon Czapek's agar, rather loosely floccose, spreading about 5 cm. in diameter and 500μ deep in ten days, margin broad white; in color white to gray or in the denser conidial area perhaps a tinge of green, consisting of trailing and branching aerial hyphae, many of them very long. Reverse uncolored; odor none; drops not seen. Penicilli either terminal on long trailing hyphae monoverticillate, or a divaricate group of two to several metula-like branchlets, or with various mixtures of branchlets and phialides in the verticil; secondary penicilli often monoverticillate; metulae or branchlets varying, mostly short; phialides up to 8–9μ long $\times$ 2μ, diverging at the tips and bearing divergent or tangled chains of conidia. Conidia mostly 2–2.5μ or 3μ in long axis, some 3.5μ

or occasionally about 4μ, elliptical to subglobose, showing connections in the chains.

From soil: Canada (15)

var. *lunzinense* Szilvinyi

From soil: Austria (131)

67. *Penicillium chrzaszczi* Zaleski

Colonies in neutral Raulin with 10 per cent gelatin, slowly growing, becoming 24–28 mm. in diameter in twelve days, and liquefying the gelatin completely, with thallus thin, irregular at margin, velvety to subfloccose, azonate, with a few (one to three) very conspicuous concentric wrinkles, with center either umbonate or depressed; margin (fimbria) white about 1 mm. wide during the growing period; in color at six days bluish-green, such as C.d.C. 371, 372, becoming 367, becoming dark orange-brown shades in age such as 173, 148, 147, 143. Reverse azonate, irregularly concentrically wrinkled, in yellow tints such as 196, 178D, 166, 157, 171. Conidiophores 300–600 × 2–3μ, straight, simple or rarely branched; penicilli 20–26μ long; metulae about 12–16 × 2.5–3μ, varying from two to eight in the verticil, with inflated apices and either symmetrical or asymmetrical in the verticil; phialides fusiform about 8–9 × 2–2.5μ, commonly eight to ten in the verticil with short tubes. Conidia 2–2.5μ or even 3μ, smooth, subglobose to globose.

From soil: Poland (168)

68. *Penicillium janthinellum* Biourge

Colonies on wort gelatin bluish-green, gray-green or bright green, azonate, with surface growth consisting of networks of hyphae and ropes of hyphae, tardily becoming reddish ("rubicante"); coremia none. Reverse yellow to ochraceous; odor weak. Conidiophores 30–40 × 2μ arising from creeping hyphae or ropes of hyphae, with all walls smooth; penicillus a single one-sided verticil of metulae with occasionally one branch from a lower node, hence short, about 15μ long or 30–50μ when branched; metulae 7–10 × 1.5–2μ mostly in threes; phialides 5.5–9 × 1.5–2μ, in pairs or threes, apparently the Eupenicillium type. Conidia globose, 2.4–3μ.

From soil: Austria (131), Canada (15), Sumatra (130)

69. *Penicillium rivolii* Zaleski

Colonies on neutral Raulin with 10 per cent gelatin in petri dishes, slowly growing becoming 27–30 mm. in diameter in twelve days, liquefying the gelatin quickly and completely, surface growth velvety, plane and zonate only in the outer area with radiate wrinkles 3–5 mm. wide and 2–3 mm. high running toward an elevated center more or less overgrown with secondary loose mycelium; margin light rosy 2–3 mm. wide formed of dense mycelium; in color conidial areas in green shades in age such as C.d.C. 141, 137, 171, 136, 137, 134; odor none. Conidiophores variable 100–600 × 2.5–3.5μ, simple or sparingly branched with walls smooth; penicilli figured as monoverticillate, divaricately branched, or as having metulae or branches in the same verticil as phialides; branches rare, distant and unequal, perhaps 20–35 × 3–4μ; metulae when present varying greatly in length and arrangement 10–20 × 3–4μ, with apices enlarged, in groups of one or two up to seven or even nine; phialides about 9–10 × 2.2–2.5μ often very variable in size in the same group, in verticils of about six to twelve, with rather long tubes. Conidia about 2.3–2.8μ, smooth globose, with connective evident.

From soil: Poland (168)

var. *lunzinense* Szilvinyi

From soil: Austria (131)

70. *Penicillium guttulosum* Gilman and Abbott

Colonies on Czapek's agar small, not spreading, surface felty; from pink in some strains to pinkish-gray or greenish-gray; often studded with exuded droplets of wine-red moisture. Reverse orange, reddish-brown to deep wine-red; color in medium same, pigment rapidly diffusing and coloring the medium a deep reddish-brown to wine-red. Conidiophores arise from aerial hyphae, 15–50μ long × 1.5–2μ, and bearing at the apex a single verticil of crowded phialides. Heads a few, short, divergent chains up to 25μ or 30μ long; phialides flask-shaped, 7–8.5 × 2–2.5μ. Conidia globose to ovate, light green, delicately echinulate, 2.5–3.5μ in diameter.

From soil: Canada (17)

United States: Texas (92, Utah (50)

71. *Penicillium gilmanii* Thom

Colonies on Czapek's agar round, slowly spreading, closely floccose; surface light gray-green; reverse colorless. On bean agar, surface gray to blue-gray; reverse colorless. Aerial hyphae often in ropes, from which the conidiophores arise as short side branches, unbranched, pitted, slightly inflated at the apex, 6–60μ long × 2–3μ in width, mode 10–25μ. The heads are loose columns of five to ten chains up to 100μ long. The apex of the conidiophores bears a single verticil of flask-shaped phialides, pitted or echinulate, 5–7 × 1.8–2.2μ. Conidia globose, light green, smooth or very delicately rugulose, 2–2.5μ in diameter.

From soil: United States: Utah (50)

72. *Penicillium glauco-roseum* Demelius

Colonies in prune gelatin floccose, gray-green (C.d.C. Nos. 347 to 78e), mycelial hyphae, conidiophores, branches, and metulae showing vesicular cells filled with rosy crystals and granules. Conidiophores 36–240 × 2–6μ, bearing a single verticil of phialides or a penicillate branching system at the apex, sometimes metulae arising from a red vesicle and showing irregular or antler-like branching with conidium formation suppressed; metulae 14.4 × 6μ; phialides 8.4–9.6 × 2μ. Conidia globose, 2.5–3.6μ, smooth.

From soil: Germany (125)

73. *Penicillium soppi* Zaleski

Colonies in neutral Raulin with 10 per cent gelatin in petri dishes, quickly growing becoming 40–45 mm. in diameter in twelve days. Liquefying gelatin slowly but completely, subfloccose, plane and indistinctly zonate only in the outer band or zone of 2–5 mm., for the most part wrinkled in the center, cerebriform becoming radiate outwardly toward the margin, margin 1–2 mm. wide; in color white then showing a conidial band near the margin yellowish green C.d.C. Nos., 271, 262, to green 347, to dark shades of orange-brown or gray such as Nos. 147, 148. Reverse in orange-yellow shades such as Nos. 191, 157; odor weak, agreeable. Conidiophores 500–800μ or 1,000 × 2.5–3.5μ with apices enlarged and with all walls smooth; penicilli commonly 18–24μ long; branches very rare; metulae enlarging from base to apex, about 10–14 × 2–4μ, in groups of five to seven; phialides about 9–10 × 2.0–2.5μ, in verticils of six to ten. Conidia about 2.2–2.5μ, smooth, mostly subglobose, occasionally globose.

From soil: Poland (168)

74. *Penicillium glauco-griseum* Sopp

Colonies at first with mycelium white, thin, wrinkled or buckled, becoming pale blue-green to blue-green with the development of conidia; hyphae very delicate. Reverse white with a yellowish tinge, odor indefinite. Conidiophores enlarged upward, very coarse, rough, long, producing a single verticil or superimposed verticils of metulae variously clavate, either bearing phialides directly or partly again proliferated to bear secondary metulae; phialides not described. Conidia elliptical to globose, 3–3.5μ long. Cleistothecia unknown.

From soil: Austria (66)

75. *Penicillium canescens* Sopp

Colonies on meat-peptone-sugar-gelatin white, fibrous ("filzig" cf.

"woolly" in *P. camemberti* with which it is compared), becoming very slightly bluish or greenish with the tardily developing conidia; hyphae delicate, fine, much finer than in *P. camemberti*; reverse white, in age brownish; substratum becoming purple-brown; odor obnoxious, suggesting mouse urine. Conidiophores short very slender, bearing few branches and phialides figured as irregularly verticillate and more or less divergent, the branches in the verticil often differing in length, and with vesicle-like apex; phialides few in the verticil, pointed, with the chains in each group figured as forming a column. Conidia bluish, smooth, globose, 2.5–3μ, in chains, figured as closely appressed (? adherent). Cleistothecia not found.

From soil: Canada (15)

76. *Penicillium jenseni* Zaleski

Colonies in neutral Raulin with 10 per cent gelatin in petri dishes, fairly rapidly growing, becoming 35–40 mm. in diameter in twelve days, liquefying the gelatin quickly and completely, velvety and zonate at the marginal areas only, variously overgrown with trailing hyphae, and thrown into wrinkles at the older areas, with center raised, umbilicate; white margin (fimbria) 4–6 mm. wide, very conspicuous; in color tardily (in about ten days) showing conidial areas in green shades about C.d.C. 322 quickly passing into orange-brown shades such as 138, 139, 143. Reverse and liquefied gelatin in pale yellow to orange shades such as 203A, 178D, 171, 166, 161; odor none. Conidiophores 10, 20–300μ, or 500 $\times$ 2–3μ, commonly arising from trailing hyphae as very short to long branches, mostly flexuous, frequently themselves branched, usually inflated at the apex; penicilli six to ten when simple, 15–25μ long when branched, with all walls smooth; metulae 8, 10–15 $\times$ 2.2–3μ, commonly enlarging from base to apex and inflated at the apex, and usually unequal in length and asymmetrically arranged in groups of two, three to six or seven; phialides about 7–8 $\times$ 2–2.5μ, in verticils of three, five, twelve, or fifteen, with short tubes, varying considerably in number and in size and shape according to the place occupied in the verticil. Conidia 2–2.5μ (or 2.8μ), smooth, globose, showing connectives.

From soil: Poland (168)

77. *Penicillium matris-meae* Zaleski

Colonies in neutral Raulin with 10 per cent gelatin in petri dishes, fairly quickly growing becoming 38–40 mm. in diameter in twelve days, liquefying the gelatin fairly quickly and completely, with surface velvety, zonate in the marginal areas, undulate or wrinkled giving a cerebriform effect in the central area and becoming radiate toward the margin; margin white, fairly wide, 2–3 mm. in the growing colony; in color conidial areas blue-green shades such as C.d.C. 371, 366, when young, later green 347, 341, 342, 348, 321, with green colors fading to leave the old colonies in pale orange shades such as 103B, 121. Reverse in orange-yellow shades such as 171, 157, 147; odor faint, agreeable suggesting *Viola odorata*. Conidiophores 300–600 × 3–4μ, straight or slightly flexuous, simple or sparingly branched, with apex dilated into a vesicle; penicilli 20–24μ, or even 28μ long, all walls smooth; metulae 11–13μ or even to 18 × 2.5–3.5μ, in groups of four to seven, mostly enlarged upward; phialides about 9–10 × 2–2.5μ, or smaller, most commonly in verticils of six to eight. Conidia about 2.5–3μ, smooth subglobose.

From soil: Austria (131), Egypt (115), Poland (168)

78. *Penicillium albidum* Sopp

Colonies on meat-peptone-sugar-gelatin, at first a white, thin mycelial growth, with conidium formation clear green, later olive-green, gray and finally brown, and commonly soon overgrown with mycelium, with surface growth spreading broadly, uneven, fibrous, rough, consisting of irregularly branching and trailing hyphae. Reverse clear reddish-yellow, or on some media yellow; odor suggestive of paraffin. Conidiophores arising as branches from trailing or ascending hyphae, from very short to fairly long, with smooth walls, enlarging to vesicle-like apices and producing either Citromyces-like verticils of phialides or divergent clavate uneven metulae each bearing a Citromyces-like cluster of phialides; phialides not described. Conidia globose, rough, 3–4μ. Cleistothecia not found.

From soil: Austria (66), Canada (15), Norway (125)

United States: Louisiana (50)

79. *Penicillium swiecickii* Zaleski

Colonies in neutral Raulin with 10 per cent gelatin in petri dishes, slowly growing, becoming 28–33 mm. in diameter in twelve days, liquefying the gelatin quickly and completely, characteristically wrinkled cerebriform in very center of the mass, thin, becoming radiate; overgrowth of secondary mycelium in the central area slight, hirsute; margin velvety azonate thin showing red from the reverse color; in color conidial area at first blue-green, such as C.d.C. 371, 372, becoming dark shades of orange-brown such as 138, 139, 134 in about two weeks. Reverse at first in yellow to orange shades such as 181, 156, 126, 106, 101; drops uncolored, small, few; odor none. Conidiophores 300, 400–600μ, even 800 × 3–3.5μ, straight, simple or rarely branched; penicilli when unbranched 20–28μ, when branched 40–56μ long, presenting anomalies of mixed verticils of metulae and phialides, and of variations in units in the verticil; branches varying from 18–35 × 2.5–3μ, common, enlarged at apex, usually unequal in length and asymmetrically placed one or two at the node; metulae varying in length in the penicillus from about 10–20 × 2.5–3.5μ, in groups of about three to six in more or less symmetrical verticils; phialides varying greatly in size but mostly about 8–9.5 × 2.2–2.8μ, in verticils of about six to ten. Conidia about 2.3–2.8μ, smooth, subglobose to globose with rather long tubes, and with connectives evident.

From soil: Poland (168)

80. *Penicillium janczewskii* Zaleski

Colonies in neutral Raulin with 10 per cent gelatin, in petri dishes, rather slowly growing, becoming 23–25 mm. in diameter in twelve days, liquefying the gelatin quickly and completely, velvety, zonate in the broad outer plane area; areas within depressed with narrow radiate furrows and wrinkles, with the center umbilicate; white margin about 1.5–2 mm. during the growing period; in color blue-green shades such as C.d.C. 367, 372, then dark yellow-green 294, 269, 274, to very dark orange-brown shades such as 115, 119, in old cultures; odor strong, suggestive of Actinomyces. Reverse in orange-yellow shades such as 161, 157, 108–110, a red-orange about 84. Conidiophores 50–200μ, or 300 × 2.3–2.8μ, frequently with short branches ten to twenty with secondary penicilli, with walls smooth; penicilli 8–25μ long, with walls smooth; metulae usually present, unequal and asymmetrically arranged, 8, 10–14μ, or even 16 × 1.5–2μ at base enlarging to 2.5–4μ at the apices, in divaricate verticils of two to five; phialides about 7–9 × 2–2.5μ, in verticils of three, six to twelve, or even fifteen, variously shaped, beaked. Conidia about 2.2–2.8μ, delicately but densely denticulate, globose, showing long slender connectives.

From soil: Poland (168)

81. *Penicillium kapuscinskii* Zaleski

Colonies on neutral Raulin with 10 per cent gelatin in petri dishes, slowly growing becoming 24–28 mm. in diameter in twelve days, liquefying the gelatin fairly rapidly, velvety, zonate in the marginal areas only, wrinkled cerebriform in center becoming radiate wrinkled toward the marginal area; white margin 1–1.5 mm. during the growing period; in color conidial area blue-green shades at first such as C.d.C. 397, 378B, then 366, 367, 373, 374, later green 322, 323, and in age dark shades of yellow such as 173, 143, 148. Reverse in shades of orange such as 146, 141, 196, 156, 191, 136; odor none. Conidiophores 100, 200–400μ, or up to 700 × 2.3–2.8μ, with apex inflated, straight or flexuous, simple or scantily branched; penicilli 8–10μ or 25–28μ long; metulae commonly unequal in length and asymmetrically arranged in the verticil 10, 12–18μ, or even 22μ long × 2.5–3μ, enlarging from base toward the apex which is usually vesicle-like; phialides about 8–9 × 2–2.5μ in verticils of six, eight to twelve, or even to sixteen, with short tubes. Conidia 2.2–2.5μ even to 3μ, smooth, mostly globose, occasionally subglobose, developed in chains showing a long slender connective.

From soil: Austria (131), Canada (15), Poland (168)

SUBSECTION v. *FUNICULOSA*

KEY TO THE SPECIES

Key	Species
a. Penicilli with metulae and branches divaricate	
b. Conidial chains not in columns	
c. Conidia rough or echinulate	
d. Colonies gray-green	
e. Reverse purple-drab	82. *P. daleae*
ee. Reverse white to bluish	83. *P. acidoferum*
dd. Colonies olive-buff	84. *P. krzemieniewskii*
cc. Conidia smooth	
d. Colonies forming broadly submerged felts	85. *P. godlewskii*
dd. Colonies lanose-funiculose	
e. White with pale green conidial areas becoming brown	86. *P. thomi* Zal.
ee. White with greenish-gray conidial areas	87. *P. intricatum*
eee. Blue-gray to brown in age	88. *P. glauco-ferugineum*
bb. Conidial chains forming compact diverging columns	
c. Reverse zonate, yellow to orange, conidia 3.5–4μ	89. *P. terrestre*
cc. Reverse sordid white to pale yellow, conidia 3.6–4.6μ	90. *P. solitum*

82. *Penicillium daleae* Zaleski

Colonies in neutral Raulin with 10 per cent gelatin in petri dishes, slowly growing becoming 24–26 mm. in diameter in twelve days, liquefying the gelatin tardily but completely, velvety or somewhat closely subfloccose, zonate only indistinctly and in the outer area, with the whole central area thrown into broad regularly radiate wrinkles, with the very center somewhat depressed and showing a few uncolored drops; white marginal zone 2–3 mm. wide; in color conidial areas at first blue-green shades such as C.d.C. 371, 372, becoming dark yellow-green shades such as 273, and later dark orange-brown such as 168, 164, 138, 139. Reverse at first in orange-yellows such as 171, 166, 157, to 133, 138, later becoming red-orange 84, 88, 92, 97; odor none or weak. Conidiophores 10–200μ or 300 × 2–2.5μ, with apex more or less enlarged or inflated, commonly unbranched, occasionally with short branches, flexuous, varying greatly in length, erect or ascending; penicilli mostly 10–12μ, less frequently 25–30μ, or 40μ long, with walls smooth, metulae 8, 10–20μ, or 24 × 2.5–3μ, in groups of two or three, commonly unequal and irregularly arranged, with apices commonly inflated; phialides about 9–10 × 2.5–3μ, commonly in verticils of three, five to ten or twelve, sometimes occurring singly. Conidia 2.5–4 × 2.5–3μ, varying considerable in size, coarsely denticulate, ovate elongated or subglobose.

From soil: Poland (168)

83. *Penicillium acidoferum* Sopp

Colonies on Czapek's or bean agar slowly spreading, cottony or closely floccose; surface olive-gray; reverse orange-buff; pigment diffuses into the medium. Aerial hyphae abundant, smooth; hyalin, creeping. Conidiophores arise from aerial hyphae as short side branches up to 40μ or 50μ long, unbranched, or once or twice branched at the apex; each branch bears a terminal head of conidial chains, which are loosely columnar, up to 75μ or 100μ long, and usually with five to ten chains in each head; apex of conidiophores slightly inflated; phialides 6.5–8.5μ long × 2–3μ thick. Conidia globose, light green, delicately rugulose under oil, 2.5–3.5μ in diameter. Elements of conidial fructifications pitted. Sopp (125) gives 3–4μ for conidia.

From soil: China (81), Norway (125)

United States: Utah (50)

This species is closely related to *P. rubens* Biourge (14). *P. acidoferum* has smaller conidia and phialides and the elements of the fructification are rough, whereas those of *P. rubens* are smooth.

84. *Penicillium krzemieniewskii* Zaleski

Colonies in neutral Raulin with 10 per cent gelatin in petri dishes, slowly growing becoming 28–30 mm. in diameter in twelve days, liquefying the gelatin slowly but completely; velvety, plane and azonate or indistinctly zonate in the outer area; areas within thrown into broad irregular wrinkles; white margin 1–1.5 mm. during the growing period; in color conidial areas green shades such as C.d.C. 347, 346, 348, becoming dark orange-brown shades such as 143, 147 in age. Reverse in light orange-yellow shades such as 171, 166, 157, 162, becoming very dark purplish at times in age; odor none. Conidiophores 50–300μ or 400 × 2.5–3μ, with apices inflated, sometimes unbranched, mostly branched more or less irregularly; penicilli when simple 10–12μ, mostly 20–30μ or 35μ long, with walls smooth with anomalies of gigantic phialides and thyrsiform masses common; metulae 10, 12, 20 × 2.5–3.5μ, commonly unequal in length and asymmetrically arranged in verticils of two, four or six; phialides about 9–11 × 2.5–3.5μ, in verticils of five to twelve or fifteen, or few or singly, straight or incurved, mostly with long slender tubes. Conidia 2.5–3.5μ or 4μ, delicately but densely denticulate, subglobose, showing distinct connectives.

From soil: Poland (168)

85. *Penicillium godlewskii* Zaleski

Colonies on neutral Raulin with 10 per cent gelatin in petri dishes; slow growing, becoming 24–26 mm. in diameter in twelve days with symmetrically radiate wrinkles 2–3 mm. wide and 1–2 mm. in height, and the center raised as a cushion or umbo often above the wrinkles, with gelatin tardily but completely liquefied; conidial area velvety or slightly floccose, azonate; margin plane, sordid white, 2–3 mm. wide, in color at first white for several days then with central conidial areas blue-green, such as C.d.C. 397, 372, 396, 371, becoming orange-brown in age 147, 148. Reverse pale yellow 0146, and 146, 153B, 128B; drops few, central; odor none. Conidiophores as short as 100μ, mostly 200–400μ (occasionally 600μ) × 2–3μ, straight or flexuous, with apices vesicle-like; penicilli from 10–25μ long, with much variation in structure; branches ordinarily absent; an occasional distant branch is indicated by the figure; metulae about 10–14 × 2.5–3μ, with apices flattened or vesicle-like, straight or incurved, absent or unequal, symmetrically or asymmetrically arranged in verticils of two to eight; phialides about 7–8 × 2.2–2.5μ, commonly six to twelve in the verticil, with short acute tubes. Conidia 2–2.5μ (even 2.8μ), smooth, globose, showing connectives and long adhering in the chains.

From soil: Poland (168)

86. *Penicillium thomi* Zaleski

Colonies in neutral Raulin with 10 per cent gelatin in petri dishes, rather quickly growing becoming 33–35 mm. in diameter in twelve days, liquefying the gelatin slowly but completely, lanose, composed of dense hyphae up to 5 mm. long, merging at the margin into a fimbriate white area 3–4 mm. wide, and including anastomosing ropes of hyphae which bear conidiophores as branches; conidial area pale greenish at first then such shades as C.d.C. 372, 322, with the green finally fading out leaving shades of orange-brown such as 137, 121, 103C. Reverse in orange-yellow and orange shades such as 171, 162, 161, 129, 109; odor none. Conidiophores from very short to 100μ or 200 × 2–2.5μ, usually flexuous, borne as branches of long (2–4 mm.) erect or trailing aerial hyphae and ropes of hyphae; penicilli 8–12μ when simple, 15–28μ when branched, with walls smooth; metulae 8, 10–16μ, or 20 × 2.5–3μ, divaricate varying in length and arrangement in groups of two to five, usually enlarging upward, or inflated at the apex; phialides about 7–9 × 2–2.5μ, very variable in size and appearance, short and inflated or straight, commonly with long tubes or beaks, verticils of five, eight to twelve, or fifteen. Conidia 2.2–2.8μ or 3μ, smooth, more or less globose, showing connectives in the chain but with the chains easily breaking down.

From soil: Poland (168)

United States: Texas (96)

87. *Penicillium intricatum* Thom

Colonies on bean agar gray, greenish-gray, when old, gray or smoky, floccose; reverse and medium colorless to yellow. Conidiophores sometimes terminal, more commonly branches of aerial hyphae, 30–50μ long. Conidial fructifications 50–140μ long, becoming longer in old cultures, usually in two stages, often only simple verticils of phialides or one to three verticils on divergent metulae; phialides 8–10 × 2–2.5μ. Conidial chains more or less divergent, frequently aggregated into a loose column. Conidia elliptical or globose, hyalin or pale green, smooth, granular within, 2.5–3μ.

From soil: Canada (15), China (81), England (36)

United States: Colorado (74), Connecticut (134), Iowa (1) (3), New Jersey (147), Texas (92)

88. *Penicillium glauco-ferugineum* Sopp

Colonies on meat-peptone-sugar-gelatin, blue-gray, in age brown, often showing margin green and center brown, with heavy mycelium wrinkled or buckled, with aerial hyphae in bundles and ropes bearing the conidiophores as branches; hyphae comparatively fine. Reverse gray-brown, to brown, or greenish-brown; odor not definite except a weak arsenic odor on potato plugs which are colored almost black. Conidiophores short, slender, often unbranched monoverticillate, again producing verticils of metulae or even an additional branch at a lower level; metulae when present with enlarged apices. Phialides very short. Conidia globose, smooth 2.5μ. Cleistothecia greenish, found once but not described, hence doubtful.

From soil: Austria (66)

89. *Penicillium terrestre* Jensen

Colonies on soil extract agar round, yellowish-green; vegetative hyphae 2–6μ in diameter; showing superficial ropes of hyphae. Conidiophores 70–375 $\times$ 2–4μ, hyalin, septate either with one or two branches near the apex or with branching limited to a terminal verticil of metulae; branches when present bearing terminal verticils of metulae; metulae 10–15μ long; phialides 7–11μ long. Conidia described as globose 2–3μ (4μ as seen by Thom in the type specimen) hyalin, in long chains.

From soil: Austria (131), Canada (15), England (36), India (21), U. S. S. R. (108)

United States (67)

90. *Penicillium solitum* Westling

Colonies in prune gelatin, somewhat floccose in central areas, bluish-green, such as C.d.C. 363, 367, then green 333, 338 to blackish green 349 in age; margin narrow, somewhat floccose, white. Reverse sordid white, or pale yellow; gelatin slowly and partly liquefied; odor weak. Conidiophores with walls smooth or nearly so, rarely rough, arising from submerged or creeping hyphae, sometimes very short but usually quite long, 300–800 $\times$ 4–6.4μ, with penicilli 60–150μ long; metulae slightly clavate 11–18 $\times$ 3.6–4.8μ; phialides 8–9.6 $\times$ 3–3.4μ. Conidia elliptical to globose or less often oval, smooth, commonly 3.6–4.6μ.

From soil: Austria (66), Greenland (96)

var. *lunzinense* Szilvinyi

From soil: Austria (131)

SUBSECTION vi. *FASCICULATA*

KEY TO THE SPECIES

a. Species producing sclerotia — 91. *P. italicum*
aa. Species not reported as producing sclerotia
  b. Colonies with simple conidiophores and fascicles mixed
    c. Colonies in blue-green shade
      d. Conidia globose — 92. *P. cyclopium*
      dd. Conidia mostly elliptical
        e. Reverse yellow to purplish nearly black — 93. *P. johannioli*
        ee. Reverse colorless to purplish or vinaceus — 94. *P. majusculum*
    cc. Colonies other than blue-green
      d. Colonies bright green or yellowish-green
        e. Margin broad, merging into a dull green
          f. Conidiophores 4–6μ in diameter — 95. *P. viridicatum*
          ff. Conidiophores 3.5–4μ in diameter — 96. *P. stephaniae*

ee. Margin narrow, bluish, then green — 97. *P. palitans*
dd. Colonies dull gray-green or glaucous; evidently zonate
e. Zones narrow (about 1 mm. intervals) — 98. *P. blakesleei*
ee. Zones broad (about 2 mm. intervals)
f. Conidia becoming subglobose to globose when ripe — 99. *P. expansum*
ff. Conidia persistently elliptical — 100. *P. polonicum*
bb. Colonies with most or all the conidiophores in fascicles or definite coremia — 101. *P. claviforme*

91. *Penicillium italicum* Wehmer

Colonies on bean agar broadly spreading, bluish-green, becoming gray-green when old. Reverse commonly brownish; color in medium none or slight. Conidiophores arise either directly from the substratum or as branches of aerial hyphae, 100–600μ long, averaging 250μ. Conidial fructification up to 300μ or more in length, usually in three stages, phialides 12–14 $\times$ 3μ. Chains of conidia loosely divergent, long; conidia 3–5 $\times$ 2–4μ, cylindrical to elliptical or slightly ovate. Numerous white sclerotia are produced upon the surface of the medium after two or three weeks' growth.

From soil: U. S. S. R. (108)

United States: California (146) (147), Iowa (3), Louisiana (2) (50), New Jersey (147), Oregon (147)

92. *Penicillium cyclopium* Westling

Colonies on Czapek's agar composed largely of coremiform masses, broadly spreading, loose; surface light blue-green. Reverse light buff to reddish-buff or orange. Conidiophores arise mostly as coremia, intertwined, directly from the substratum, up to 1 mm. or more in length, unbranched or dichotomously branched, each branch being also once or twice dichotomously branched near the apex. Heads long, columnar masses up to 350μ in length; fructification usually in three stages, sometimes four, elements closely appressed, metulae oblong 8–10 $\times$ 1.8–2.5μ; phialides 4–6 $\times$ 1.5–2μ. Conidiophores and elements of fructification delicately spinulose. Conidia globose to ovate, smooth, 2–3μ in diameter.

Westling (153) gives 9.5–14 $\times$ 3.2–4.4μ for metulae, 8–9 $\times$ 2.2–2.8μ for phialides and 2.6–3.2μ for conidia.

From soil: Canada (16), China (81), England (37)

United States: New Jersey (146) (147), Oregon (147), Porto Rico (147), Utah (50)

93. *Penicillium johannioli* Zaleski

Colonies in neutral Raulin with 10 per cent gelatin quickly growing becoming 4.5–5 cm. in diameter in twelve days, liquefying the gelatin very rapidly beginning about the tenth day, thin, velvety or with some funiculose masses or ropes of aerial hyphae, either entirely plane or slightly plicate, definitely zonate throughout with zones narrow (dense?) with an elevated point or umbo in center, with an overgrowth of scattered delicate gray aerial hyphae; margin during the growing period 2–3 mm. wide; in color blue-green when young, C.d.C. 363 to 358, 362, and later 339, 340, 345. Reverse showing crowded zones, plane or with a few radiate but delicate wrinkles with the mycelium giving a typically radiate appearance; in color pale yellow 171 to orange about 108; liquefied gelatin in yellow or orange color; drops uncolored, few; odor weak. Conidiophores 400–500 × 3–4μ, straight, simple or occasionally branched; penicilli mostly 40–55μ long; branches (rami) 18–32 × 3–4μ, in number two or three, in groups of four to six, commonly vesicle-like at the apex, figured as uneven in length; phialides about 9–10 × 2–2.2μ, commonly six to eight in the verticil, straight or slightly incurved. Conidia 2.5–4 × 2.2–3.5μ, mostly subglobose, some ovate, smooth, not persistently adherent in chains or masses.

From soil: Poland (168)

94. *Penicillium majusculum* Westling

Colonies in prune gelatin, at first blue-green, such as C.d.C. 362, 363, then dark green 334 and related, becoming brown in age (one month or more); not floccose, with narrow sterile margin; reverse pallid yellow; gelatin slowly and partly liquefied; odor weak, moldy. Conidiophores arising from creeping hyphae, smooth at first then commonly slightly verruculose, up to 550μ, commonly 150–300 × 4–6.5μ, with penicilli 45–195μ, commonly 90–150μ long, with or without branches preceding the formation of verticils of metulae; metulae 12–20 × 4–6.5μ, occasionally one-septate, hence two-celled; phialides 10.5–15 × 3–3.6μ. Conidia at first elliptical then oval-globose to globose, smooth, 4.5–6μ, swelling in germination to 7.5–9.6μ. Sclerotia or cleistothecia not reported.

From soil: U. S. S. R. (108)

95. *Penicillium viridicatum* Westling

Colonies on Czapek's or bean agar velvety, slowly spreading; surface bright leaf-green, sometimes with shades of blue-green. Reverse colorless to buff or brown shades. Aerial sterile mycelium not abundant, usually warty and rough; colonies consist mostly of conidiophores and heads. Conidiophores usually arise from the substratum, but also from aerial mycelium, 75–250 × 4–6μ. Heads vary from loose, almost radiate masses of chains, to loose columns. Fructification in three stages, usually with one primary branch arising laterally, a second primary branch being the prolongation of the conidiophore through the center of the head. Primary branches variable in length, 17–30 × 3–4μ; metulae 13–20 × 3.5–4μ; phialides 7.5–10.5 × 2.5–3μ. Some heads have only metulae and phialides. Conidia smooth, globose, light green, 3–4μ in diameter.

Westling's (153) measurements are: conidiophores 600 × 4.4–6.5μ; metulae 10.5–12 × 4–5.6μ; phialides 8–9.6 × 3.2–3.5μ; conidia 3–3.8μ.

From soil: Canada (15), England (37), U. S. S. R. (108)

United States: Colorado (75) (76), Idaho (106), Iowa (50), New Jersey (146) (147), Porto Rico (147)

96. *Penicillium stephaniae* Zaleski

Colonies on neutral Raulin with 10 per cent gelatin in petri dishes, slowly growing, becoming 2.5–2.8 cm. in diameter in twelve days, with liquefaction of gelatin beginning at the fourteenth day and proceeding rapidly, outer areas rather thin, velvety, buckled giving the central area either cushion-like elevated or depressed (concave), indefinitely zonate (*parum distincte*), with all or more commonly the outer areas radiate wrinkled; margin (fimbria) 1–2 mm. in the growing period; in color white for the first six days becoming green shades, such as C.d.C. 346, 347, 342, 343, 348, with ripening conidia and very dark shades of orange-brown, such as 173, 174, in old cultures. Reverse at first in yellows 166, 157, later becoming orange 127, 128, 129, 130, and related shades; drops uncolored, small, abundant in the outer areas, less frequent in the central area; odor weak, suggesting rotting potatoes. Conidiophores about 300–500 × 3.5–4μ, straight or somewhat flexuous, with apices vesicle-like; penicilli 40–50μ (or even 60μ) long, with walls smooth, figured as with branches uneven in length and phialides borne at various levels; branches straight or incurved about 20–32 × 3–3.5μ, in groups of two to three; metulae about 12–14 × 3–3.5μ, in groups of about five to seven, clavate or with apices vesicle-like; phialides 10–11 × 2.2–2.5μ, in verticils of about six to ten. Conidia smooth globose or subglobose 2.5–3μ, persisting in masses.

From soil: Egypt (115), Poland (168)

97. *Penicillium palitans* Westling

Colonies in prune gelatin, green, such as C.d.C. 333, 313, 329, 309, in age dark gray-green, not floccose, with narrow white granular scarcely woolly sterile margin consisting of creeping hyphae from which the conidiophores arise. Reverse uncolored, or very pale yellow; gelatin slowly or only in part liquefied; hyphae rather coarse, 3–6.5μ even to 8μ in diameter; odor moldy. Conidiophores arising from creeping hyphae, with range of 50–600μ, but usually 90–300μ in length × 4.4–6.5μ, rarely up to 8μ, with walls smooth when young, often verrucose in old cultures, more often branched than in other species; penicilli 60–175μ long, figured as (1) branch or verticil of branches fairly appressed, (2) metulae, (3) phialides; metulae 12–16 × 6.5μ, with walls smooth or "uneven;" phialides about 9–11.5 × 3.2–4μ. Conidia at first pear-shaped, oval, or oblong (langlich), then becoming globose, to resume the broadly oval form when ripe, smooth 4–4.7 × 3.6–4.3μ, becoming 7–8μ in germination.

From soil: Canada (16), Egypt (115)

98. *Penicillium blakesleei* Zaleski

Colonies upon neutralized Raulin with 10 per cent gelatin in petri dishes, quickly growing and becoming 6 cm. or more in diameter in twelve days, with gelatin at first softened then partly liquefied; thin, plane, velvety, definitely zonate throughout, with center raised like an umbilicus and loosely overgrown with secondary gray aerial mycelium; margin (fimbria) white, loose, 1–2 mm. wide in the growing colony; in color at six days green. Reverse plane, zonate; at margin shades, such as C.d.C. 342, then to 338 and in center becoming 335 by the tenth day, passing 334, to dark fuliginous shades, such as 140, 156, and 215, under various conditions in age; reverse zonate, in pale yellow-greens, and yellows, to dirty orange-browns; drops few small uncolored and developing mostly in the newer areas; liquefied gelatin almost uncolored; odor strong (intolerable). Conidiophores straight or slightly flexuous, occasionally branched, 300–400μ or even 600μ long, mostly 4–5μ in diameter, with all walls smooth; penicilli mostly 50–60μ long; branches straight or incurved 22–32 × 4–5μ, commonly one or two in number; metulae cylindrical or somewhat enlarged at base and apex, about 14–22 × 3–4μ in groups of four to six; phialides 10–13 × 2.5–3μ, in verticils of five to eight. Conidia smooth, globose or subglobose, at first about 3.5μ mostly 4–4.5μ occasionally 5μ, in chains which break up in mounting.

From soil: Egypt (115), Poland (168)

99. *Penicillium expansum* (Link) Thom

Colonies on bean agar green or gray-green, broadly spreading, becoming brown with age, floccose, often with concentric zones. Soil strains rarely produce coremia in artificial culture. Reverse brown; color in medium, none to brownish. Conidiophores either very short lateral branches of aerial hyphae or very long, 1 mm. or more, arising singly or sometimes grouped to form coremia. Conidial fructification typically in three stages, 130–200 × 50–60μ, consisting of one to three primary branches bearing verticils of metulae supporting crowded whorls of phialides; phialides 8–10 × 2–3μ. Conidia elliptical to globose, 2–3.3μ or 3–3.4μ, green.

From soil: Canada (16), England (11) (37), Egypt (115), Germany (6), Greenland (96), U. S. S. R. (108)

United States: California (147), Colorado (74) (75) (76), Idaho (106), Iowa (1) (3), Louisiana (2), New Jersey (146) (147), New York (67), North Dakota (147), Utah (50)

100. *Penicillium polonicum* Zaleski

Colonies on neutral Raulin with 10 per cent gelatin in petri dishes, thin, plane, quickly growing, becoming more than 5 cm. in diameter in twelve days, with the gelatin becoming strongly liquefied about the tenth day; appearance velvety, entirely plane, regularly concentrically zonate, overgrown in the raised central area with loose white rather short mycelium; margin (fimbria) of the growing colony 2, 3 or 4 mm. wide, in color blue-green, such as C.d.C. 371, 372, at margin, 359, 360 within, becoming very dark shades, such as 250, 224, in age; reverse in pale to brighter orange-yellow tints, such as C.d.C. 196, 146, 171, drops small, uncolored appearing in the marginal area; odor strongly fetid.

Conidiophores commonly varying from 300–600μ long, 4–5μ or even 6μ in diameter, straight or slightly flexuous unbranched or rarely with one short branch figured as perpendicular, with walls sparingly asperulate; branches two, three or four in number commonly 20–30μ long enlarging from base to the capitate apex with a range of 3–5μ, with walls somewhat rough; metulae enlarging slightly from base to apex, 12–14 × 2.5–3.5μ enlarging to 4–5μ, usually in groups of four or five; phialides straight, rather long, cylindrical 10–11 × 2.3–2.8μ, commonly five to eight in the verticil. Conidia smooth, subglobose, or occasionally ovate, 2.5–3.5μ, more rarely 4μ, showing connections between the conidia and the chain.

From soil: Poland (168)

101. *Penicillium claviforme* Bainier

Colonies on Czapek's agar, white or gray with more or less loose floccose hyphae bearing scattered inconspicuous simple penicillate conidial masses among the bases of conspicuous coremia with stalks compact, fibrous, white to flesh color or rose, up to 1–2 cm. long, simple or branched bearing well-differentiated heads olive-green in color and producing long conidial chains massed into columns often up to 1–3 mm. long and splitting variously with increasing length. Reverse brown in age especially at bases of the stalks; odor strong, penetrating; drops abundant during the growing period, colorless. Simple penicilli sparingly branched bearing verticils of few phialides 9–10 × 2μ; heads composed of complex hymenium-like masses covered with phialides, crowded, radiating from the surface. Conidia elliptical 4–4.6 × 3–3.3μ.

From soil: England (11)

SECTION III. BIVERTICILLATA-SYMMETRICA

KEY TO THE SPECIES

a. Species producing cleistothecia
  b. Ascospores described — 102. *P. spiculisporum*
  bb. Ascospores not described — 103. *P. parasiticum*
aa. Species not producing cleistothecia
  b. Species producing erect, definite coremia bearing the conidia — 104. *P. duclauxi*
  bb. Species without erect coremia
    c. Colonies with green conidial areas on a mycelium usually showing yellow shade
      d. Colonies with superficial net works of hyphae or ropes
        e. Conidiophores long and coarse, up to 2,000 × 8μ — 105. *P. elegans*
        ee. Conidiophores shorter and finer
          f. Colonies not deeply floccose
            g. Colonies olive-gray with yellow-green margin — 106. *P. humicola*
            gg. Colonies yellow-green — 107. *P. pinophilum*
            ggg. Colonies deep green with yellow margin — 108. *P. funiculosum*
          ff. Colonies deeply floccose — 109. *P. herquei*
      dd. Colonies with ropiness reduced or absent
        e. Colonies with dense dark green conidial areas more or less bordered, mixed with or overgrown by yellow to orange hyphae
          f. Reverse orange — 110. *P. rugulosum*
          ff. Reverse corinthian red — 111. *P. crateriforme*
        ee. Colonies not dark green
          f. Reverse purple-red
            g. Colonies zonate, deep green to olive-green — 112. *P. sanguineum*

gg. Colonies not zonate
  h. Colonies restricted, velvety — 113. *P. rubrum*
  hh. Colonies more floccose
    i. Colonies showing yellow-green and yellow to red colors — 114. *P. variabile*
    ii. Colonies floccose with green conidial areas and deep red reverse — 115. *P. purpurogenum*
ff. Reverse yellow to orange
  g. Yellow area on marginal zone — 116. *P. aureo-limbum*
  gg. Yellow area not confined to marginal zone
    h. Conidial areas gray to green on pale yellow mycelium — 117. *P. luteum*
    hh. Conidial areas yellow-green to blue-green — 118. *P. citricolum*
cc. Colonies lacking the yellow to red mycelium
  d. Colonies never green — 119. *P. braziliense*
  dd. Colonies some shade of green
    e. Colonies velvety
      f. Conidia 2–3 × 1.5–2μ — 120. *P. namylowskii*
      ff. Conidia 2.5–3μ — 121. *P. tardum*
    ee. Colonies floccose
      f. Conidia subglobose, 4.5 × 3.5–4.5μ — 122. *P. lagerheimii*
      ff. Conidia subglobose 2–2.5 × 3μ
        g. Colonies yellow-green — 123. *P. miczynskii*
        gg. Colonies gray-green — 77. *P. matris-meae*

102. *Penicillium spiculisporum* Lehman

Colonies on potato or bean agar white or gray or with few small greenish conidial areas, becoming cream, yellowish or pinkish with the ripening of the cleistothecia; aerial mycelium floccose, hyphae 2–3.5μ in diameter.

Cleistothecia subglobose, 0.4–2 mm. in diameter, indehiscent, with wall of interwoven hyphae, persistently white or becoming cream, pink and yellow with age. Asci elliptical, pyriform to globose 7–10.8 × 6.3–7.7μ, hyalin, six- to eight-spored, with evanescent walls. Ascospores ovate or elliptical 7.5–4 × 1.8–2.8μ, minutely spiny.

Conidiophores scattered 10–50 × 2–2.5μ (usually less than 20μ long), bearing a single verticil of two to five phialides, 11–16μ long, uneven in length, with long acuminate points. Conidia 2.5–4 × 1.8–2.5μ.

From soil: United States: Iowa

103. *Penicillium parasiticum* Sopp

Colonies on meat-peptone-sugar-gelatin with mycelium intensively yellow, with clear yellow-green conidial areas produced only when parasitic upon an insect larva and only cleistothecia upon sulphur-yellow mycelium in cultures. Conidiophores moderately large, figured as bearing one or two side branches at different levels, with penicilli consisting of a verticil of six to twelve metulae, phialides and diverging chains of conidia; phialides few in the verticil. Conidia smooth, oval, $3 \times 4\mu$, or occasionally globose. Cleistothecia with a thin wall of dark green almost black cells, and central mass yellow were developed in large clusters and ripened spores only once in culture in which the ascospores were smooth, oval, yellow and larger than the conidia.

From soil: Austria (66)

104. *Penicillium duclauxi* Delacroix

Colonies on bean agar clear deep green to olive-green, often brown when old, strict, consisting of short crowded conidiophores arising for the most part singly from the substratum. Reverse and medium yellow to red. Conidiophores short, 10–50μ long, conidial fructifications simple, 100μ or 160μ in length, consisting of a terminal whorl of phialides or with both metulae and phialides; phialides 10–12μ long. Conidia elliptical, green, smooth when young, rugulose when mature, $3.6\text{–}4 \times 2\text{–}2.5\mu$.

From soil: Canada (15), Japan (132)

United States: Colorado (74), Iowa (3)

105. *Penicillium elegans* Sopp

Colonies broadly zonate, deeply blue-gray, becoming progressively more yellowish-green, with a heavy white marginal zone; mycelium white in reverse, in age greenish-yellow; hyphae coarse; odor strong, from petroleum-like to ether-like. Conidiophores very long, uniform in diameter, coarse, septate, somewhat rough, branched, figured as producing a long partly divergent branch at some distance below the penicillus which bears a separate penicillus; penicillus figured as the main stalk with one branch, then verticils of metulae; phialides very numerous, short, cylindrical. Conidia 3.5–4μ yellowish-green. Cleistothecia not found.

From soil: U. S. S. R. (108)

106. *Penicillium humicola* Oudemans

Colonies on Czapek's agar cottony or floccose, not broadly spreading, with bright yellow aerial mycelium and olive to gray-green fruiting areas. Young colonies are bright yellow shades, becoming green as fruiting areas develop, reverse nearly colorless to orange or reddish. Colonies consist of densely woven aerial hyphae and conidiophores, the latter arising both directly from the substratum and from aerial hyphae up to 135μ long. Heads loosely penicillate and straggling, breaking up easily; fructification in two stages, a single verticil of oblong metulae bearing the elongate and slightly pointed phialides; metulae 9.5–11.5 × 2–3μ; phialides 6.5–8 × 1–2μ. Some heads have phialides only. Conidia ovate to globose, smooth, olive-green, 2–3 × 1.5–2μ.

From soil: China (81), Holland (100), Japan (132)

United States: Colorado (75) (76), Idaho (106), Iowa (50), Louisiana (50), Michigan (51), New York (67)

107. *Penicillium pinophilum* Hedgcock

Colonies on Czapek's agar cottony or very closely floccose, surface from green through various shades of mixed green, yellow, and red; reverse and medium red. Conidiophores arise as short side branches of aerial hyphae, the latter often in ropes, up to 200μ long. Conidial fructifications in two stages up to 120 long, chains parallel but not in columns; metulae 10–16 × 2–2.5μ; phialides acuminate, 13–15 × 2–2.5μ. Conidia elliptical, smooth, pale green or yellow, 3–3.6 × 2μ.

From soil: India (21)

United States: Illinois (129), Iowa (1) (3), Louisiana (2), New Jersey (146) (147)

108. *Penicillium funiculosum* Thom

Colonies on bean agar deep green, broadly spreading, surface closely floccose; reverse and medium red or purple to almost black. Conidiophores arise laterally from aerial hyphae, the latter commonly in ropes; occasionally arise directly from the substratum, 20–80μ or 100μ long. Conidial fructification in three stages, up to 160μ long, columnar; phialides 10–14 × 2–3μ, in dense, parallel verticils. Conidia elliptical, 3–4 × 2–3μ, green, smooth.

From soil: Canada (15)

United States: Colorado (74), Iowa (3), Louisiana (2)

109. *Penicillium herquei* Bainier and Sartory

Colonies on Czapek's agar with mycelium more or less floccose with some trailing hyphae and ropes of hyphae, white, then golden-yellow (hyphae covered first with drops then crusted with yellow granules) becoming yellow-green with the development of conidial areas. Reverse and agar yellow, such as C.d.C. 206, with color extending rapidly beyond the limits of the colony; vegetative hyphae closely felted 1–3μ in diameter; odor none. Conidiophores up to 500μ, even to 1,000μ long × 3–3.5μ, septate, unbranched, but occasionally anastomosing with adjacent conidiophores, with walls heavily incrusted with yellow granules (? pitted), producing at the apex a crowded, compact verticil of metulae; metulae 8–10μ long; phialides about 7–9μ long, tapering rather abruptly to a narrow conidia-bearing tube. Conidia subfusiform, delicately roughened, from 3 $\times$ 2μ to 3.5–4 $\times$ 2.5μ, occasionally doubled in size, very slightly colored, in loosely divergent and tangled chains. Cleistothecia not found.

From soil: Canada (16), Egypt (115)

United States: Illinois (129)

var. *lunzinense* Szilvinyi

From soil: Austria (131)

110. *Penicillium rugulosum* Thom

Colonies on bean agar yellowish-green, then green, and finally dark green; surface growth of densely crowded conidiophores with few aerial hyphae; reverse yellow to bright orange; medium slightly colored. Conidiophores arise separately or as branches of aerial hyphae, up to 200 $\times$ 2.5–3μ. Conidial fructifications typically in two stages, (sometimes three), up to 150μ long, conidial chains divergent; phialides 9–12 $\times$ 2μ. Conidia 3.4–3.8 $\times$ 2.5–3μ, elliptical, green, verrucose when mature.

From soil: Austria (131), Canada (15), England (36), Greenland (96)

United States: Iowa (3), New Jersey (146) (147), Porto Rico (147), Texas (92)

var. *atricolum* (Bainier) Thom

From soil: Canada (15)

United States: Iowa (3), New Jersey (146)

var. *lunzinense* Szilvinyi

From soil: Austria (131)

111. *Penicillium crateriforme* Gilman and Abbott

Colonies on Czapek's agar small, round, restricted, velvety, consisting of conidiophores and heads with little aerial mycelium; surface deep dark green to dull blackish-green (Ridgway XLI). Reverse colorless, or reddish in old cultures. Conidiophores densely crowded, arising from the substratum or from surface hyphae, up to 250μ long, mostly 100–150μ. Fructification a loose column, not divergent, not densely packed, up to 225μ long; in two stages, metulae oblong, 8.5–11.5 $\times$ 3–4μ; phialides flask-shaped, 6.5–9.5 $\times$ 2–3μ. Conidia globose to slightly ovate, smooth, light green, 2.5–3μ.

From soil: China (81)

United States: Louisiana (50)

112. *Penicillium sanguineum* Sopp

Colonies upon Czapek's agar showing well differentiated conidial zones often with sterile separating zones, with conidial masses a shade of dark yellow-green near Andover-green; zonation often obliterated in age and sometimes wanting at any stage; overgrowth of orange hyphae more or less evident in many cultures in age. Reverse a shade of orange-red, near Corinthian-red without showing yellow shades. Conidiophores 50–100 × 4μ, with walls smooth; penicilli biverticillate, sometimes with the third verticil more or less asymmetrical, with long chains of conidia parallel, or the chains from the single verticils of phialides more or less adherent into columns and the columns forming more or less continuous masses up to 400μ thick which break off in crusts when the container is tapped or struck; metulae up to 10–12 × 2–3μ, in compact verticils; phialides acuminate up to 10–12 × 2–2.5μ, in compact verticils. Conidia elliptical to subglobose, 3–2.5–3.5 × 3μ or 3–3.5μ, rarely 4μ when subglobose, green, becoming dark green in mass.

From soil: Austria (66)

113. *Penicillium rubrum* Stoll

Colonies on bean agar from green through ochraceous to ochraceous-red with varying conditions; consisting of green conidia with yellow mycelium in sugar media; aerial portion velvety or very closely floccose. Reverse yellowish to red; coloring medium in old cultures. Conidiophores arise directly from substratum or as very short lateral branches of aerial hyphae, 15–30 × 3–3.5μ, slightly swollen at the apex. Conidial fructification in two stages, usually massed into a heavy column with a broad triangular base, 100–200μ in length; metulae slightly swollen at the apex; phialides 10–13 × 2–3μ. Conidia elliptical to globose, 3.4 × 2μ or 2.4–3.3μ, yellowish-green to green, smooth.

From soil: United States: Louisiana (2)

114. *Penicillium variabile* Sopp

Colonies on meat-peptone-sugar-gelatin showing a wide range of color in red and green shades, and in reverse from light to very dark red almost black; mycelium orange to rose or carmine-red, on some substrata in shades of green and yellow-green; substratum from pink-red to deep blood-red; hyphae comparatively fine or delicate; odor suggestive of the bark of aspen (*Populus tremuloides*?). Conidiophores rather coarse, stiff, septate, brownish, thick-walled, swollen somewhat at apex and producing a verticil of two to six metulae often with one or more superposed verticils produced by prolongation of the main stalk; phialides narrowly cylindrical ("needle-like"), eight to twenty in the verticil. Conidia elliptical, fusiform $3 \times 4\mu$, with a tendency to become adherent in a ball, like Gliocladium, figured as chains more or less adherent into columns. Cleistothecia not found.

From soil: Austria (66), Canada (15)

115. *Penicillium purpurogenum* Stoll

Colonies on Czapek's agar slowly spreading, very closely floccose to almost velvety, white at first, becoming yellow to pinkish shades, and finally light gray-green; reverse and medium colored deep red to purple. Conidiophores arise from aerial mycelium, up to 100μ or 300μ long. Conidial fructifications consist of long, divergent chains, up to 100μ long, in two stages; metulae $10–16 \times 2–2.5\mu$; phialides $11–12 \times 2.5\mu$. Conidia elliptical, $3.4–3.8 \times 2–2.5\mu$, smooth, pale green.

From soil: Canada (15) (147), China (81)

United States: Colorado (75) (76), Iowa (3), Louisiana (2), New Jersey (146) (147)

116. *Penicillium aureo-limbum* Zaleski

Colonies in neutral Raulin with 10 per cent gelatin in petri dishes, slowly growing, becoming 26–30 mm. in diameter in twelve days, liquefying the gelatin quickly and completely, with central area characterized by cerebriform wrinkles 1–2 mm. high and 1–2 mm. wide, which become radiate in the outer areas; marginal area velvety, zonate with the outer zone or "fimbria" 1 mm. in width; in color conidial areas at first blue-green, such as C.d.C. 378B, to 396, 372, 366, becoming later green 346, 347, 322, 318. Reverse in orange-yellow shades such as 171, 166, 151, 126; odor rather fetid, strong, similar to rotting potatoes. Conidiophores varying 100μ, commonly $150–300\mu$, less commonly $500 \times 2.5–3\mu$, straight or flexuous, simple or rarely branched, with apices usually inflated, with all walls smooth; penicilli about $18–25\mu$ in length; metulae about $12–15\mu$ or even $18 \times 2.5–3\mu$, in groups of three to six, with apices variously enlarged or inflated, equal or unequal in length and symmetrically or asymmetrically arranged in the verticil; phialides about $8–9 \times 2.2–2.5\mu$, in verticils of six to ten with short tubes. Conidia $2–2.5\mu$ or even 2.8μ, smooth, more or less globose, showing distinct connectives in the chains.

From soil: Poland (168)

117. *Penicillium luteum* Zukal

Colonies on bean or Czapek's agar white at first, then yellow, with few pale green conidial areas, and with abundant bright yellow to red sclerotia. Colonies commonly consist almost entirely of sclerotial areas with scant production of conidia. Reverse of colonies red. Conidiophores scantily produced, mostly as lateral branches of aerial hyphae, 20–100 × 3μ. Conidial fructification in two stages, up to 80μ in length, phialides 13–16 × 3–4μ. Conidia elliptical to fusiform, 2.4–2.3μ, greenish, smooth.

From soil: Canada (15), Egypt (115), Greenland (96)

United States: Iowa (1) (3) (147), Louisiana (147), New Jersey (146) (147), Texas (152)

var. *lunzinense* Szilvinyi

From soil: Austria (131)

118. *Penicillium citricolum* Bainier and Sartory

Colonies on licorice sticks with conidial areas yellowish-green, such as C.d.C. 313, then 263, 267, 257, on some media blue-green. Reverse and medium citrine-yellow. Conidiophores 2μ in diameter, figured as septate, more or less sinuous, rather long and arising from creeping or ascending hyphae; penicillus consisting of a verticil of four to six closely crowded metulae about 8μ long each bearing three to six phialides. Conidia oval 2μ in diameter (apparently in short axis—C.T.).

From soil: Austria (131)

119. *Penicillium braziliense* Thom

Colonies white or very faintly tinged, velvety, spreading broadly, not over 200μ deep, slowly and incompletely zonate in age. Reverse (four weeks) yellow to olive-buff or a dirty yellow-orange mixture; drops not seen; odor none. Conidiophores 100–200μ long ascending rather than erect, × 3μ or 4μ, with walls pitted or rough; penicilli variously branched only partly biverticillate; metulae 16–20μ enlarged at apex unequal in the verticil; chains tangled—breaking up in mounts; phialides 13–16μ. Conidia about 3μ, very pale with more or less internally granular appearance, thin walls.

From soil: Canada (15)

var. *lunzinense* Szilvinyi

From soil: Austria (131)

120. *Penicillium namylowskii* Zaleski

Colonies in neutral Raulin with 10 per cent gelatin in petri dishes, slowly growing, becoming 24–28 mm. in diameter in twelve days, liquefying the gelatin only slightly, with surface uneven, velvety, rarely indistinctly zonate near the margin; central area showing deep cerebriform wrinkles which become radiate outwardly; margin 1–1.5 mm. wide during the growing period; in color conidial areas green, such as C.d.C. 346, 347, becoming dark orange-brown shades such as 148 in age. Reverse, yellow-green, to yellow shades such as 0296, 277, 171; odor strong, offensive. Conidiophores 70–150μ or even 300 × 3.5–4μ, with apices enlarged usually to 5–6μ, straight or flexuous, simple or rarely branched, with walls smooth; penicilli 20–30μ when simple, 40–50μ when a branch is present; branches rare, when present 20–25 × 3–4μ, with apex enlarged to 6μ; metulae about 14–20 × 3–4μ with apices vesicle-like, equal or unequal in length, in symmetrical or asymmetrical verticils of three to six; phialides about 9.5–11 × 3–3.5μ, commonly 6–10μ, or even 15μ in the verticil, with short heavy tubes, crowded in the verticils and straight or incurved. Conidia 2.5–4 × 2.5–3.0μ, even to 3.5μ, smooth, long, ovate to subglobose, showing distinct connectives in the chains.

From soil: Poland (168)

121. *Penicillium tardum* Thom

Colonies on Czapek's agar, slow and restricted in growth, green, with a submerged zone 3–4 mm. broad. Reverse and agar not colored; odor none; vegetative hyphae slender. Conidiophores slender, 300–400 × 2–2.5μ, bearing a symmetrical verticil of metulae and phialides characteristic of the group; metulae unequal in the verticil but about 8–10 × 2μ; phialides up to 8 × 2μ. Conidia elliptical mostly up to 2.5–3μ in long axis, less often subglobose, in slightly diverging or fairly closely arranged but not adhering chains, forming masses up to 160μ long and broadening to 100μ at the apex.

From soil: Canada (16)

122. *Penicillium lagerheimii* Westling

Colonies in wort gelatin, deeply lanose, yellowish-white, coremia none. Reverse orange-yellow, odor indefinite. Conidiophores 1.2–1.5μ diameter slender, with wall encased in crystalline material; penicillus 25–40μ, even 75μ long, smooth, (or ? incrusted or sheathed with crystalline material), figured as a main axis and terminal verticil of phialides or with one or more branches or metulae from the next node; either short or much longer than the continuation of the main axis; metulae 24–30μ or rarely 60 × 1.8–2.5μ; phialides 7.5–11 × 2.8–3.5μ, one, two, or rarely three together. Conidia oblong (?) subglobose 4.5 × 3.5–4.5μ, few or scantily produced.

From soil: Austria (131)

123. *Penicillium miczynskii* Zaleski

Conidia in neutral Raulin with 10 per cent gelatin in petri dishes, slowly growing, becoming 32–35 mm. in diameter in twelve days, liquefying the gelatin quickly and completely; surface velvety-sublanose, azonate, with marginal area plane or slightly elevated, the center cerebriform wrinkled and depressed, passing outward as radiate wrinkles; surface growth consisting of a dense lanose mass of white mycelium tardily showing a faint bluish-green and later green in the central area, tending again to be overgrown with white mycelium, only locally showing gray-green conidial areas. Reverse and liquefied gelatin in yellow shades such as C.d.C. 221, 191, 178A, 216, 211; odor weak or none. Conidiophores 300, 400–800μ, or even 1,000 × 2.5–3.5μ commonly enlarged at the apices, simple or rarely branched, straight or flexuous, all walls smooth; branches rarely encountered, usually unequal, distant, 15–20–30μ, or even 36 × 2.5–3μ, sometimes two or three at the node; penicilli 18, 22–28μ, or even 35μ long; metulae about 12–18μ, or even 24 × 2.5–3.5μ, commonly enlarged upward, usually four to eight in the verticil in which the units are unequal in length and symmetrically or rarely asymmetrically arranged; phialides about 8.5–9.5 × 2.0–2.5μ, commonly six to ten in the verticil, with short tubes. Conidia 2–2.5μ, or even 3μ, smooth, subglobose or occasionally globose.

From soil: Poland (168)

var. *lunzinense* Szilvinyi

From soil: Austria (131)

SECTION IV. POLYVERTICILLATA-SYMMETRICA

A single species treated.

124. *Penicillium albicans* Bainier

Conidiophores figured as short perpendicular branches from trailing hyphae, much larger in diameter than the sterile hyphae, consisting of one or two swollen cells, bearing a penicillus figured as regularly two to three verticillate with elements coarse, short, vesiculose rather than tubular, successively smaller in diameter, and with occasional branches directed backward from the upper cell of the conidiophore or the first verticil of branches. Conidia oval, at first white then slowly fawn to reddish in age.

From soil: U. S. S. R. (108)

Species of uncertain position:

Penicillium aequabile Szilvinyi
Penicillium brunneo-viride Szilvinyi
Penicillium cavum var. *lunzinense* Szilvinyi
Penicillium fusco-glaucum var. *lunzinense* Szilvinyi
Penicillium gracile Szilvinyi
Penicillium griseo-viride Szilvinyi
Penicillium hermanni Szilvinyi
Penicillium huberi Szilvinyi
Penicillium impar Szilvinyi
Penicillium internascens Szilvinyi
Penicillium kuhnelti Szilvinyi

Penicillium lanceolatum Szilvinyi
Penicillium lanoso-viride var. *lunzinense* Szilvinyi
Penicillium lunzinense Szilvinyi
Penicillium luteo-viride var. *lunzinense* Szilvinyi
Penicillium martensii var. *lunzinense* Szilvinyi
Penicillium multiforme Szilvinyi
Penicillium paecilomyceforme Szilvinyi
Penicillium pavoninum Szilvinyi
Penicillium ruttneri Szilvinyi
Penicillium schmidtii Szilvinyi
Penicellium subviride Szilvinyi
Penicillium varians Szilvinyi
Penicillium viride-albo Szilvinyi
Penicillium virido-brunneum var. *lunzinense* Szilvinyi
Penicillium wallandi Szilvinyi
Penicillium zaleskii Szilvinyi
From soil: Austria (131)
Penicillium hyphomycetis Saccardo
From soil: England (11)
Penicillium silvaticum Oudemans
From soil: Holland (100)
Penicillium candidum Link
From soil: Japan (132)
United States (51)
Penicillium javanicum Szilvinyi
Penicillium sumatrense Szilvinyi
Penicillium victoriae Szilvinyi
From soil: Sumatra (130)
Penicillium desiscens Oudemans
From soil: United States: California (146), Maine (147), New Jersey (146) (147), New York (67), Porto Rico (147), Texas (147)
Penicillium ochraceum Raillo
Penicillium pigmentaceum Raillo
Penicillium salmonicolor Raillo
From soil: U. S. S. R. (108)

(11) **Scopulariopsis** (Bainier) Thom

Colonies never green, with aerial hyphae, partly at least, in trailing and anastomosing ropes or fascicles. Conidiophores very short or wanting, commonly borne along the fasiculate hyphae; conidial apparatus Penicillium-like or consisting of varying aggregations of branches and phialides, at times reduced to single phialides scattered along aerial hyphae; phialides more or less specialized, tapering gradually from a basal tubular section or even the base itself toward a conidium-bearing apex, or narrowly tubular without tapering, cutting off conidia from the apex by cross-walls. Conidia more or less pointed at the apex and truncate at the base with a more or less thickened basal ring surrounding a basal germinal pore, with walls usually thickened and often variously marked or roughened.

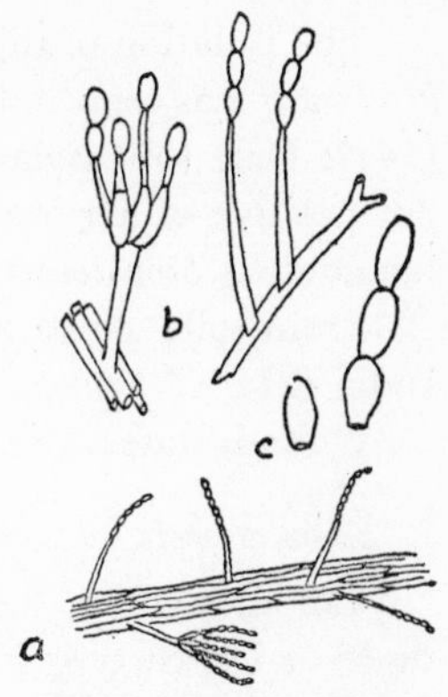

FIG. 71. **Scopulariopsis.** *a* conidial rope; *b* conidiophores; *c* conidia (after Thom).

KEY TO THE SPECIES OF THE GENUS SCOPULARIOPSIS

a. Conidia in mass light brown to chocolate — 1. *S. brevicaulis*
aa. Conidia in mass cream to yellow
  b. Conidia, 6–8 × 3.5–6.5μ — 2. *S. costantini*
  bb. Conidia, 5.6–11 × 3.6μ — 3. *S. communis*

1. *Scopulariopsis brevicaulis* Bainier
   Syn. *S. rufulus* Bainier
   S. repens Bainier
   Acaulium nigrum Sopp

Colonies on sugar gelatin, white at first, then yellowish-brown or chocolate, consisting of short closely crowded conidiophores making powdery areas overgrown by loose trailing floccose hyphae and ropes of hyphae, with broadly spreading, indeterminate margin. Conidiophores short, 10–30μ long, arising directly from the submerged hyphae, or numerously and irregularly borne as lateral and perpendicular branches of trailing aerial hyphae and ropes of hyphae. Conidial fructifications either simple chains, terminating unbranched or sparingly branched conidiophores in young colonies, or verticillately and irregularly twice verticillately branching systems bearing numerous divergent chains, often 150μ in length in old colonies. Phialides continuous with conidiophores, 12–15 × 4μ, tapering at the apex. Conidia somewhat pear-shaped, slightly tuberculate at the apex, with broad base, 6.5–7.5 × 7.5–9μ, in mass light brown to chocolate, smooth at first, then with thick tuberculate walls.

The isolation of this species from the soil under the name *Scopulariopsis brevicaulis* was reported by Waksman in New Jersey (146) (147) and Texas (147); Dale (37) isolated *S. repens* Bainier and *S. rufulus* Bainier, but was not positive of her identifications, and it is probable that both should be included as *Scopulariopsis brevicaulis* Bainier.

From soil: Austria (131), Canada (15), China (81), Germany (104), India (21)

United States: New Jersey (146) (147)

2. *Scopulariopsis constantini* Bainier

Colonies white, then dirty white, sometimes yellowish, in old cultures more or less dirty golden-yellow. Conidial heads 30–50μ long, irregular; fructification without metulae, or rather deformed, 7–8 $\times$ 3–5μ; phialides 9–16μ or 20–25 $\times$ 2.5–5μ. Conidia 6–8 $\times$ 3.5–6.5μ.

From soil: England (37)

3. *Scopulariopsis communis* Bainier

Colonies producing prominent ropes and networks of hyphae almost perpendicular to the substratum, and bearing abundant conidial fructifications as branches; penicillus very short (one-celled) stalk, bearing a verticil of phialides directly or a mixed verticil of metulae and phialides, or more complex branching. Conidia more or less oval, with truncated base and pointed apex, about 5.6–11.2 $\times$ 3.6μ, almost colorless at first becoming cream when ripe.

From soil: England (37)

var. *lunzinense* Szilvinyi

From soil: Austria (131)

Species of uncertain position:

Scopulariopsis alba Szilvinyi
Scopulariopsis argentea Szilvinyi
Scopulariopsis lilacea Szilvinyi
Scopulariopsis olivacea Szilvinyi
Scopulariopsis olivacea var. *parva* Szilvinyi
Scopulariopsis polychromica Szilvinyi
Scopulariopsis rosacea Szilvinyi

From soil: Austria (131)

Scopulariopsis roseum

From soil: England

(12) **Gliocladium** Corda

Conidiophores erect, simple or branched, septate, producing at the apex a fructification composed of successive verticils of primary branches, secondary branches, metulae, and phialides, or in some cases without secondary branches; primary branches often arise laterally on the conidiophores below the main head. Conidial heads enveloped in slime, conidia in chains, or held together in a mass of slime in which chains are not distinguishable.

KEY TO THE SPECIES OF THE GENUS GLIOCLADIUM

a. Mature colonies never green	
b. Mature colonies pure white to cream	1. *G. penicilloides*
bb. Mature colonies pink or rose shades	
c. Colonies pink to rose	
d. Conidia remaining in chains	2. *G. vermoeseni*
dd. Conidia established in slime balls	3. *G. roseum*
cc. Colonies light ochraceous salmon	4. *G. salmonicolor*
aa. Mature colonies green	
b. Conidia definitely in chains which form dense columns	5. *G. catenulatum*
bb. Conidial heads round, enveloped in slime in which chains are not distinguishable	
c. Colonies pure white, with green fruiting areas	6. *G. fimbriatum*
cc. Colonies always dark green, never white	
d. Slime production very abundant, conidiophores hyalin, rough	7. *G. deliquescens*
dd. Slime production not abundant, conidiophores olivaceous, smooth	8. *G. atrum*

1. *Gliocladium penicilloides* Corda

Colonies on Czapek's agar broadly spreading, floccose, surface pure white to pale cream in fruiting areas; reverse colorless. Aerial mycelium abundant, from which the conidiophores arise as side branches, erect, septate, 50–100μ long $\times$ 3μ in diameter, pitted or rough. Fruiting heads enveloped in slime, columnar. Fructification in three stages, primary branches 15–25μ long $\times$ 3.2μ in diameter; metulae 10–15 $\times$ 2.5μ; phialides 10–14 $\times$ 1.5μ. Conidia in definite chains, elongate elliptical to bacillate, smooth, hyalin, 3.5–4 $\times$ 2μ.

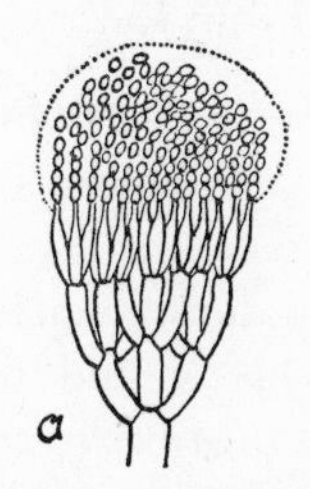

FIG. 72. **Gliocladium.** *a* conidiophore and conidia.

From soil: Canada (15), China (81), England (37)
United States: Iowa (3), Louisiana (50)

2. *Gliocladium vermoeseni* (Biourge) Thom

Colonies on wort gelatin, producing numerous salmon-colored coremia 10 mm. in height or more. Conidiophores about 5μ in diameter; metulae 7–15 $\times$ 2.5–5μ, irregularly borne, irregular in number or none; phialides 10–20 $\times$ 2.5–3.5μ, in groups of two to five, or even seven. Conidia elliptical 4–6 $\times$ 3–4μ.

In soil: Austria (130), Egypt (115)

3. *Gliocladium roseum* (Link) Thom
Syn. *Penicillium roseum* Link

Colonies on potato agar loose floccose, with simple hyphae and ropes of hyphae, surface white to pink or salmon in fruiting areas; reverse colorless. Produces dense irregular pinkish masses or sclerotia up to 1 mm. or more in diameter in old cultures. Conidiophores borne as branches of aerial hyphae, 45–125μ long. Conidial fructification enclosed in slime, up to 140μ long, in two or three stages, phialides 12–17 $\times$ 2–3μ bearing conidia in gelatinous balls or masses. Conidia colorless (pink or rosy in mass), elliptical, 5–7 $\times$ 3–5μ, slightly apiculate, smooth, appearing granular within.

From soil: Canada (15), Japan (132)
United States: Colorado (74), Illinois (129), Iowa (101)

4. *Gliocladium salmonicolor* Raillo

Colonies on rice light ochraceous salmon. Conidiophores branched terminally forming heads, 54–94μ long; phialides 28.6–29 $\times$ 3μ. Conidia hyalin, oval, smooth 5.4 $\times$ 3μ, seldom 8.4 $\times$ 4μ.

From soil: U. S. S. R. (108)

5. *Gliocladium catenulatum* Gilman and Abbott

Colonies on Czapek's agar pure white, spreading, floccose, becoming olive-green to bright green in the center as fruiting areas develop, and clear dark green in old cultures; fruiting areas are usually confined to center of colony and one or two concentric zones separated by sterile mycelium; reverse colorless to yellowish. Aerial mycelium abundant, simple or in ropes, from which the conidiophores arise as branches. Conidiophores often once and sometimes twice branched, coarse, pitted or rough, 50–125μ long. Heads are composed of conidial chains in long, close columns, enveloped in slime, up to 150μ long. Fructification in three stages, elements of fructification pitted or rough; primary branches 15–20 $\times$ 3.5–4μ; metulae 7–9 $\times$ 15–25μ; phialides 10–20μ long. Conidia elliptical, smooth, pale green, 4–7.5 $\times$ 3–4μ.

From soil: Canada (15)
United States: Utah (50)

6. *Gliocladium fimbriatum* Gilman and Abbott

Colonies on Czapek's agar broadly spreading, orbicular, pure white at first, with zones of dark leaf green fruiting areas appearing near the center of the colony. Conidiophores arise from aerial hyphae, smooth, up to 25μ long; several from one point, stolon-like hyphae usually present at point of origin. Heads enveloped in round balls of slime in which chains are not distinguishable; fructification in two stages, with divergent branchlets or metulae which bear elongate flask-shaped, appressed phialides, or with conidia borne directly on a few finger-like phialides which arise irregularly from the conidiophore; in most heads one or more branchlets arise laterally from the conidiophore some distance below the main head; metulae elongate, extremely variable in size, phialides usually 10–20μ long, from flask-shaped to irregular elongate. Conidia elliptical or elongate, ovate, smooth, pale green, 6.5–9.5 $\times$ 2.5–4μ.

From soil: China (81)

United States: Iowa (50), Louisiana (50), Texas (92)

7. *Gliocladium deliquescens* Sopp

Growth not abundant on Czapek's agar. On bean agar, broadly spreading, producing a thin, transparent growth of sterile hyphae over the entire medium, from which the dark green fruiting areas soon develop; surface deep, dark green to blackish-green; reverse colorless. Aerial mycelium scant, colony consisting almost entirely of conidiophores and slimy heads. Conidiophores arise from submerged and surface hyphae, several from one point; both aerial and submerged stolons present at these points; conidiophores 100–225 $\times$ 8–10μ. Fructification typically in four stages, consisting of three to five primary branches arising from the apex of the conidiophore; these bear a verticil of secondary branches, and these verticils of metulae; phialides closely crowded on the metulae, club-shaped; primary and secondary branches and metulae elongate oblong, slightly inflated at the apex. Primary branches 15–20 $\times$ 3–3.5μ; secondary branches 13–15 $\times$ 3μ; metulae 8–10 $\times$ 1.5–2μ; phialides 6–8 $\times$ 1–1.5μ. Conidia elliptical, greenish, smooth, granular within, 3–3.8 $\times$ 2–2.5μ. Hyphae, conidiophores, and elements of fructification coarse and pitted, or rough. Slime production very abundant, usually enveloping the entire colony.

From soil: Norway (125)

United States: Louisiana (50)

8. *Gliocladium atrum* Gilman and Abbott

Colonies on Czapek's brown-green, small, slowly spreading, largely submerged; aerial mycelium olivaceous, scanty, aerial growth consisting mostly of conidiophores; colonies moist with slime which envelops the heads. On bean agar considerable aerial mycelium is produced. Conidiophores arise mostly from submerged hyphae, olivaceous, thick-walled, smooth, septate, often slightly flexuous, 75–300 × 3–4μ. Conidial heads enveloped in slime, round, chains not distinguishable; fructification typically in three stages, sometimes in two or four. Primary branches oblong, 8.5–9.5 × 3–3.5μ; metulae oblong, 7.5–9.5 × 3μ; phialides flask-shaped, 7.5–10 × 1.5–2.5μ. Conidia oval to ovate, smooth, light green to almost hyalin, 2.5–4 × 2–2.5μ.

From soil: Canada (15)

United States: Louisiana (50)

There is sufficient color in the conidiophores of this fungus to place it with the Dematiaceae. However, the morphological structure is that of the genus Gliocladium, and it was placed in this genus because of its evident relationship to the other species included in this group.

(13) **Acremonium** Link

Hyphae forming a turf, branched, septate, prostrate, possessing side branches which become erect and serve as conidiophores. Conidia single on the conidiophores, terminal, hyalin or bright colored, usually ovate, small. Differs from Sporotrichum by the erect, unbranched laterals which bear a single conidium at their tips.

A single species treated.

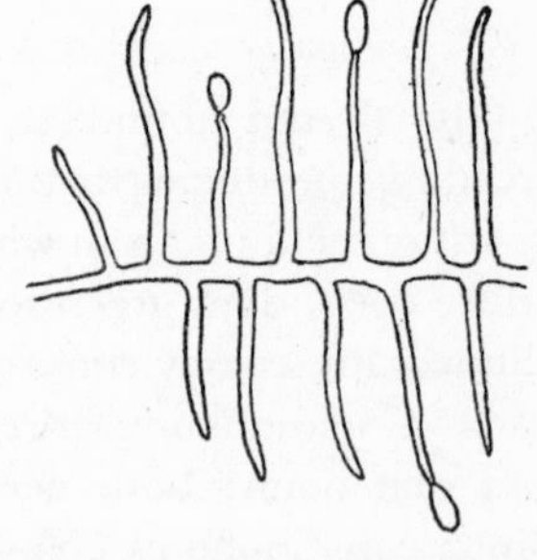

FIG. 73. **Acremonium.** conidiophore and conidia (after Lindau).

1. *Acremonium vitis* Cattaneo

Arachnoid, white; hyphae prostrate, variously branched, hyalin, delicate, obscurely septate, broadly extended and loosely aggregated. Conidiophores awl-shaped, verticillate, often in fours, with a single conidium at each apex. Conidia ovate, one-celled, hyalin, 3–4μ, persistent.

From soil: U. S. S. R. (108)

(14) **Sporotrichum** Link

Hyphae creeping, irregularly branched, but never in whorls, branches repeatedly branched. Conidiophores not formed or only as projections from the side branchlets. Conidia borne laterally and terminally on the hyphae or the branches, usually very numerous, sessile or on small phialides, ovate or globose, hyalin, or brightly colored, usually small.

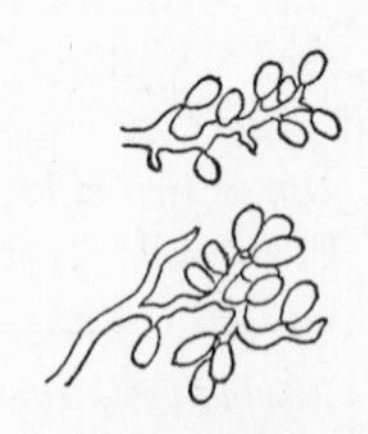

FIG. 74. **Sporotrichum.** conidiophores and conidia (after Lindau).

KEY TO THE SPECIES OF THE GENUS SPOROTRICHUM

a. Conidia globose, 2–4μ in diameter
  b. Colonies olive — 1. *S. olivaceum*
  bb. Colonies white, conidia 2–3μ in diameter — 2. *S. epigaeum* var. *terrestre*

aa. Conidia elliptic to elongate
  b. Colonies reddish, conidia 4 × 3μ with an oil drop
    c. Colonies reddish with an oil drop — 3. *S. roseum*
    cc. Conidia 3–5 × 3–4μ without guttulae — 4. *S. roseolum*
  bb. Colonies white, conidia 9.5 × 5.5–7.5μ
    c. Conidia elliptical — 5. *S. pruinosum*
    cc. Conidia obovate — 6. *S. laxum*
  bbb. Colonies green — 7. *S. chlorinum*

1. *Sporotrichum olivaceum* Fries

Mycelium white, with close hyphae, forming an olive-gray turf. Hyphae irregularly branched, septate, 2–6μ in diameter. Conidiophores alternate, with secondary branches carrying two to three phialides at the same point, which bear single spores at their tips. Conidia, olive, globose, to slightly elliptic, 2–4μ in diameter.

From soil: Switzerland (38)

2. *Sporotrichum epigaeum* Brunard var. *terrestre* Daszewska

Mycelium white forming a velvety turf, short, becoming gray with conidial formation. Hyphae septate, little branched, hyalin, slender, 2.5μ in diameter. Conidia on very slender branchlets, which are forked or trifurcate, each branch budding from its tip a gray globose conidium, 2–3μ, rarely forming short chains.

From soil: Switzerland (38)

3. *Sporotrichum roseum* Link

Colonies broadly spreading, red. Hyphae creeping, sparsely septate. Conidiophores arising as short side branches, unbranched with two to three phialide-like branches at the tip. Conidia terminal, ovate, reddish, 4 × 3μ, with an oil drop.

From soil: Canada (15), China (81), England (37)

United States: Louisiana (50), North Dakota (147), New Jersey (147), Texas (147)

4. *Sporotrichum roseolum* Oudemans and Beijerinck

Turf pale rose, widely extended. Hyphae prostrate, irregularly branched, rarely septate, very delicate, with erect simple or branched laterals which serve as conidiophores. Conidia terminal, globose or oval, single hyalin, in mass pale rose, 3–5μ in diameter or 4–5 × 3–4μ, without vacuoles or guttulae.

From soil: Egypt (115)

5. *Sporotrichum pruinosum* Gilman and Abbott

Colonies on Czapek's agar pure white broadly spreading, cottony; on bean agar low growing, dusty or powdery; reverse colorless; consisting of branched, hyalin, often roughened, aerial hyphae from which the conidiophores arise as branches; sterile hyphae often roped, up to 10μ thick. Conidiophores freely branched, oppositely or irregularly, up to 25μ long, bearing terminal conidia, oval or lemon-shaped, 9.5–13.5 × 6–10μ; appearing grayish.

From soil: Canada (16)

United States: Colorado (75), Iowa (50), Louisiana (50)

6. *Sporotrichum laxum* Nees

Turf at first small, about 1 mm. in diameter, then combining into larger colonies 5 cm. or more in diameter, very delicate, from creeping floccose, diffusely branched hyphae. Conidiophores tapered, occurring as side-branches. Conidia white, obovate, few.

From soil: England (11)

7. *Sporotrichum chlorinum* Link

Turf rather thick, extended, olive-green, floccose; hyphae rather thick, branched, smooth. Conidia obovate, on the tips of the branches, 4–6 × 2–3.5μ, greenish with one oil drop.

From soil: U. S. S. R. (108)

Species of uncertain position:

Sporotrichum flavissimum var. *lunzinense* Szilvinyi

Sporotrichum parvum Szilvinyi

From soil: Austria (131)

Sporotrichum pulviniforme Thom

From soil: United States: New Jersey (146)

(15) **Monosporium** Bonorden

Sterile hyphae creeping, septate, branched, forming a turf. Conidiophores erect, septate or nonseptate, branched in a tree-like form usually with two or more erect or horizontal branches occurring above one another, which may often branch again at the tip into two or three short branches. Conidia on the final branchlets of the conidiophore, terminal, seldom one borne laterally and sessile, or less frequently with a short stipe, hyalin or bright colored, smooth, one-celled, thin walled, usually rather large, ovate or spherical.

Differs from Sporotrichum by the erect conidiophores; from Verticillium by the complete absence of whorled branches; from Sepedonium by the smooth spores.

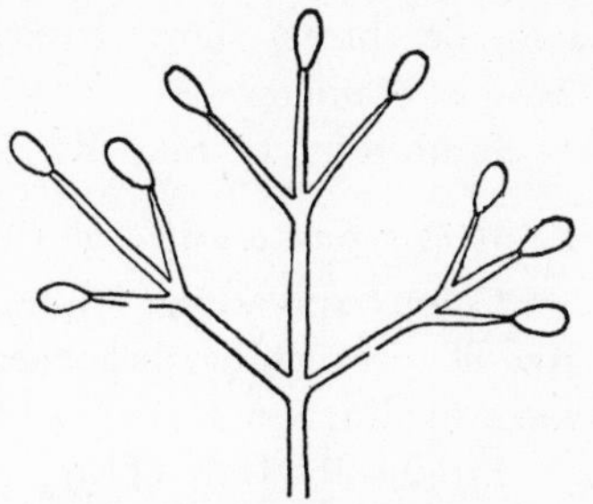

FIG. 75. **Monosporium.** conidiophore and conidia (after Mason).

KEY TO THE SPECIES OF THE GENUS MONOSPORIUM

a. Colonies white	
b. Spores small, 3–2 μ in diameter	
c. Spores obovate	1. *M. silvaticum*
cc. Spores globose	2. *M. minutissimum*
bb. Spores larger, 4–6 × 3–4 μ	
c. Conidia colorless 5–6 × 3 μ	3. *M. acuminatum*
cc. Conidia green, 4–6 × 3–4 μ	4. *M. ellipticum*
aa. Colonies some shade of green	
b. Mycelium forming a pellicle	5. *M. glaucum*
bb. Mycelium forming a turf	
c. Conidia large, oval, 6–10 × 3–4 μ	6. *M. olivaceum*
cc. Conidia smaller, globose to elliptic	
d. Turf powdery	7. *M. viridescens*
dd. Turf floccose	
e. Turf yellow, becoming green	8. *M. flavum*
ee. Turf green from first	
f. Spores small 2–4 μ	9. *M. subtile*
ff. Spores larger 3–4 μ	
g. Hyphae with characteristic swelling	10. *M. humicolum*
gg. Hyphae without swelling	11. *M. reflexum*

1. *Monosporium silvaticum* Oudemans

Colonies orbicular, white; vegetative hyphae creeping, branched, hyalin. Conidiophores erect, continuous, hyalin, dendroidly branched, with ultimate branches commonly two-, rarely three-forked. Conidia single, acrogenous, obovate, 3 × 2 μ.

From soil: Holland (100)

2. *Monosporium minutissimum* Rivolta

Turf white. Conidiophores irregularly branched, final branchlets in twos or threes, short, truncate. Conidia acrogenous, globose, hyalin, 2–3μ in diameter.

From soil: U. S. S. R. (108)

3. *Monosporium acuminatum* var. *terrestre* Saccardo

Colony spreading, white. Conidiophores erect, slightly septate with tree-like branching, branches erect, simple or forked, pointed. Conidia long, hyalin, 5–6 × 3μ.

From soil: Italy (116)

Its presence in soil is uncertain. Reported from moist soil by Saccardo.

4. *Monosporium ellipticum* Daszewska

Mycelium white, floccose, forming conidia only on cellulose media. Hyphae hyalin, septate, 3–4μ in diameter, branching alternately. Conidiophores arise in twos or threes at the same point and bear single terminal conidia. Conidia oval, green, 4–6 × 3–4μ. Chlamydospores numerous, 6–14μ in diameter.

From soil: Switzerland (38)

5. *Monosporium glaucum* Daszewska

Mycelium white forming a green pellicle. Hyphae irregularly branched, septate, 3–6μ in diameter. Conidiophores bearing two to three slightly swollen phialides arise at single points. Conidia oval, single, 4–8 × 3–4μ. Chlamydospores numerous, in chains, 8–12μ in diameter.

From soil: Switzerland (38)

6. *Monosporium olivaceum* Cooke and Massee var. *major* Daszewska

Mycelium white, floccose. Conidiophores erect, forming an olive-green turf, branching irregular; phialides alternate or opposite, at times united into groups of two or three. Conidia terminal, single, oval, pale green, 6–10 × 3–4μ.

From soil: Switzerland (38)

7. *Monosporium viridescens* Bonorden

Mycelium white, floccose, forming a short dark green powdery turf. Hyphae hyalin, septate with nonseptate fertile branchlets; branching dichotomous. Conidiophores dichotomous, with two very slender phialides, which bud globose conidia; rather dark green, 2–4μ in diameter.

From soil: Switzerland (38)

8. *Monosporium flavum* Bonorden

Mycelium white, floccose, forming a short yellow turf, becoming green with age. Hyphae much branched, septate, 2–6μ in diameter, ranching opposite or alternate. Terminal branchlets slender, bearing a ingle round spore, yellow with a single oil drop, 2–4μ in diameter.

From soil: Switzerland (38)

9. *Monosporium subtile* Daszewska

Mycelium white, forming a very short, fine, dark green turf. Conidia produced on acid media. Conidiophores branching irregularly forming bushes. Terminal branches (phialides) budding off at their tips, globose to ovate conidia, green, 2–4μ in diameter.

From soil: Switzerland (38)

10. *Monosporium humicolum* Daszewska

Mycelium white, floccose, forming a short green turf, becoming yellow with age. Hyphae hyalin, branched, septate, with characteristic swellings. Conidiophores dichotomously branched, bushy. Phialides slender and slightly curved, at all times undulant. Conidia globose, to slightly elliptic, hyalin, 3–4 $\times$ 3μ, gathered in masses of a green tint.

From soil: Switzerland (38)

11. *Monosporium reflexum* Bonorden var. *viride* Daszewska

Mycelium white, forming a short turf, green at first becoming brown with age. Hyphae hyalin, septate, branching irregularly, 2–4μ in diameter. Conidiophores alternate, dividing into two to three recurved branchlets which bear the globose, hyalin, (green in mass) conidia at their tips. Conidia 3–4μ in diameter.

From soil: Switzerland (38)

Species of uncertain position:

Monosporium guttulans Szilvinyi

From soil: Austria (131)

(16) **Botrytis** Micheli

Hyphae creeping. Conidiophores simple or frequently branching in an irregular dendroid arrangement, erect. Branches either thin or thicker and narrowing to a point, truncate or with swollen warts on the tips or toothed comb-like. Conidia frequently on the tips of the branches, but not uniformly in heads, globose, ellipsoid or long, hyalin or bright colored, one-celled. The genus contains a large number of forms which have nothing in common,

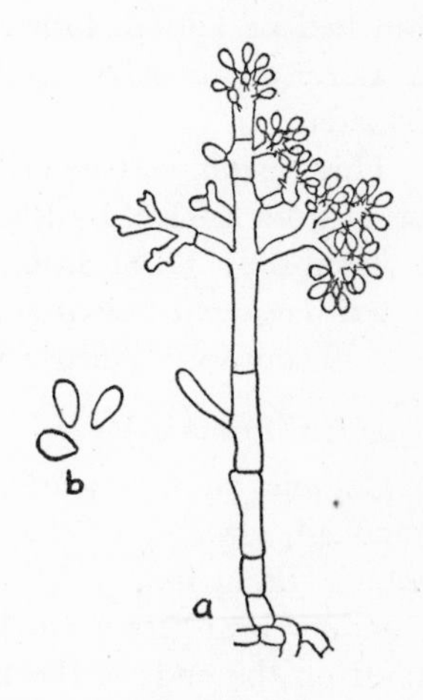

FIG. 76. **Botrytis.** *a* conidiophore; *b* conidia (after Lindau).

one with the other, except a certain superficial similarity. In some species the genus forms mycelial sclerotia which are very similar to those of the genus Sclerotinia.

KEY TO THE SPECIES OF THE GENUS BOTRYTIS

a. Turf white
  b. Conidia ovate 5–7 × 3–3.3μ — 1. *B. pyramidalis*
  bb. Conidia globose 2–3μ — 2. *B. bassiana*
aa. Turf gray
  b. Conidia small, 2.5–3 × 3–4μ — 3. *B. terrestris*
  bb. Conidia larger, 9–12 × 6.5–10μ — 4. *B. cinerea*

1. *Botrytis pyramidalis* (Bonorden) Saccardo
Syn. *Botryosporium pyramidale* Costantin

Turf quite white. Conidiophores long, septate, many times dichotomously branched and branches relapsing. At the axis of the conidiophores there occur in great numbers short nonseptate branchlets, which are cut off from the axis by a septum. The tips of the branchlets are swollen and on the vesicles occur three to six short branchlets which are swollen club-shaped at their tips and carry many tiny phialides on the ends. Conidia single on the phialides, long egg-shaped, round at the tip and having at the base a fine papilla, 5–7μ long, 3–3.3μ wide.

From soil: England (11)

2. *Botrytis bassiana* Balsamo

Turf extended, felt-like, white. Conidiophores erect, white, unbranched or seldom forked with short side-branches. Conidia, globose on lateral phialides, or on the sides of the conidiophores in heads; 2–3μ in diameter.

This fungus was described as a parasite on insect larvae, but is reported from the soil by Raillo.

From soil: U. S. S. R. (108)

var. *lunzinense* Szilvinyi

From soil: Austria (131)

3. *Botrytis terrestris* Jensen

Colonies at first white, later gray; sterile hyphae creeping, hyalin, branched, septate, 1.5–3μ in diameter. Conidiophores erect, ascending, septate, branched, 2–3.5μ in diameter, 50–200μ high; primary and secondary branches verticillate, dichotomous, or alternate. Conidia produced on the ends of the branches, forming a more or less compact triangular cluster that averages 20–25μ, obovate, somewhat angled, uniform 2.5–3 × 3–4μ, hyalin to light gray. Clusters of conidia separate very easily.

From soil: Canada (15), China (81)
United States: New York (67)

4. *Botrytis cinerea* Persoon

Colonies diffuse, gray, gray-green, dark olive-green to brown-black, seldom brown or reddish-green, dusty from the conidia, loose or dense, up to 2 mm. high. Conidiophores erect, unbranched or seldom branched, septate, 11–23μ thick, wall blackish-brown, toward the tip almost hyalin, with several (three and more) projections at the tip from which the conidia are formed singly on very fine warts. The point of the conidiophore grows between the warts, thereby pressing them back, usually some distance from one another, and they become lateral. The conidia stand so thickly on the projections that thick heads are produced which soon fall off. Conidia ovate or elliptical to almost globose, finely apiculate at the base, 9–12 $\times$ 6.5–10μ, with almost hyalin, slightly brownish wall.

From soil: Canada (15), Denmark (68), England (11) (36) (37), Japan (132), Switzerland (67)

United States: Louisiana (50), New Jersey (146) (147), Porto Rico (146)

The following species of Botrytis are listed by Jensen (67) as soil fungi, but they were isolated from leaves and sticks in contact with the soil, and not from the soil itself:

Botrytis fulva Link
Botrytis dichotoma Corda
Botrytis geophila Bonorden
Botrytis epigaea Link
Botrytis purpureospadicea (Fuckel) Lindau

(17) **Cylindrophora** Bonorden

Hyphae forming a turf, prostrate. Conidiophores erect, with or without septa, with simple or forked branches occurring on one or both sides, carrying single conidia at their tips. Conidia cylindric, with rounded ends, hyalin.

A single species treated.

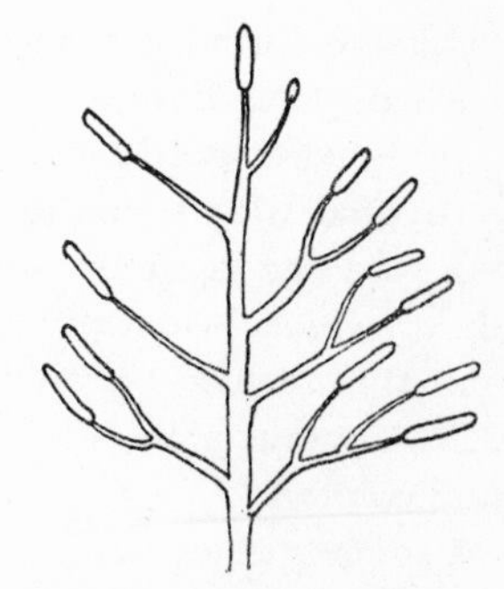

Fig. 77. **Cylindrophora.** conidiophore and conidia (after Lindau).

1. *Cylindrophora hoffmanni* Daszewska

Mycelium white, branching. Hyphae very long, little branched, colorless, septate and filled with oil drops. Conidiophores short, alternate with single spores at their tips. Conidia cylindric; rounded at both ends, hyalin, 6–12 × 2–4μ.

From soil: Switzerland (38)

(18) **Sepedonium** Link

Hyphae creeping loosely branched, carrying conidia at the tips of the final branchlets. Conidia single or two or three, terminal, warty, globose, or ovate, hyalin, or bright colored. Sometimes simple ovate conidia occur on the upper branches of the conidiophore. Then as in Stephanoma the warty spores are known as chlamydospores, the small egg-shaped forms as conidia.

A single species treated.

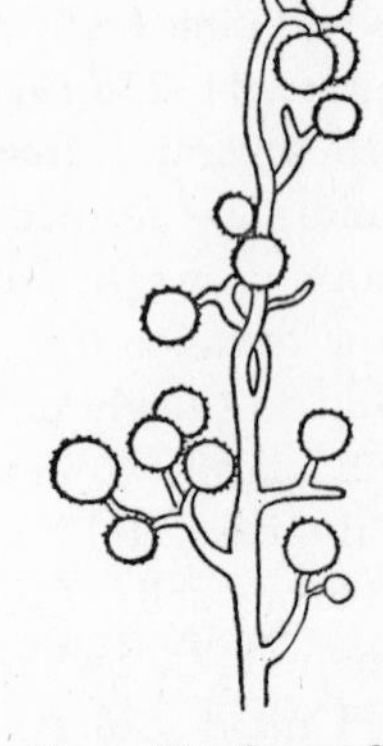

FIG. 78. **Sepedonium.** condiophore and conidia (after Lindau).

1. *Sepedonium chrysospermum* (Bulliard) Fries
   Syn. *Ozonium croceum* Persoon

Hyphae widespread forming a thick white, then golden-yellow turf creeping in and on the substrate, septate, branched, tolerably thick, bearing lateral short simple or clustered branches on the tips of which the spores occur. Chlamydospores single, acrogenous, formed in large numbers, globose, warted, yellow or golden-yellow, 13–17μ in diameter, with rather thick wall.

From soil: England (37)

United States: Colorado (76), New Jersey (147)

(19) **Pachybasium** Saccardo

Hyphae forming a turf, creeping, septate, branched. Conidiophores erect, branched; primary branches sterile, ending in long, curved, thin hyphal tips; secondary branches alternating or standing in almost opposite whorls, on the ends of which occur many short flask-shaped terminal branchlets on which are formed the conidia. Conidia globose or elongate, hyalin or bright colored.

A single species treated.

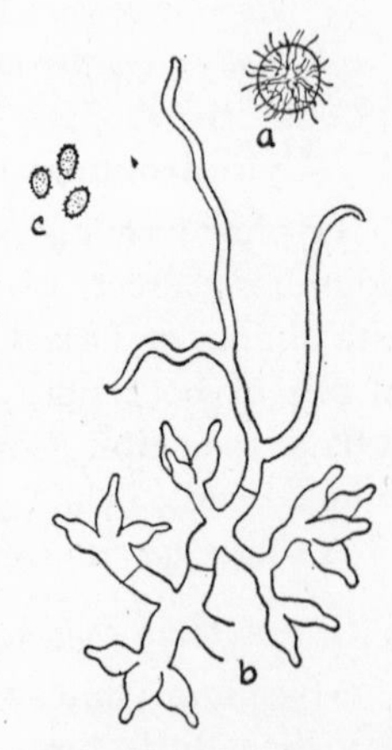

FIG. 79. **Pachybasium.** *a* habit; *b* conidiophore; *c* conidia (after Lindau).

1. *Pachybasium hamatum* (Bonorden) Saccardo

Turf more or less extended, white or straw-colored, formed of floccose, curved, septate, branched hyphae, up to 16μ in diameter. Conidiophores erect, simple or many times dichotomously divided, septate and the primary branches sterile, ending in long tips bending back and forth. Side branches short, alternating or whorled, forming at their tip two to four flask-shaped branchlets which are elliptical in outline and narrowed to a phialide-like tip, 1–2μ long, the whole being 10–12μ long and 7μ broad. Conidia single, terminal on the branchlet, ovate, 7μ long × 4.3μ wide, hyalin or slightly colored.

From soil: United States: Michigan (52)

(20) **Verticillium** Nees

Sterile hyphae creeping, septate, branched, hyalin or lightly colored. Conidiophores erect septate, branched. Branches of the first order whorled, opposite or alternate; branches of the second order whorled, dichotomous or trichotomous on the branches of the first order; further branching similar; terminal branchlets usually flask-shaped and distinctly pointed at the apex. Conidia always borne singly on the branchlets, soon falling away. Round, elliptical, ovate, inverted egg-shaped, or short spindle-shaped, hyalin or slightly colored.

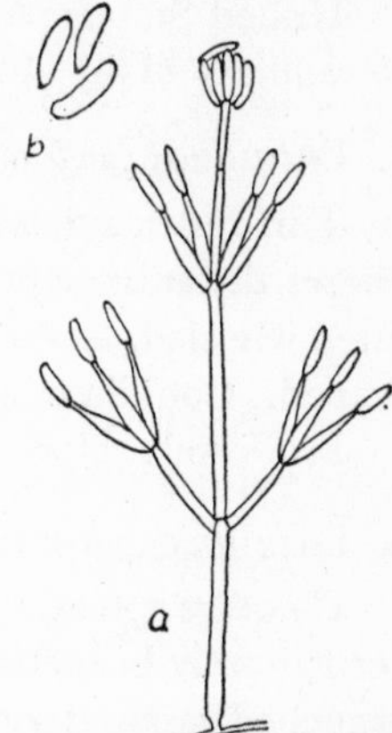

FIG. 80. **Verticillium.** *a* conidiophore; *b* conidia (after Lindau).

KEY TO THE SPECIES OF THE GENUS VERTICILLIUM

a. Colonies white	
b. Colonies spreading, floccose	1. *V. terrestre*
bb. Colonies restricted, velvety	2. *V. candelabrum*
aa. Colonies other colors than white	
b. Colonies green	
c. Colonies blue-green	3. *V. glaucum*
cc. Colonies pale green	
d. Spores 4–6 × 3–4μ	4. *V. cellulosae*
dd. Spores 2–12 × 3μ	5. *V. albo-atrum*
bb. Colonies not green	
c. Colonies yellow	
d. Colonies brownish-yellow	6. *V. effusum*
dd. Colonies pale yellow	7. *V. sulphurellum*
cc. Colonies red	
d. Colonies brick red	8. *V. lateritium*
dd. Colonies pale rose	9. *V. puniceum*

1. *Verticillium terrestre* (Link) Lindau

Colonies pure white, spreading, floccose, consisting of dense, cobwebby, branched hyphae. Conidiophores erect, septate, usually with four whorls of branchlets, branchlets rarely again verticillately branched. Conidia formed singly at the tips of the branchlets, globose to elliptical, hyalin 4.4–5.0 × 3.5–4.5μ.

From soil: Canada (15)

United States: Alaska (147), Colorado (74), Iowa (3), Louisiana (2), New Jersey (146) (147)

2. *Verticillium candelabrum* Bonorden

Turf restricted, white, confluent and then extended, velvety. Conidiophores distantly septate, sparingly branched at the tip; branchlets with three-whorled, secondary branchlets, final branchlet short, inverted club-shaped. Conidia ovate, 4–6 × 3μ, hyalin.

From soil: U. S. S. R. (108)

3. *Verticillium glaucum* Bonorden

Colonies spreading, blue-green. Conidiophores erect, 100 × 3μ, twice verticillately branched, sparsely septate; branches usually trichotomously branched, secondary branches with three branchlets at the apex. Conidia globose, 2.5μ in diameter, almost hyalin.

From soil: Canada (15), Egypt (115), Switzerland (38), U. S. S. R. (108)

United States: New Jersey (146) (147)

4. *Verticillium cellulosae* Daszewska

Mycelium white, forming a loose pale green turf about 0.5 cm. in height. Branches opposite, hyphae hyalin, septate, 3–6μ in diameter. Conidiophores branched, the final branches carrying swollen phialides, three or four at the same level, with an oval spore budding from the tip of each, 4–6 × 3–4μ, slightly greenish. The spores collect in spherical masses.

From soil: Switzerland (38)

5. *Verticillium albo-atrum* Reinke and Berthold

Mycelium spreading, brownish. Conidiophores erect, simple, dark colored, paler at the apex, with up to eight whorls, three to five branches in the whorl; branches sparsely septate, simple or further branched in whorls, terminal branchlets thickened at the base and narrowed at the apex, erect. Conidia elongate egg-shaped, hyalin, then brownish, 5–12 × 3μ.

From soil: England (36)

Dale (36) reported the isolation of this species from soil in England. It is doubtful whether her species was *Verticillium albo-atrum*, however, since she stated that her fungus was pure white in all stages. *Verticillium albo-atrum* is slightly olivaceous.

6. *Verticillium effusum* Otth

Mycelium white, turf yellow-brown, short, uniform. Hyphae 3–6μ in diameter. Conidiophores with opposite branches, bearing short bulbous phialides, three or four in each verticil, 6–10μ long. Conidia single, globose to slightly elliptic, 3.5μ in diameter.

From soil: Switzerland (38)

7. *Verticillium sulphurellum* Saccardo

Heads effuse, pale yellow, cottony. Conidiophores erect, septate, 3μ in diameter, sub-verticillately branched. Conidia ovate-oblong, $3 \times 1\mu$, sub-hyalin.

From soil: Austria (131)

8. *Verticillium lateritium* Berkeley

Turf delicate, extended, evanescent, brick-red, with white margin. Conidiophores many-times dichotomously branched, rarely whorled, yellow-red, 2–3μ in diameter. Conidia ovate, $2\text{–}3 \times 2\mu$, very small and very numerous, brick-red.

From soil: U. S. S. R. (108)
United States: Illinois (129)

9. *Verticillium puniceum* Cooke and Ellis

Colonies compact, heads semi-globose, elliptic or confluent; hyphae delicate, septate, branched. Conidiophores verticillately branching, short, rosy. Conidia elliptic, small, one-celled, hyalin, $4 \times 2\mu$.

From soil: Austria (131)

Species of uncertain position:
Verticillium chlamydosporium Goddard (52)
Verticillium nubilum

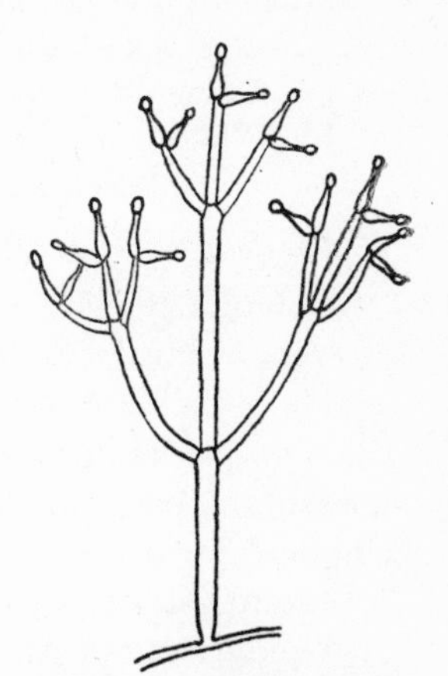

FIG. 81. **Verticilliastrum.** conidiophore and conidia (after Daszewska).

(21) **Verticilliastrum** Daszewska

Conidiophores erect, delicate, with opposite, alternate or three-branched whorls. The fertile branches end in two clavate phialides which are produced at right angles to each other and carry a conidium on each tip. Conidia globose, hyalin.

1. *Verticilliastrum glaucum* Daszewska

Mycelium white, turf green about 0.5 cm. high. Hyphae colorless, septate branched. Conidiophores branch dichotomously, budding-off a spherical, green conidium, 3–4μ in diameter. Chlamydospores intercalary on the hyphae, 8–16μ in diameter.

From soil: Switzerland (38)

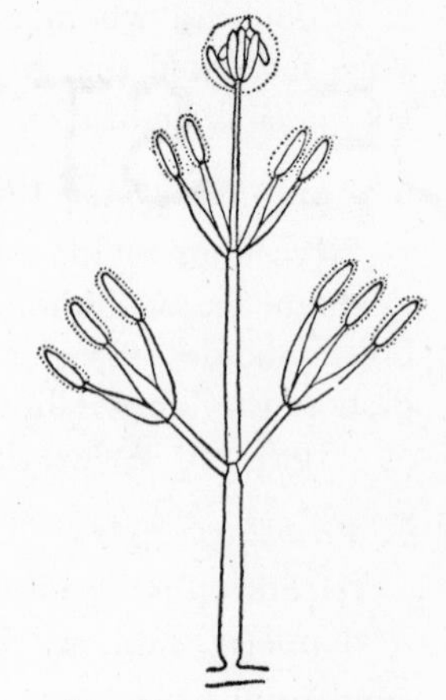

FIG. 82. **Acrostalagmus.** conidiophore and conidia (after Lindau).

(22) **Acrostalagmus** Corda

Hyphae creeping, septate, branched. Conidiophores erect, septate, usually branched in whorls. Conidia borne at the points of the branchlets, produced successively but not catenulate, forming a head held together by slime; conidia hyalin, egg-shaped to elliptical.

KEY TO THE SPECIES OF THE GENUS ACROSTALAGMUS

a. Colonies floccose, pure white to creamy — 1. *A. albus*
aa. Colonies not floccose
  b. Orange to avellaneous — 2. *A. cinnabarinus*
  bb. White — 3. *A. fungicola*

1. *Acrostalagmus albus* Preuss

Colonies spreading, floccose, pure white or creamy. Sterile hyphae creeping, indistinctly septate, sparingly branched. Conidiophores arise as side branches of aerial hyphae, erect, up to 200μ or 220μ long, sometimes simple, but usually with one or two whorls of branchlets; branchlets nonseptate, pointed, each bearing conidia on the point. Conidia hyalin, elliptical 3.0–3.5 $\times$ 1.0–1.5μ.

From soil: Austria (131), Canada (16), China (81), England (11)

United States: Alaska (147), Illinois (129), Iowa (3) (100), Louisiana (2), New Jersey (146) (147), North Dakota (147), Texas (147)

var. *varius* Jensen

Colonies effused, thin, subfloccose, white; vegetative hyphae hyalin branched, septate, 2–3.5μ. Conidiophores creeping, ascending, or erect branched, 15–75 × 2–3.5μ; usually simple but occasionally alternately branched, verticillate, alternate toward apex, slightly curved at the summit producing a head of conidia, 15–36 × 2–3μ. Conidia hyalin, oblong, 3.3 × 1.5μ.

From soil: China (81)

United States: New York (67)

2. *Acrostalagmus cinnabarinus* Corda var. *nana* Oudemans

Colonies orbicular, orange mixed with red; vegetative hyphae septate. Conidiophores septate, with two or three series of opposite branchlets, branches terminated by three-rayed verticils, with each ray in the form of a tenpin, 36–45μ long, bearing the conidia. Conidia elliptical or oblong 5–8 × 3–5μ, formed in a head enveloped by slime.

From soil: Austria (131), Canada (15), China (81), Denmark (68), England (11), Holland (100)

United States: Hawaii (146), Michigan (52), North Dakota (146) (147), New Jersey (146) (147), Oregon (146), Porto Rico (146) (147)

3. *Acrostalagmus fungicola* Preuss

Colonies effuse; mycelium prostrate, branched, septate. Conidiophores erect, hyalin, septate, branching above, branches continuous, verticillate, bearing globose white heads of conidia at their apices. Conidia elongate, hyalin.

From soil: U. S. S. R. (108)

(23) **Acrocylindrium** Bonorden

Hyphae prostrate; conidiophores erect, branching in whorls, seldom dichotomous, with tapering tips. Conidia short cylindrical, hyalin, quickly falling. Differs from Verticillium by its longer spores.

A single species treated.

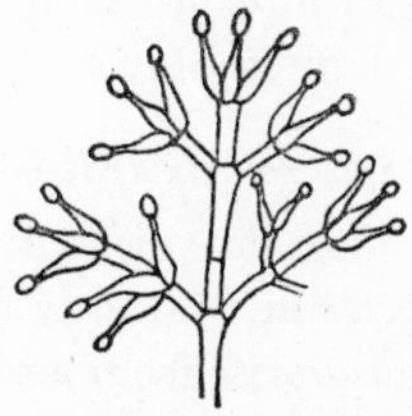

Fig. 83. **Acrocylindrium.** conidiophore and conidia (after Lindau).

1. *Acrocylindrium granulosum* Bonorden

Turf extended, white dusty. Conidiophores erect, with opposite branches, laterals in three whorls, pointed. Conidia cylindric, straight or slightly curved, rounded at both ends, 5μ long × 1.7–2μ wide, hyalin.

From soil: Denmark (68)

(24) **Spicaria** Harting

Conidiophores erect, septate, usually freely branched, branching often in whorls but also irregular; each branchlet bears a terminal fructification composed of a verticil of divergent metulae on which are borne a verticil of divergent phialides; heads divergent and seldom penicillate; conidial chains usually long. Conidia hyalin, round ovate, elliptical, or elongate.

KEY TO THE SPECIES OF THE GENUS SPICARIA

a. Conidiophores smooth
  b. Colonies gray-green, conidia elliptical, 6–12 × 4–6μ — 1. *S. silvatica*
  bb. Colonies gray, sometimes with rosy tints
    c. Conidia globose, 2–3μ — 2. *S. simplicissima*
    cc. Conidia ellipsoid 2.5 × 1.5μ — 3. *S. griseola*
  bbb. Colonies brownish-olive to olive-buff — 4. *S. divaricata*
  bbbb. Colonies bright violet or lavender — 5. *S. violacea*
aa. Conidiophores spiny or echinulate, colonies white — 6. *S. elegans*

1. *Spicaria silvatica* Oudemans

Colonies orbicular, light gray-green; vegetative hyphae creeping, hyalin, septate, with forked branching. Conidiophores erect, sparsely branched; branches alternate; variable in length, simple or forked at the tip, bearing phialides 20–25μ long, cylindrical and somewhat curving Conidia in long chains, elliptical or oblong, hyalin, smooth, 6–12 × 4–6μ.

From soil: Holland (100), India (21)

United States: Rhode Island (110)

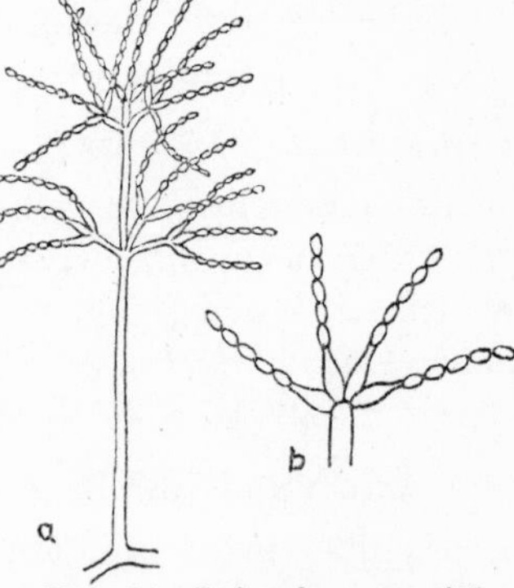

Fig. 84. **Spicaria.** *a* conidiophore; *b* conidial head (after Lindau).

2. *Spicaria simplicissima* Oudemans

Colonies orbicular, with alternating zones of cream-yellow sterile mycelium and gray fruiting areas, occasionally with rosy tints. Conidiophores arise from aerial mycelium, 40μ long, septate, usually unbranched, bearing metulae and whorls of phialides only; phialides 8–12μ long, verticillate. Conidia in short chains, globose, 2–3μ in diameter.

From soil: Holland (100)

United States: Colorado (74), Iowa (3), New York (67)

3. *Spicaria griseola* Saccardo

Turf extended, velvety, gray. Conidiophores erect, almost nonseptate 80–90μ long, 4μ broad, with four to six whorls or branches at the tip which in turn are forked three or four times. The final branchlets are inverted club-shape carrying a conidial chain at their tip. Conidia ellipsoid, 2.5μ long $\times$ 1.5μ wide, at first hyalin, then gray.

From soil: England (11)

4. *Spicaria divaricata* (Thom) Gilman and Abbott
   Syn. *Paecilomyces varioti* Bainier
   Penicillium divaricatum Thom

Colonies on Czapek's agar broadly spreading, low growing, felty, with scattered, floccose aerial mycelium; surface olive, olive-buff or brownish-olive, never true green; reverse colorless. Conidiophores arise from aerial or submerged mycelium, freely and irregularly branched, conidiophores up to 325μ long. The conidial fructification is typically in two stages the branches of the conidiophore bearing a terminal verticil of divergent metulae, with divergent phialides. Metulae extremely variable in length, phialides 10–25 $\times$ 2.5–4μ. Conidial chains very long divergent, seldom more than five or six in a head. Conidia elliptical, smooth, 4.5–6 $\times$ 2.5–4μ.

From soil: China (81), Greenland (96)
United States: Illinois (129), Iowa (50), Utah (50)

5. *Spicaria violacea* Abbott

Colonies on Czapek's agar floccose, spreading, surface white at first, becoming bright lavender or violet when mature; reverse colorless. Aerial mycelium abundant, consisting of a dense network of interwoven hyphae. Conidiophores arise as branches of aerial mycelium, erect up to 100μ long, usually once or twice branched, but often short and unbranched. Conidial chains very long, up to 700μ or more in length; fructification a divergent head with both metulae and phialides or with phialides only; phialides 6.5 $\times$ 2μ. Conidia elliptical, smooth, hyalin, 3–3.5 $\times$ 2–2.5μ.

From soil: United States: Iowa (3), Louisiana (2)

6. *Spicaria elegans* (Corda)

Colonies somewhat spreading, white, velvety; vegetative hyphae creeping, hyalin, septate. Conidiophores erect, septate, with two to four circles of opposite, or three to four verticillate branches; branches short, fusiform, each divided at the tip into a verticil of three branchlets; branchlets lageniform, swollen at the tip. Conidia ovate-fusiform, united to form long chains, 4.5–7 $\times$ 3.5–4μ, hyalin.

From soil: United States: Iowa (101)

Species of uncertain position:

Paecilomyces austriacus Szilvinyi
Paecilomyces subflavus Szilvinyi
Paecilomyces viride Szilvinyi
From soil: Austria (131)
Spicaria decumbens Oudemans
From soil: Holland (100)

25) **Nematogonum** Desmazieres

Hyphae creeping. Conidiophores erect with sterile and fertile cells. Sterile cells thickened on both sides, fertile cells globosely swollen, smooth. Conidia formed singly, ovate, bright colored.

A single species treated.

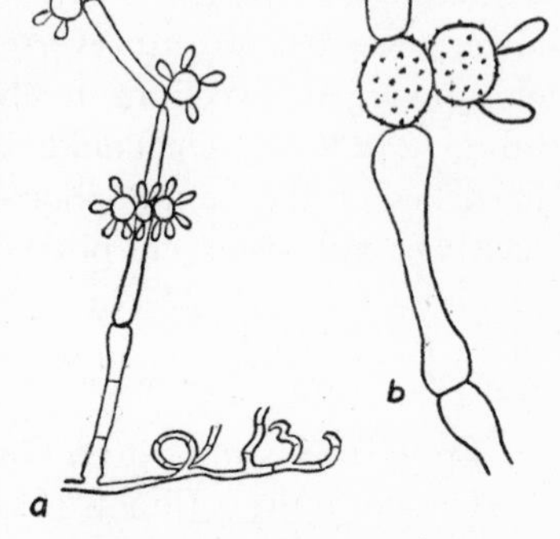

FIG. 85. **Nematogonum.** *a* conidiophore; *b* conidia (after Lindau).

1. *Nematogonum humicola* Oudemans

Turf circular, ribbon-like, at first white, then bright gray, finally cream colored. Conidiophores erect, 2.2–3.3μ thick, hyalin, septate, unbranched, with longer cells not swollen at both ends and shorter smooth-walled fertile cells swollen at both ends. Conidia globose 3–4μ broad, sessile, almost hyalin.

From soil: England (36), Holland (100)

(26) **Trichothecium** Link

Hyphae creeping. Conidiophores erect, septate, unbranched. Conidia terminal, single, two-celled, hyalin or bright colored.

A single species treated.

1. *Trichothecium roseum* Link

Syn. *Cephalothecium roseum* Corda

Turf forming a powdery case, widespread, mold-like or arachnoid, white, finally pink, formed of creeping, branched, septate, white hyphae. Conidiophores erect, little or nonseptate, usually unbranched and scarcely swollen at the tip. Conidia acrogenous, single, one after another, but remaining attached and forming a head by apical growth, pear-shaped, two-

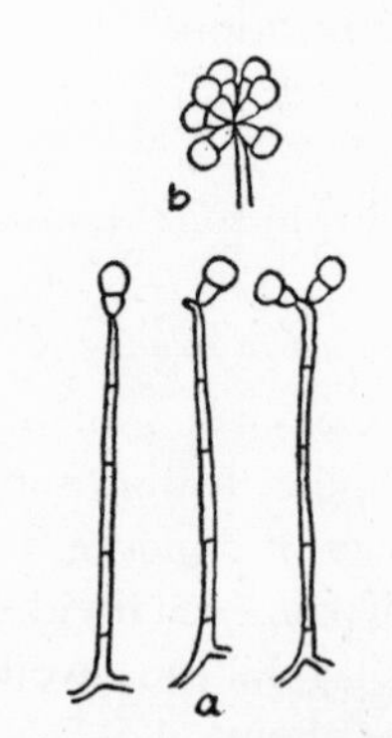

FIG. 86. **Trichothecium.** *a* conidiophores; *b* conidial head (after Lindau).

celled, the apical cell being larger, hyalin, then pink, 12–18μ long × 8–10μ broad.

From soil: China (81), England (36)

United States: Hawaii (146), North Dakota (146), New Jersey (147), New York (67), Porto Rico (147)

(27) **Mycogone** Link

Hyphae branched, interwoven. Conidiophores short, occurring laterally. Conidia single on the tips of the conidiophores, dissimilar, two-celled, the upper cell larger, usually warty, bright colored, the lower cell pale. As in Sepedonium there are found here conidia and chlamydospores.

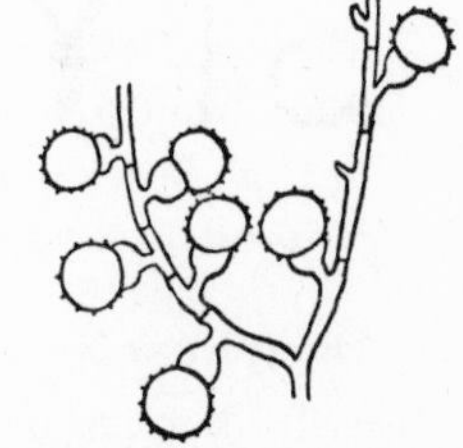

Fig. 87. **Mycogone.** *a* conidiophore and conidia (after Lindau).

KEY TO THE SPECIES OF THE GENUS MYCOGONE

a. Turf with yellow tint; upper cell of conidium, dark brown	
b. Conidia 12–15μ in diameter	1. *M. nigra*
bb. Conidia 8–8.4μ in diameter	1a. *M. nigra* var. *minor*
aa. Turf white; upper cell of conidium, hyalin	2. *M. alba* var. *minor*

1. *Mycogone nigra* (Morgan) Jensen

Colonies at first hyalin, later showing yellowish tint, and finally becoming black-brown and zonate. In rapidly growing colonies, the hyphae near the margin are aerial as well as immersed and show a distinct yellow tint. Mycelium branched, septate, with numerous fertile branches bearing a single spore at the apex, 2.5–4μ thick. Conidiophores varying from scarcely none to a length of 30μ, width 2–3μ, ascending or erect. Conidia uniseptate, upper cell dark brown, smooth, thick-walled, globose, 12–15μ in diameter, lower cell hyalin to slightly colored, smooth, hemispherical, 8–10 × 9–12μ. Intercalary cells are often formed. Variations in which the lower cell is not cut off, and again when a second small cell is formed, occur in culture. All conidia may probably be considered as chlamydospores.

From soil: Canada (15), Denmark (68), U. S. S. R. (108)

United States: New York (67)

1a. *Mycogone nigra* Morgan var. *minor* Raillo

Differs from *M. nigra* in size of spores; conidia one-celled: 8–8.4μ in diameter, two-celled: upper cell brown, globose, 8–8.4μ in diameter, lower cell hyalin, oval, 5.4 × 4μ.

From soil: U. S. S. R. (108)

2. *Mycogone alba* Persoon

Turf white, extended, woolly, formed of interwoven septate hyphae, 4μ in diameter. Conidia two-celled, hyalin, entire length 36μ, width $15–21\mu$; upper cell 21μ long, $15–20\mu$ wide, warty; lower cell 8μ long, 7μ wide, punctate or smooth.

var. *minor* Raillo

Differs from *M. alba* in size of spores; upper cell globose $13.5–16.8\mu$ in diameter; lower oval $8.4–10.8 \times 6.7–8.4\mu$.

From soil: U. S. S. R. (108)

(28) **Dactylium** Nees

Sterile hyphae forming a turf, prostrate, branching, septate, hyalin. Conidiophores ascending or erect, simple or branching in many whorls. Conidia acrogenous, single with two or more cross-walls, hyalin.

A single species treated.

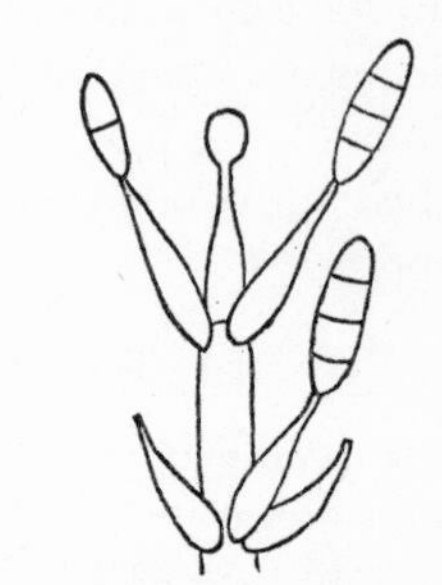

Fig. 88. **Dactylium** *a* conidiophore and conidia (after Lindau).

1. *Dactylium dendroides* (Bulliard) Fries

Turf extended, white, mold-like. Conidiophores erect, septate, with approximately opposite branches, which divide above into usually three-pointed branchlets. Conidia elongate, pointed at the base, with three septa, scarcely constricted at the septa; $26–32 \times 10–13\mu$, hyalin.

From soil: Canada (15)

b. *DEMATIACEAE*

Hyphae septate, usually prostrate, seldom short, dark to black, seldom bright colored or hyalin and in that case having dark conidia. Conidiophores either not sharply differentiated from the mycelium or differentiated, erect, simple or much branched, usually dark or bright colored. Conidia of various forms, dark or hyalin, in the latter case the conidiophores and hyphae dark.

KEY TO THE GENERA OF THE DEMATIACEAE

a. Conidia and conidiophores one or both dark colored
  b. Conidia one-celled
    c. Conidia and conidiophores both dark colored

d. Conidiophores not differentiated from the mycelium (Conisporieae)
  e. Conidia single (1) **Papularia**
  ee. Conidia in chains or clustered
    f. Conidia in chains
      g. Conidia budding from dark colored septate mycelial threads (2) **Pullularia**
      gg. Conidia formed by rounding up of mycelial cells
        h. Conidia remaining attached to each other (3) **Hormiscium**
        hh. Conidia easily separating from one another (4) **Torula**
    ff. Conidia in clusters on tips of short side branches (5) **Echinobotryum**

dd. Conidiophores differentiated from the mycelium
  e. Conidia not in chains
      g. Conidia in terminal heads (Periconieae)
        h. Conidia irregularly distributed at the tip of the conidiophore, or raised on small delicate points
          i. Conidiophores inflated at tip; conidia globose (6) **Periconia**
          ii. Conidiophores not inflated at tip; conidia elongate (7) **Synsporium**
        hh. Conidia produced at the tip of the conidiophores from rather long phialides
          i. Conidia not enveloped in slime (8) **Stachybotrys**
          ii. Conidia enveloped in slime (9) **Gliobotrys**
      gg. Conidia not in terminal heads (Trichosporieae)
        h. Conidia sessile, both terminal and lateral (10) **Trichosporium**
        hh. Conidia single on branching conidiophores (Monotosporeae)
          i. Conidia on lateral branchlets of the mycelium

j. Conidiophores inflated at tip (11) **Nigrospora**

jj. Conidiophores not inflated at tip (12) **Acremoniella**

ii. Conidia on truly branched conidiophores (13) **Monotospora**

ee. Conidia in chains (Haplographieae)

f. Conidiophores with terminal whorl of phialides (14) **Haplographium**

ff. Conidiophores with branching chains of conidia, without phialides (15) **Hormodendrum**

cc. Conidia hyalin or bright colored, conidiophores always dark

d. Conidia terminal, single or in chains, not in heads (Chloridieae)

e. Conidia on lower lateral branches of the conidiophore; upper branches sterile, conidia ovate (16) **Mesobotrys**

ee. Conidiophores at base of sterile hyphae, conidia globose (17) **Botryotrichum**

dd. Conidia in heads (Stachylidieae) (18) **Stachylidium**

bb. Conidia more than one-celled, dark

c. Conidia, two-celled, dark colored

d. Conidia ovate to elongate

e. Conidiophores little differentiated from the mycelium (Bisporeae)

f. Conidia borne singly, on simple branchlets (19) **Dicoccum**

ee. Conidiophores differentiated from mycelium

f. Conidia smooth walled, not in heads (Cladosporieae)

g. Conidia in chains

h. Chains of conidia short (20) **Cladosporium**

hh. Chains of conidia long (21) **Diplococcium**

gg. Conidia not in chains; on long thread-like branches (22) **Scolecobasidium**

cc. Conidia more than two-celled, dark; ovate, cylindric or vermicular

d. Conidia with cross walls only

e. Conidiophores very short or lacking (Clasterosporieae) (23) **Clasterosporium**

ee. Conidiophores well-formed and differentiated

f. Conidia formed singly, either terminal or lateral, not in chains (Helminthosporieae)
  g. Conidia terminal and lateral, not in whorls
    h. Conidia four- or five-celled (24) **Curvularia**
    hh. Conidia more than five-celled (25) **Helminthosporium**
  gg. Conidia in whorls (Acrotheciae)
    h. Whorls lateral (26) **Spondylocladium**
    hh. Whorls only terminal (27) **Acrothecium**

dd. Conidia muriformly divided (Dictyosporae)
  e. Conidiophores differentiated from mycelium
    f. Conidia not in chains
      g. Conidia cruciately divided into four cells, warted (28) **Tetracoccosporium**
      gg. Conidia with more than four cells
        h. Conidiophores decumbent (29) **Stemphylium**
        hh. Conidiophores erect (30) **Macrosporium**
    ff. Conidia in chains (31) **Alternaria**

(1) **Papularia** Fries

Mycelium at first plentiful and white, later becoming gray and then black. Submerged hyphae branched, septate, at first hyalin, later yellow-brown. Conidiophores hyalin, short, collapsing. Conidia lenticular, black by reflected light, yellow-brown by transmitted light, with a hyalin rim around the periphery.

FIG. 89. **Papularia.** conidiophores and conidia (after Mason).

A single species treated.

1. *Papularia sphaerosperma* (Persoon) von Höhnel
   Syn. *Coniosporium arundinis* Saccardo
   Periconia lanata Gilman and Abbott

Colonies spreading, floccose woolly white on the surface, reverse at first white, later yellow to brown, spotted by occurrence of dark brown to black sclerotia. Conidiophores prostrate or ascending, very variable in length up to 100μ, hyalin.

Conidia borne at the apex in an irregular head, on short phialides with inflated bases, lenticular $3 \times 9\mu$ usually $5–7\mu$ in diameter, smooth, black by reflected light, yellow-brown by transmitted light with a hyalin rim around the periphery.

From soil: Canada (15)
United States: Louisiana (50)

(2) **Pullularia** Berkhout

Hyphae dark colored with age. Blastospores ovate and hyalin, occasionally absent. Hyphae composed of chains of dark, thick-walled cells, connected by strands of lighter colored thin-walled cells.

A single species treated.

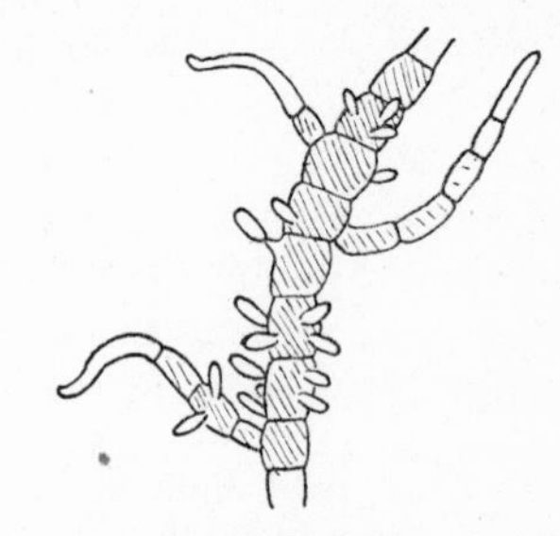

FIG. 90. **Pullularia.** fruiting habit (after Corda).

1. *Pullularia pullulans* (de Bary) Berkhout
   Syn. *Dematium pullulans* de Bary

Conidia as oval to elongate hyalin cells, budding from brown, branching, and septate mycelial threads, both terminally and laterally. After abstriction the conidia may continue to multiply by budding and abstriction. Mycelial cells later divide into a number of isodiametric cells with rounded sides and thick double wall, filled with oil drops.

From soil: Egypt (115)
United States (146)

(3) **Hormiscium** Kunze

Sterile mycelium either entirely lacking or very sparse, usually the entire thallus is formed of threads whose members arose as buds. Conidia or buds remain attached to one another, not separating, dark colored.

A single species treated.

1. *Hormiscium stilbosporum* (Corda) Saccardo

FIG. 91. **Hormiscium.** fruiting habit (after Lindau).

Turf erumpent, dusty, confluent, black. Conidial chains various, branched, forked or simple, curved. Conidia almost cubical, of similar size, brown, $7–8\mu$ in diameter.

From soil: U. S. S. R. (108)

(4) **Torula** Persoon

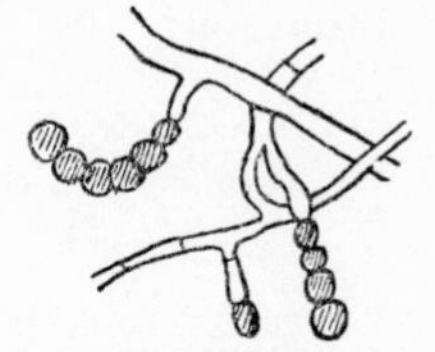

FIG. 92. **Torula.** conidiophores and conidia (after Lindau).

Sterile hyphae lacking or also spreading mold-like, branched, septate, hyalin or dark colored. Conidiophores either entirely lacking or formed as short lateral branchlets. Conidia either formed by the entire thread breaking up into bud-like spores, or budding irregularly on the mycelium and occurring as bud-colonies by further multiplication or as long chains of spores produced by growth of the basal cells of mycelial branches. Conidia, hence, bound in chains which break apart in single cells or in short pieces, usually black, brown, olive-green or gray, globose or elongate or ovate or almost spindle-shaped, smooth or rough to warty.

KEY TO THE SPECIES OF THE GENUS TORULA

a. Conidia small, globose 2.5–3μ in diameter — 1. *T. convoluta*
aa. Conidia larger, more than 8μ in diameter
  b. Conidia dark green, 10–22 × 8–10μ — 2. *T. lucifuga*
  bb. Conidia brownish-black, 14μ in diameter — 3. *T. allii*

1. *Torula convoluta* Harz

Mycelium prostrate, richly branching, septate. Conidiophores erect, very short, unbranched or branched, septate, terminating in long chains of conidia which are produced basipetally, with enrolled ends. Conidia globose, black, translucent, 2.5–3μ in diameter.

From soil: Canada (15)

2. *Torula lucifuga* Oudemans

Turf, cushion-like, at first straw colored, later by the occurrence of irregular fruiting areas, becoming mottled above; reverse greenish-black to black. Hyphae prostrate, hyalin, nonseptate, bent and curved, branched, finally divided into chains by increasingly numerous septa. Conidial chains made up of dark green, globose to ellipsoid or elongate, conidia with granular interiors, 10–22 × 8–10μ.

From soil: England (11)
United States: New York (67)

3. *Torula allii* (Harz) Saccardo

Hyphae prostrate; septate, richly branched, with, here and there, short hyalin lateral branchlets, on which the conidial chains occur. Conidia from five to ten in slightly curved chains, becoming larger at the tip, 14μ in diameter, globose, hyalin at first, becoming brownish-black.

From soil: England(11)

Species of uncertain position:
Torula grisea Szilvinyi
Torula lanosa Szilvinyi
Torula rubefaciens Szilvinyi
From soil: Austria (131)

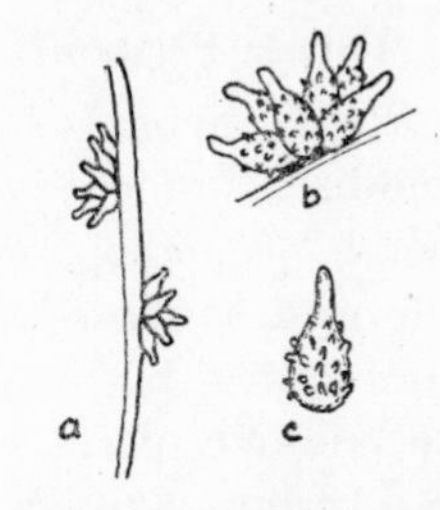

Fig. 93. **Echinobotryum.** *a* habit; *b* conidial mass; *c* conidium (after Lindau).

(5) **Echinobotryum** Corda

Sterile hyphae very little developed, continuous or little branched, hyalin. Conidia ovate or flask-shaped, smooth or somewhat spiny, formed at the tips of mycelial threads or short lateral branches, clustered or almost in heads, black.

KEY TO THE SPECIES OF ECHINOBOTRYUM

a. Conidia 10–12μ long	
b. Conidia on very short stalks on the conidiophore	1. *E. laeve*
bb. Conidia sessile on the conidiophore	2. *E. atrum*
aa. Conidia 4μ long	3. *E. subterraneum*

1. *Echinobotryum laeve* Saccardo

Turf loose, extended, black. Conidia-carrying threads unbranched or with very short branches, slightly septate, hyalin. Conidia at the ends of the branches in loose, head-like clusters, ovate or almost spindle-shaped, narrowed toward the tip and somewhat pointed, truncate at the base, with very short hyalin stalks, smooth, smoke-colored, paler at the tip, 12 × 6–7μ.

From soil: England (11)

2. *Echinobotryum atrum* Corda

Turf broad, black. Conidiophores of short branched hyphae not evident, septate, bright brown. Conidia inverted pear-shaped, drawn to a point at the tip, gathered into star-shaped knots, finely echinulate, brown, paler at the tip, 10–12 × 6–8μ.

From soil: England (11)

3. *Echinobotryum subterraneum* Raillo

Mycelium hyalin, densely floccose, sparsely septate, branched. Conidiophores slightly differentiated from vegetative hyphae. Conidia single or rosettes, pear-shaped with smooth walls, 3–4 × 4μ.

From soil: U. S. S. R. (108)

(6) **Periconia** (Tode) Bonorden

Sterile hyphae creeping, abundant, scarcely transparent. Conidiophores erect or reclining, unbranched, brown, more or less swollen at the apex where the conidia are borne, seldom with short branchlets at the apex. Conidia borne singly, globose or ovate, brown.

KEY TO THE SPECIES OF THE GENUS PERICONIA

a. Colony brown — 1. *P. byssoides*
aa. Colony gray-green — 2. *P. felina*

1. *Periconia byssoides* Persoon

Conidiophores grouped to form a colony, thread-like, rather rigid, septate, brown, light colored at the apex, about 1 mm. in height. Conidial heads globose, firm. Conidia globose, rather large, finely echinulate, dark brown, 5–7μ in diameter.

From soil: Canada (16), China (81)
United States: Idaho (106)

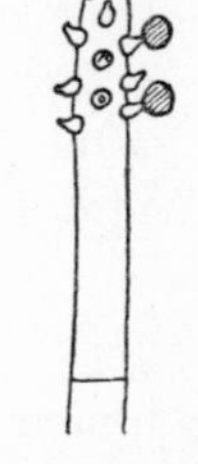

Fig. 94. **Periconia.** conidiophore and conidia (after Lindau).

2. *Periconia felina* E. Marchal

Turf thick, wide-spread, indefinite, at first white, then gray-green. Sterile hyphae prostrate, loosely branched. Conidiophores erect, somewhat curved, often buhsy, unbranched or dichotomous at the base and then with one or two septa, thickened at the tip, 40–60 $\times$ 2–3μ. Conidia in a single head 12–18μ in diameter, held in a mucus, ovate, brown, slightly translucent, with dark content, 4–6 $\times$ 3–3.4μ.

From soil: Canada (16) England (11)

(7) **Synsporium** Preuss

Sterile hyphae creeping. Conidiophores forming a turf, erect, septate, branched. Conidia oblong, borne in heads at the apex of the conidiophores, brown, nonseptate. (The genus is Acrotheca with branched conidiophores).

A single species treated.

1. *Synsporium biguttatum* Preuss

Colonies spreading, at first dark, then coal black. Conidiophores creeping, then erect, branched, brown. Conidia large, ovate, at first hyalin, then black-brown, usually with an oil drop at each end.

From soil: Austria (66), England (37)

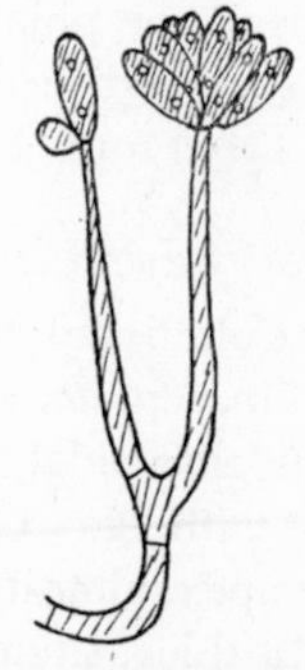

Fig. 95. **Synsporium.** conidiophores and conidia (after Lindau).

(8) **Stachybotrys** Corda

Mycelium creeping, spreading over the substratum, septate, branched, hyalin, or slightly colored. Conidiophores arise as branches of the mycelium, erect, variously branched, septate, dark colored or almost hyalin, bearing at the apex of the main stalk and branches small sterigma-like cells (phialides), which are nonseptate, hyalin or slightly dark colored, and either borne in whorls or arise irregularly below the point of the branch, appearing singly or more or less grouped. Conidia borne singly on the points of the phialides round or elongate, black, smooth or echinulate.

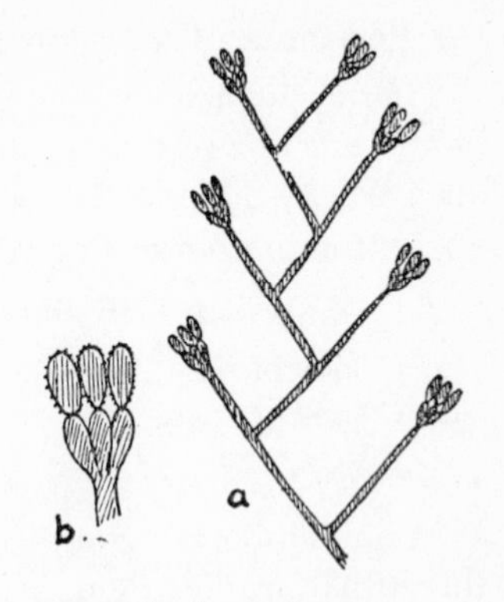

FIG. 96. **Stachybotrys.** *a* conidiophore; *b* conidial head (after Lindau).

KEY TO THE SPECIES OF THE GENUS STACHYBOTRYS

a. Branching of conidiophores regularly alternate 1. *S. alternans*
aa. Branching of conidiophores not regularly alternate
  b. Conidiophores short, up to 75 μ long, conidia smooth 2. *S. atra*
  2a. (*S. cylindrospora*)
  bb. Conidiophores up to 1 mm. long, conidia echinulate 3. *S. lobulata*

1. *Stachybotrys alternans* Bonorden

Sterile hyphae creeping, branched, sparsely septate, black-brown, 3–5 μ thick, with abundant papillae. Conidiophores erect, gray or almost hyalin, 3.5 μ thick, mostly unbranched, branching when present regularly alternate, not swollen at the apex, with crowded, inverted egg-shaped or club-shaped phialides, gray or hyalin, 10 × 4–5 μ. Conidia borne at the ends of the phialides, elliptical to ovate, with or without two oil drops, black, roughened, 8–12 × 5–7.5 μ.

From soil: United States: Porto Rico (147)

2. *Stachybotrys atra* Corda

Colonies spreading, at first hyalin, becoming black with age; mycelium hyalin, septate, 5–6 μ thick, with branches almost at right angles, and with oval, ellipsoidal or globose chlamydospores up to 12 μ in diameter; articulate with age. Conidiophores arise from aerial mycelium fuliginous near the apex, almost hyalin near the base, branched, septate, 65–74 μ long × 2–4 μ thick, slightly alternate toward the apex, bearing on the summit a whorl of papillate phialides; phialides 10–12 × 4.5–5 μ. Conidia single, smooth, elliptical, usually with acute ends and mostly with two oil drops, slightly colored when young to fuliginous and black when mature.

From soil: Egypt (115)

United States: Illinois (129), New York (67)

2a. *Stachybotrys cylindrospora* Jensen

Jensen's description of *S. cylindrospora* so closely resembles that of *S. atra*, it seems probable that they should be considered synonymous.

Colonies round, thin, diffuse, becoming black with age; mycelium branched, septate, hyalin, 0.5–3μ thick. Conidiophores hyalin at base, fuliginous toward apex, branched, septate, attenuate toward tip, 40–65μ high, bearing on the summit from three to nine phialides; phialides subclavate, with or without short papillae, 8–11 × 4–5μ; conidia borne singly, smooth, subcylindrical to sometimes ovate, 6–16 × 3.8–5μ, hyalin when young, becoming fuliginous with age.

From soil: United States: Colorado (74), Iowa (101), New York (67)

3. *Stachybotrys lobulata* Berkeley

Colonies broadly spreading, black, dense; hyphae creeping, almost hyalin, septate. Conidiophores arise from aerial mycelium, erect, up to about 1 mm. long × 3–4μ thick, septate, almost hyalin at the base, darker toward the apex, with few branches, 30–35μ long, which are granular within. Phialides borne at the apex of the branches, usually three to five, black, finely warty, 11–12 × 6μ, each bearing a conidium. Conidia black, finely warty or echinulate, round to elliptical, 9–12 × 7–8μ.

From soil: Egypt (115)

United States: Iowa (3)

Species of uncertain position:

Stachybotrys lunzinense Szilvinyi

From soil: Austria (131)

(9) **Gliobotrys** von Höhnel

Sterile hyphae sparse, creeping. Conidiophores hyalin, erect, scarcely swollen at the tip, carrying a thick crown of short hyalin, simple branchlets placed in a whorl. Conidia olive-green, ellipsoid, incased in slime and forming a round head.

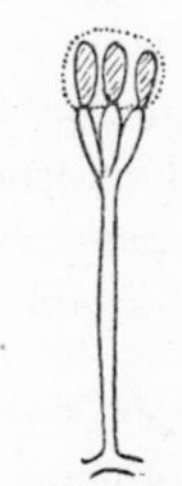

Fig. 97. **Gliobotrys.** conidiophores and conidia.

A single species treated.

1. *Gliobotrys alboviridis* von Höhnel

Conidiophores hyalin with one to five crosswalls, usually unbranched,

12μ long, 5–8μ thick, a little thicker at the tip, carrying at the end five to eight simple nonseptate cylindric branches, 10–12μ long. Conidia egg-shaped, bright olive or green, 4–6μ long, 3–4.5μ thick, incased in slime and forming a spherical head.

From soil: Canada (16)

United States: Colorado (74), Iowa (101)

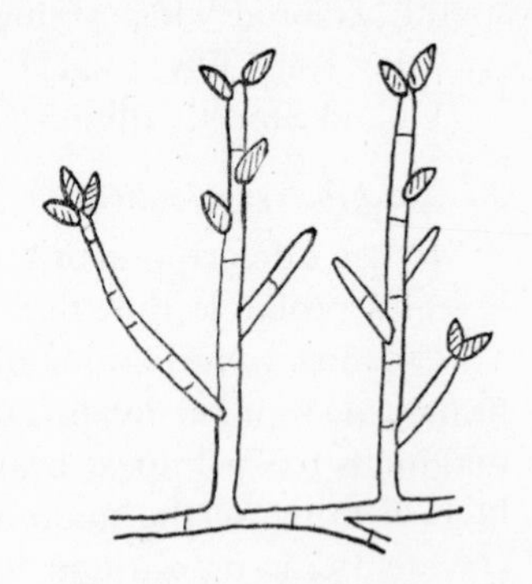

Fig. 98. **Trichosporium.** conidiophores and conidia (after Bonorden).

(10) **Trichosporium** Fries

Hyphae prostrate, irregularly branched, brown or pale in color. Conidia terminal or lateral on the hyphae, globose or ovate, smooth or slightly rough, brown or seldom almost hyalin.

KEY TO THE SPECIES OF THE GENUS TRICHOSPORIUM

a. Conidia globose	1. *T. nigricans* *f. lignicola*
aa. Conidia ovate	
b. Conidia greenish-brown 10–12 × 8μ	2. *T. murinum*
bb. Conidia brown, 8–11 × 6–7μ	3. *T. fuscum*

1. *Trichosporium nigricans* Saccardo

Effuse black, sub-velvety or somewhat dusty; hyphae prostrate, anastomosing at the base simple or furcate, distinctly septate, thread-like, very slightly inflated at the apex, smoky-gray. Conidia inserted at the tips, abundant, globose, 6.5–8μ in diameter, smoky-black with a paler guttula.

form *lignicola*

Differs from the typical species by its slightly larger conidia, 10–12μ in diameter, black, guttulate.

From soil: United States: Illinois (129)

2. *Trichosporium murinum* (Ditmar) Saccardo

Producing a turf of widely spreading, floccose, gray-green, finally brownish-black hyphae dichotomously or irregularly branched, septate, greenish-brown. Conidia occurring at the tip of the branches, almost spike-like, ovate, pointed at their ends, with one oil drop, green-brown, 10–12 × 8μ.

From soil: England (11)

3. *Trichosporium fuscum* (Link) Saccardo

Hyphae brown, forming a rather thick felt, much branched, septate. Conidiophores lateral, short, tapering. Conidia terminal, loosely racemose, brown, 8–11 × 6–7μ.

From soil: India (49)

(11) **Nigrospora** Zimmermann

Hyphae creeping, at first hyalin, later dark, ultimate branchlets bearing jar-shaped conidiophores either laterally or terminally. Conidia solitary, subglobose, smooth. This genus approaches the genus Pachybasium among the Botrytideae and Rhinocladium among the Trichosporieae.

A single species treated.

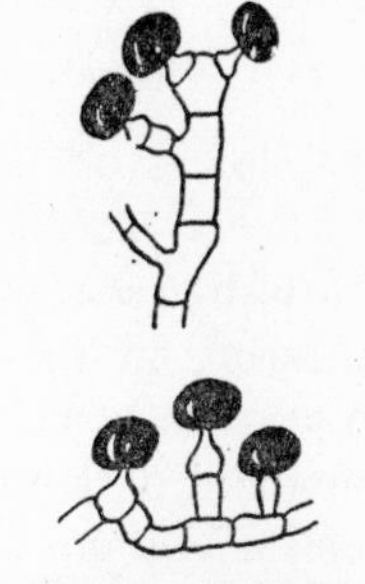

FIG. 99. **Nigrospora.** conidiophores and conidia (after Durrell).

1. *Nigrospora sphaerica* (Saccardo) Mason

Both sterile and fertile hyphae creeping, at first hyalin, then dark; sterile hyphae septate 18μ in diameter; fertile hyphae septate, 4μ in diameter, much branched bearing swollen jar-like cells terminally and laterally on which are borne single the subspherical, smooth, black conidia. Conidia one-celled, 11–14μ in diameter.

From soil: England (36)

United States: New Jersey (147)

(12) **Acremoniella** Saccardo

Hyphae creeping or slightly ascending, unbranched or branched, hyalin or dark colored, bearing here and there short side branches, which bear conidia on the points. Conidia round or ovate, borne singly, brown, one-celled.

A single species treated.

FIG. 100. **Acremoniella.** conidiophore and conidia (after Mason).

1. *Acremoniella fusca* Kunze var. *minor* Corda

Colonies spreading, greenish-brown. Hyphae yellowish-brown, forked, with irregularly placed side-branches. Conidia round, pale ochre color, transparent, finally olive-green.

From soil: Switzerland (38)

United States: Alaska (147), Iowa (50), North Dakota (147)

(13) **Monotospora** Corda

Sterile hyphae creeping, branched, septate, usually plain. Conidiophores erect, straight, septate, unbranched, rather long, brown. Conidia single, apical, globose or subglobose, brown, one-celled.

KEY TO THE SPECIES OF THE GENUS MONOTOSPORA

a. Conidia 6.5–9.5μ ... 1. *M. brevis*
aa. Conidia 9–16μ ... 2. *M. daleae*

1. *Monotospora brevis* (Gilman and Abbott) Mason
Syn. *Acremoniella brevis* Gilman and Abbott

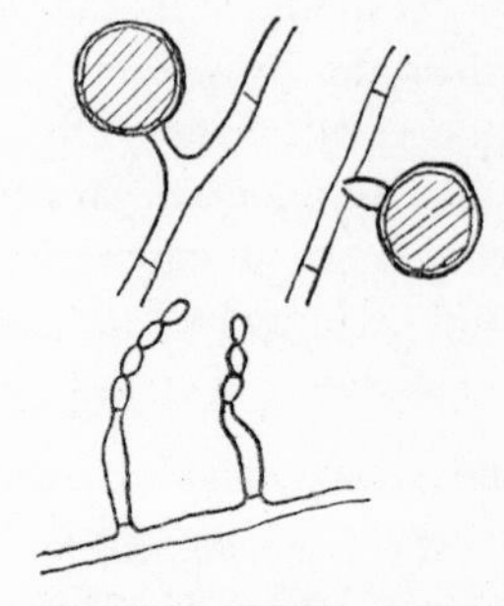

FIG. 101. **Monotospora.** conidiophores and conidia (after Mason).

Colonies on Czapek's agar spreading, felty or closely floccose; surface dark grayish to brownish-green, with whitish superficial hyphae appearing in the center of old colonies. Reverse green-black to black. Colonies consist of abundant, dark colored, verrucose, multiseptate aerial mycelium, bearing very short scattered conidiophores, 2–15μ long, verrucose, dark colored. Conidia borne terminally, pyriform to subglobose, delicately rugulose, dark brown, 6.5–9.5 × 5–6μ, one-celled, sometimes almost sessile. Chlamydospores common.

From soil: United States: Louisiana (50)

2. *Monotospora daleae* Mason

Hyphae septate, hyalin, 4μ in diameter, bearing masses of yellow-brown conidia. Conidiophores (short lateral branches) subcylindrical or dilated upwards, with a single apical spore. Spores typically globose with double wall, granular, yellow-brown, black by reflected light, 9–16μ in diameter.

From soil: England (36)

(14) **Haplographium** Berkeley and Broome

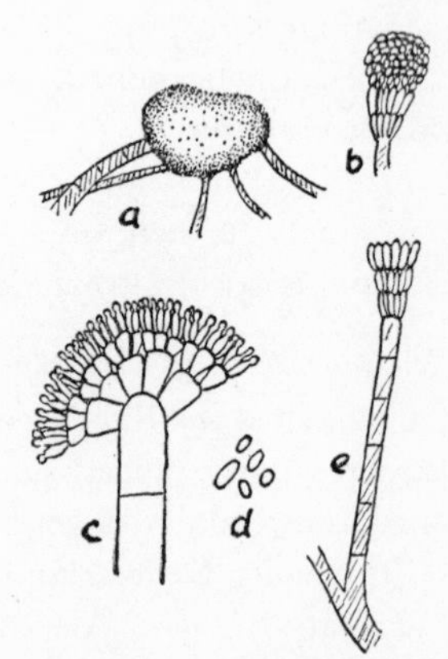

FIG. 102. **Haplographium.** *a* habit; *b*, *c*, *e* conidial heads; *d* conidia (after Mason).

Sterile hyphae creeping and inconspicuous. Conidiophores erect, unbranched, septate, brown, and at its apex forming a head of very small irregular buds from which the spores are abstricted. Conidia globose to elongate, green, brown or almost hyalin, one-celled.

KEY TO THE SPECIES OF HAPLOGRAPHIUM

a. Conidia olive-green	1. *H. chlorocephalum*
aa. Conidia honey-yellow	2. *H. bicolor*
aaa. Conidia hyalin	3. *H. fuscipes*

1. *Haplographium chlorocephalum* (Fresenius) Grove

Conidiophores gregarious, erect, simple or with one or more tolerably thick phialides at the tip, 210–250 × 8–9μ. Conidia terminal in chains, globose, elliptic, sometimes, somewhat angular, olive-green, 4–6μ in diameter.

From soil: U. S. S. R. (108)

2. *Haplographium bicolor* Grove

Conidiophores widely scattered, forming a turf, erect, stiff, septate, somewhat swollen below, brownish-black, opaque, paler above and rounded at the apex, 250–300μ long, 8μ in diameter. Phialides numerous at the top, arranged radially, in three series, forming a head up to 25μ long. Conidia elongate or ovate, somewhat pointed, 4–5μ long, enveloped in mucus, and forming an irregular ovate mass, pale honey-yellow.

From soil: Canada (16)

3. *Haplographium fuscipes* (Preuss) Saccardo

Turf delicate, scarcely visible, indefinite, brown. Mycelium in a black-brown layer. Conidiophores erect, unbranched, septate, below black-brown almost opaque, above translucent. Phialides very short, simple, hyalin, united into a brush-like head. Conidial chains equal, mealy, white. Conidia globose, hyalin.

From soil: Canada (15)

(15) **Hormodendrum** Bonorden

Sterile hyphae creeping, branched, septate. Conidiophores erect, septate, brown, variously branched or only little branched. Conidial chains acrogenous on the branches (often all the branches are borne on a single main stipe). Conidia globose or ovate, olive-green or brown, one-celled.

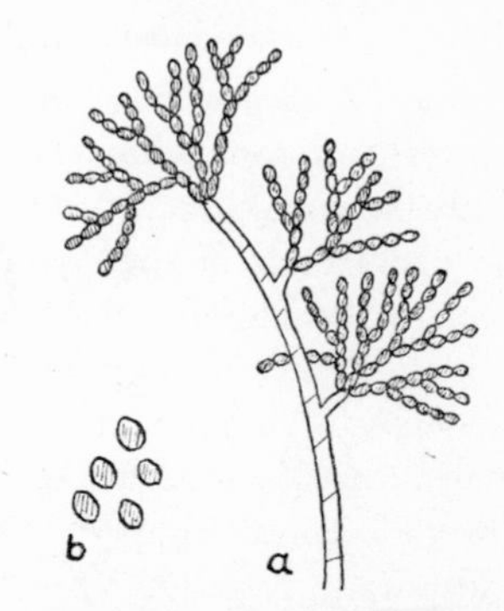

FIG. 103. **Hormodendrum.** *a* conidiophore; *b* conidia (after Lindau).

KEY TO THE SPECIES OF THE GENUS HORMODENDRUM

a. Colonies olive-green	
b. Conidia less than 7μ in length	
c. Conidia $3–6 \times 2.5–3.6\mu$	1. *H. cladosporioides*
cc. Conidia $5.5–7 \times 3.5–4.5\mu$	2. *H. resinae*
bb. Conidia more than 7μ in length	
c. Conidia $8–12 \times 4–5\mu$	3. *H. olivaceum*
cc. Conidia $4–12 \times 2–4\mu$	4. *H. viride*
aa. Colonies not olive-green	
b. Colonies brown	5. *H. hordei*
bb. Colonies gray	6. *H. pallidum*
bbb. Colonies black	7. *H. nigrescens*

1. *Hormodendrum cladosporioides* (Fresenius) Saccardo

Colonies dark olivaceous-green, round, dense. Conidiophores erect, branched, $100–200\mu$ long, olivaceous, toward the apex gradually attenuate, ultimate branches copiously dividing with predominant tendency to dichotomy, septate, articulate above. Conidia cylindrical to broadly oval, olivaceous, smooth, $3–6 \times 2.5–3.6\mu$, continuous or inferior ones rarely septate.

From soil: Austria (66), Canada (15), China (81), Czechoslovakia (91), Denmark (68), Egypt (115), England (11) (36) (37), U. S. S. R. (108),

United States: California (147), Colorado (75) (76), Illinois (129), Iowa (3) (101), Louisiana (147), Michigan (51) (52), New Jersey (147), New York (67), Oregon (146) (147), Texas (147)

2. *Hormodendrum resinae* Lindau

Turf effuse, woolly, brownish-green, rather floccose. Sterile hyphae prostrate or somewhat raised, $4–4.5\mu$ in diameter, little branched, distantly septate, greenish-brown, transparent. Conidiophores formed at the distal ends of the hyphae or infrequently as long lateral branches, erect, with closely placed septa, sometimes articulate, green-brown, scarcely lighter toward the tips and scarcely tapered, about 4μ in diameter below, 3.5μ above; branching confined to the tip of the conidiophores, alternating or numerous to form a head, conidia in short chains. Conidia ellipsoid spindle-shaped, bluntly pointed at both ends, green-brown, transparent, $5.5–7 \times 3.5–4.5\mu$.

From soil: Austria (131)

3. *Hormodendrum olivaceum* (Corda) Bonorden

Colonies olive-green, spreading. Conidiophores erect, unbranched except at the apex, olive-green, 75–200μ long, borne as lateral branches of the sterile hyphae. Conidial chains short. Conidia elliptical to short cylindric, 8–12 × 4–5μ, with intermediate cells of the chain swollen and terminal cells often much smaller and globose.

From soil: China (81)

United States: Iowa (101), Texas (92)

4. *Hormodendrum viride* (Fresenius) Saccardo

Colonies gray-green, small. Conidiophores arising from prostrate mycelium, erect, septate, branched at the tip and ending in forked conidial chains. Conidia long or egg-shaped, frequently with two oil drops, green, smooth, 4–12 × 2–4μ (Lindau (107) gives the conidial measurements as 7–8μ long).

From soil: Canada (16)

United States: Iowa (101)

5. *Hormodendrum hordei* Bruhne

Colonies brown at maturity, circular, dense; mycelium brown, septate, branched, 3–6μ thick; conidiophores simple septate, ascending or erect, 50–100μ. Conidia various, some cylindrical with ends rounded, truncate, or subattenuate, others ellipsoidal, ovate, or subglobose, regular or somewhat angular; with age many become once septate and verrucose, 4–14 × 3–5μ, chains of conidia short.

From soil: Austria (131)

United States: Alaska (147), New York (67)

6. *Hormodendrum pallidum* Oudemans

Colonies orbicular, gray, not plainly zonate. Conidiophores erect, very light gray, upward dendroidly branched; primary and even secondary branches decussate, each succeeding branch and branchlet shorter than preceding, consisting of single cells, constricted at septa. Conidia variable in size, 12–20 × 5–8μ.

From soil: China (81), Holland (100)

United States: Colorado (74)

7. *Hormodendrum nigrescens* Paine

Colonies somewhat elevated, at first hyalin, becoming olive-green, and finally black beneath with white surface; the hyalin mycelium appears slightly floccose; margin 2 mm. or more wide, hyalin. Sterile hyphae arise at the apex of the colony as fine bristle-like tufts above the conidiophores. Conidiophores originating in the substratum, smoky,

300–400μ long $\times$ 4.5–5μ in thickness, dendroidally branched, erect. Conidia green, subspherical to ellipsoidal or spindle-shaped, seldom pointed at the ends, 4–10 $\times$ 2.5–4μ.

From soil: United States: Iowa (101)

Species of uncertain position:
Hormodendrum bergeri Szilvinyi
Hormodendrum cleophae Szilvinyi
From soil: Austria (131)

(16) **Mesobotrys** Saccardo.

Conidiophores erect, dark colored, with whorls of fertile branchlets arising in the middle portion; apex sterile. Conidia borne terminally on the branchlets; ovate, hyalin.

FIG. 104. **Mesobotrys.** *a* conidiophore and conidia.

1. *Mesobotrys simplex* Gilman and Abbott

Colonies on Czapek's agar dark olive to brown-green slowly spreading, velvety to subfloccose, consisting of both submerged and aerial hyphae; reverse greenish-black. Conidiophores arise in groups from aerial hyphae, with aerial and submerged stolons surrounding the point of origin, swollen at the base, and tapering to a pointed apex; brown in color; sparsely and indistinctly septate, bearing on the middle portion whorls of fertile branchlets, which may also be once branched. Conidiophores 150–350μ long $\times$ 5–6μ in diameter near the base; branches up to 100μ long. Conidia borne singly and terminally on the branches, oval to ovate, light brown-green, smooth, 3.5–5 $\times$ 3–3.5μ.

From soil: United States: Louisiana (50)

(17) **Botryotrichum** Saccardo and Marchal

Sterile hyphae loosely bush-like, ascending, unbranched, septate, gray. Conidiophores produced at the base of the sterile hyphae, irregularly branched, hyalin. Conidia acrogenous, globose, hyalin, one-celled.

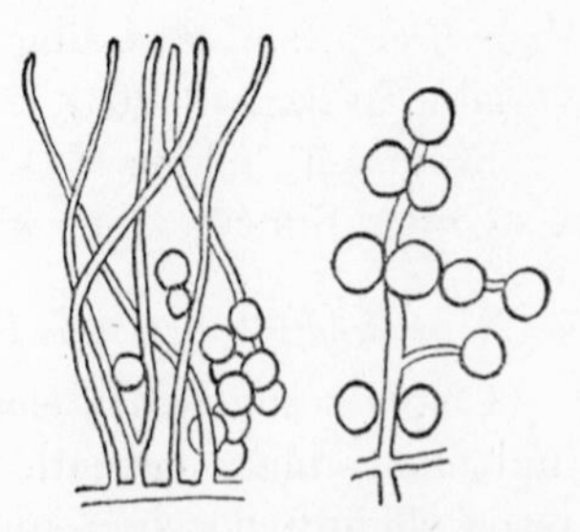

FIG. 105. **Botryotrichum.** conidiophores and condidia (after Marchal).

KEY TO THE SPECIES OF THE GENUS BOTRYOTRICHUM

a. Reverse of colonies yellow-brown; conidia 11–14μ — 1. *B. piluliferum*
aa. Reverse of colonies nearly black; conidia 13–22μ — 2. *B. atrogriseum*

1. *Botryotrichum piluliferum* Saccardo and Marchal

Colonies dark gray above, yellow-brown below. Sterile hyphae turf-like, bushy, slightly curved, smooth or somewhat roughened, slightly thickened at the base, 200–250 × 3.5–5μ. Fertile hyphae branched, growing between the sterile hyphae, prostrate. Conidia terminal, globose, hyalin, 11–14μ in diameter.

From soil: Canada (15)

2. *Botryotrichum atrogriseum* van Beyma

Colonies dark gray on surface, nearly black below. Spores globose, hyalin, 13–22μ in diameter.

From soil: Canada (15)

(18) **Stachylidium** Link

Sterile hyphae prostrate, sparingly present. Conidiophores erect, branching approximately in whorls, dark colored. Conidia formed terminally on the branches, and held in a head, globose to ovate, usually hyalin.

A single species treated.

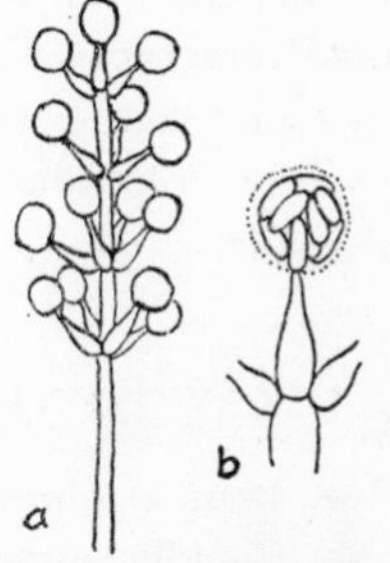

Fig. 106. **Stachylidium.** *a* conidiophore; *b* conidial head. (after Saccardo).

1. *Stachylidium extorre* var. *majus* Saccardo

Widespread, turf-like, brown-gray. Conidiophores erect, cylindric, slightly thickened at the base, tapering toward the tip, 50–200 × 4–5μ, septate, sooty-gray, branching in whorls to the end. Branches in three- to six-membered whorls, smaller toward the tip, unbranched, at times branching in whorls, paler than the primary conidiophores. Conidia terminal, and at first, forming heads enveloped in mucus, transparent, 6–10μ in diameter, later dissolving in moisture, elongate, hyalin 3–4 × 1.5μ.

From soil: England (11)

(19) **Dicoccum** Corda

Hyphae creeping, branched, septate, dark colored. Conidia terminal on short, erect side branches, elongate or short clavate, two-celled, sometimes biscuit-form, dark colored.

A single species treated.

Fig. 107. **Dicoccum.** conidiophores and conidia (after Lindau).

1. *Dicoccum asperum* Corda
Syn. *Trichocladium asperum* Harz

Colony floccose, white at first, becoming dark to almost black. Mycelium consists of hyalin, branched, sparsely septate, yellowish hyphae, 2.6–3.5μ thick. Conidiophores arise as short side branches, 3–12μ long. Conidia at first hyalin, then brown to black, oval, two-celled; upper cell spherical, brown, thick-walled, spiny; lower cell usually smaller, spiny; 20–22 × 9–13μ.

From soil: Canada (15), U. S. S. R. (108)
United States: Colorado (147), New Jersey (146)

(20) **Cladosporium** Link

Hyphae creeping, septate, on the surface or in the substrate. Conidiophores almost erect, branched, and floccose, often forming a turf, olive colored. Conidia globose and ovate, at first one-celled, then usually with a cross-wall, usually greenish, terminal and then pressed to the side.

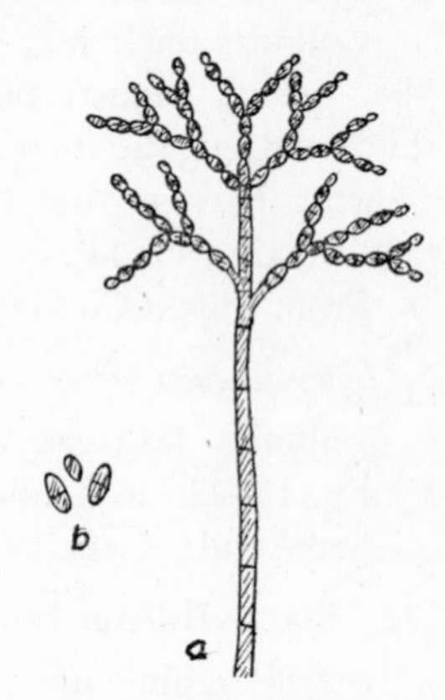

FIG. 108. **Cladosporium.** *a* conidiophore; *b* conidia (after Lindau).

KEY TO THE SPECIES OF THE GENUS CLADOSPORIUM

a. Conidiophores pale green — 1. *C. epiphyllum*
aa. Conidiophores brown or olive-green
b. Conidia, two-celled, dark brown — 2. *C. lignicola*
bb. Conidia, one- to four-celled, smoky brown — 3. *C. herbarum*

1. *Cladosporium epiphyllum* Persoon

Colonies greenish-black, large, thick; conidiophores at first erect, then falling, pale green; conidia very numerous, soon falling from the chain, at first one-celled, then two- to more-celled, olive-green, 10–22μ long × 4–6μ thick.

Waksman (146) gives: conidia one- or two-celled, 10–14 × 3.8–5.2μ.

From soil: England (36) (37)
United States: California (146), New Jersey (146), Texas (147)

2. *Cladosporium lignicola* Corda

Turf extended, felty, black, 1–1.5 cm. in diameter. Conidiophores scarcely branched, brown. Conidia long, ellipsoid, usually in chains, two-celled, dark brown and usually opaque, 8–10 × 5–6μ.

From soil: England (11)

3. *Cladosporium herbarum* (Persoon) Link

Turf matted, yellow-green, later black-green; conidiophores erect, little branched, septate, brown or olive-green, 5–10μ in diameter; or various heights up to ⅓ mm. Conidia terminal, by extension of the tip falsely lateral, on short knee-like swellings, single or at times in chains, of various shapes, elongate, oval, and then usually one-celled, or cylindrical ellipsoid and then with one- to four-septa, smoky brown or olive-green, slightly constricted at the septa, with a finely granulate or spiny wall, of very different diameter and length.

From soil: Austria (66) (131), Canada (15), England (36) (37), India (21) (49), U. S. S. R. (108)

United States: Iowa (3), Louisiana (2), New Jersey (146), Texas (152)

(21) **Diplococcium** Grove

Hyphae creeping, thread-like, branched, septate, dark. Conidiophores erect or ascending, septate, branched like the mycelium, dark. Conidia in terminal, long chains, usually biscuit-shaped, sometimes longer, dark, two-celled.

A single species treated.

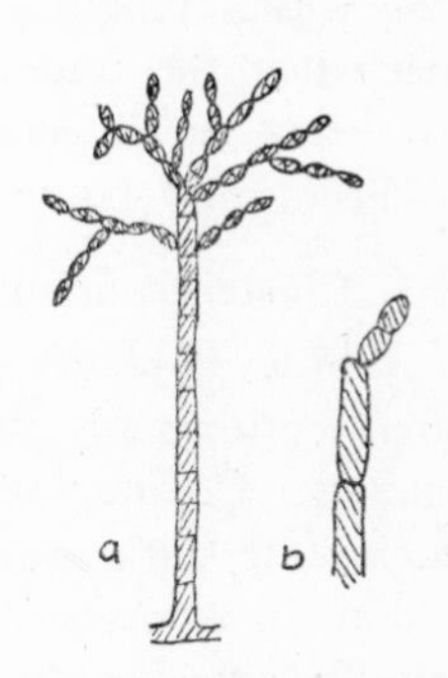

Fig. 109. **Diplococcium.** *a* conidiophore; *b* conidium (after Lindau).

1. *Diplococcium resinae* (Corda) Saccardo

Turf extended, usually somewhat floccose. Mycelium rather sparingly branched, septate, transparent, brown, 3.5–4.5μ in diameter, black. Conidiophores breaking up into spores at their tip, short. Conidia in chains, sometimes forming a head, ellipsoid, brown, two-celled, constricted, smooth or warty, 6.5–9 × 4.5–5μ.

From soil: England (11)

(22) **Scolecobasidium** Abbott

Hyphae creeping, septate. Conidiophores arising as short side branches from aerial hyphae, not erect, nonseptate. Conidia elongate, two-celled, smooth, light olivaceous to almost hyalin, borne single on short, terminal, thread-like phialides; one to three phialides on each conidiophore.

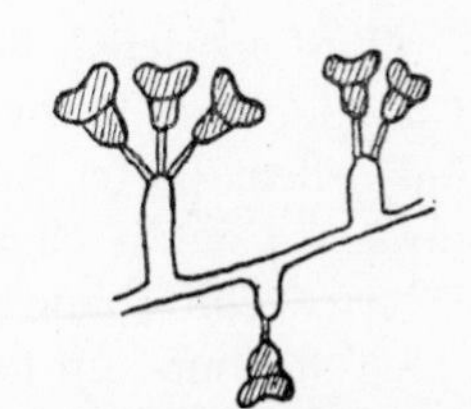
Fig. 110. **Scolecobasidium.** conidiophore and conidia (after Abbot).

KEY TO THE SPECIES OF THE GENUS SCOLECOBASIDIUM

a. Conidia T- or Y-shaped — 1. *S. terreum*
aa. Conidia oval, constricted at center — 2. *S. constrictum*

1. *Scolecobasidium terreum* Abbott

Cultivated on dextrose bean agar, colonies round, 2–3 cm. in diameter; surface velvety, olivaceous; reverse greenish-black. Hyphae light olivaceous, septate. Conidiophores 5.0–8.0μ long × 2.0–2.5μ wide. Phialides 0.5–1.0μ long. Conidia T- or Y-shaped, two-celled, light olivaceous to almost hyalin, smooth, 4.0–12.0μ long × 2.0–2.5μ wide. Perithecia or sclerotia not observed.

From soil: United States: Louisiana (4)

2. *Scolecobasidium constrictum* Abbott

Cultivated on dextrose bean agar, colonies round, seldom more than 3 cm. in diameter; surface fuscous, olivaceous, reverse greenish-black. Hyphae light olivaceous, septate. Conidiophores 5.0–8.0μ long × 2.0–2.5μ wide. Phialides 0.5–1.0μ long. Conidia two-celled, slightly constricted at the center, smooth, light olivaceous, 6.0–12.0μ long × 2.5–4.0μ wide. Perithecia or sclerotia not observed.

From soil: United States: Louisiana (4)

(23) **Clasterosporium** Schweinitz

Sterile hyphae prostrate, septate, branched, often entirely or almost entirely lacking, dark colored. Conidiophores erect, short, septate or nonseptate, dark colored. Conidia terminal, single, seldom in a bush, ovate, elongate, cylindrical or spindle-shaped, with two or more cross-walls and slightly or not constricted at the septa, dark colored, often not all the cells of like-colors.

A single species treated.

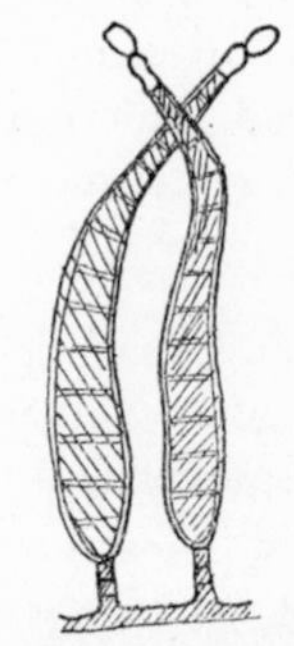

Fig. 111. **Clasterosporium** conidiophores and conidia (after Lindau).

1. *Clasterosporium carpophilum* (Leveille) Aderhold

Turf restricted, black, on a submerged stroma. Conidiophores bushy, simple or branched, usually one-celled, shorter than the spores, hyalin or yellow-brown. Conidia elongate, rounded at their ends, seldom club-shaped or inverted club-shaped, brownish-yellow to brownish-black, usually with three- or four-, up to seven septa, not or slightly constricted at the septa in age, 23–62 × 12–18μ.

From soil: England (11)

(24) **Curvularia** Boedijn

Mycelium branched, septate, subhyalin or brown; conidiophores brown, thread-like, unbranched, septate. Conidia acrogenous, verticillate or spirally arranged, olivaceous or brown, ellipsoid or cylindrical, curved or bent, (rarely straight) three- or four-septate, one of the central cells being distinctly larger and darker than the terminal cells; germination bipolar.

KEY TO THE SPECIES OF THE GENUS CURVULARIA

a. Spores symmetrical
  b. Conidia obclavate — 1. *C. subulata*
  bb. Conidia tapering toward both ends
    c. Conidia dark olivaceous — 2. *C. tetramera*
    cc. Conidia transparent brown — 3. *C. interseminata*
aa. Spores unequilateral or curved
  b. Conidia four-septate — 4. *C. geniculata*
  bb. Conidia three-septate — 5. *C. lunata*

1. *Curvularia subulata* (Nees) Boedijn

Syn. *Helminthosporium subulatum* Nees

Colonies floccose. Conidiophores usually unbranched, straight, 117μ thick at base, 6–7μ thick at the apex. Conidia cylindric-ellipsoid, rounded at the apex, often attenuated at the base, with three- to four-septa, black- brown, 22–26 × 9–11μ.

From soil: Japan (132)

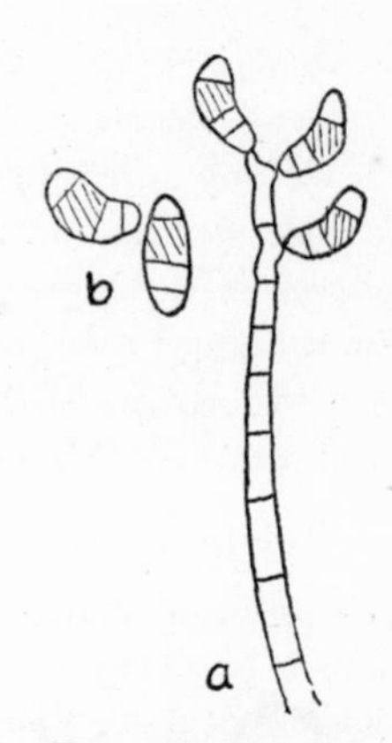

FIG. 112. **Curvularia.** *a* conidiophore; *b* conidia.

2. *Curvularia tetramera* (McKinney) Boedijn

Syn. *Helminthosporium tetramera* McKinney

Conidiophores dark olivaceous to brown, very irregular; simple or compound, septa 5–50μ apart. Conidia produced at irregular distances from the bases, chiefly four-celled, borne in clusters of 2 or 3 to 50 or more; dark olivaceous to brown, rather symmetrical in shape, tapering toward the rounded ends, 20.4–40.8 × 8.5–20.4μ. Long, simple or branched stromata produced in cultures.

From soil: Canada (15), India (49)

3. *Curvularia interseminata* (Berkeley and Ravenel)

Syn. *Helminthosporium interseminatum* Berkeley and Ravenel

Colonies broadly spreading, brown-black, velvety. Conidiophores grouped thickly together, erect, also sometimes reclining, unbranched or branched, multiseptate, bent or with geniculations, brown, transparent, sometimes swollen and with knobs at the apex, up to 500μ long × 3.5–4μ thick. Conidia borne terminally or laterally on the geniculations, slender elongate, rounded at both ends, almost always with three septa, more seldom with two or four, cells of the same size, brown, transparent, 15.5–23 × 5.5–7.5μ.

From soil: England (36)

4. *Curvularia geniculata* (Tracy and Earle) Boedijn

Syn. *Helminthosporium geniculatum* Tracy and Earle

Mycelium septate, richly branched, subhyalin to brown; hyphae 2.5–7μ in diameter. Conidiophores brown, septate, at times narrower at the base, lighter colored near the tip; 340–900μ, basal cells 2–2.5μ, at the tip 3.5–5μ in diameter. Conidia usually in a dense panicle, boat-shaped, of unequal sides or more or less strongly curved, with four septa; the third cell much larger and darker colored than the others; 19–45 × 7–14μ.

From soil: Canada (15)

5. *Curvularia lunata* (Walker) Boedijn

Syn. *Acrothecium lunatum* Walker

Colony spreading, subfloccose, dark olive-gray, reverse bluish-black; hyphae septate and much branched, olive, 3–3.6μ in diameter. Conidiophores erect, more than 100μ long × 3.6μ in diameter, unbranched. Spores borne more or less in a whorl at tip of conidiophore, three-septate, curved, brown, 18–29 × 10–8μ.

From soil: Czechoslovakia (97), Egypt (115), India (21) (49)

(25) **Helminthosporium** Link

Colonies consist of conidiophores, loose or dense, regularly or irregularly velvety, brown to black, with strict or spreading margin. Conidiophores usually arise in groups, erect and straight, sometimes reclining, usually unbranched, only seldom with small side branches, septate, geniculate at points below the conidia, brown, green-brown to black, transparent or non-transparent. Conidia terminal or lateral on the geniculations, elongate, cylindrical, clavate or obclavate, smooth, mostly rounded at both ends, or sometimes pointed at the base or at both ends, straight or bent, with more than four cross-wall, dark brown, green-brown to black, often with the end cells lighter colored.

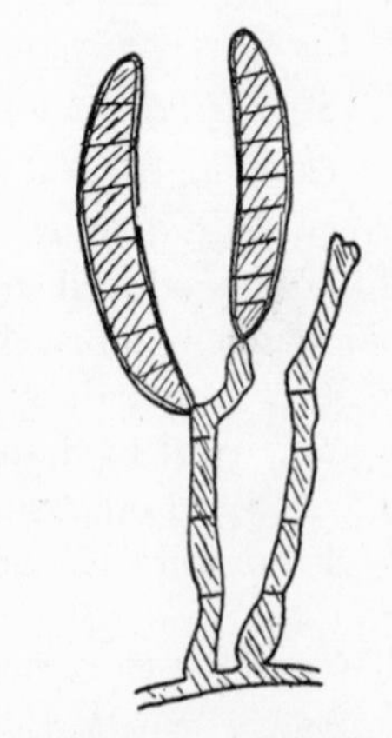

FIG. 113. **Helminthosporium.** conidiophores and conidia.

KEY TO THE SPECIES OF THE GENUS HELMINTHOSPORIUM

a. Colonies floccose. Conidia, 90–130 × 15–20μ ... 1. *H. sativum*
aa. Colonies velvety. Conidia, 40–90 × 10–15μ ... 2. *H. anomalum*

1. *Helminthosporium sativum* Pammel, King and Bákke

Colonies at first white becoming brown with spore production. Conidiophores fasciculate 150–180μ long, 6–10μ in diameter, dark reddish-brown. Conidia straight or curved, tapering toward the ends, ends rounded, olivaceous, 90–130 × 15–20μ with seven to fourteen cells.

From soil: Canada (15), Egypt (115), India (49)

2. *Helminthosporium anomalum* Gilman and Abbott

Colonies on Czapek's agar slowly but broadly spreading, at first consisting largely of submerged hyphae, but later developing aerial hyphae and conidiophores; velvety; surface greenish-black to black, reverse black; aerial mycelium dark brown, submerged mycelium dark brown, to almost black, multiseptate. On bean agar colonies become floccose and are dark brown-green in color. Conidiophores arise usually from submerged hyphae, more or less bent, and bearing a terminal group of conidia, with lateral conidia borne singly and irregularly on the geniculations; mostly 150-400μ long, brown. Conidia elongate, straight, rounded at both ends, five to twelve times septate, mostly seven, when mature 40–90μ long × 10–15μ broad.

From soil: United States: Iowa (50), Utah (50)

Species of uncertain position:
Helminthosporium lunzinense Szilvinyi.
From soil: Austria (131)

(26) **Spondylocladium** Martius

Hyphae creeping, septate. Conidiophores erect, unbranched, slightly rigid. Conidia borne in lateral whorls, spindle-shaped, usually three-celled, dark colored.

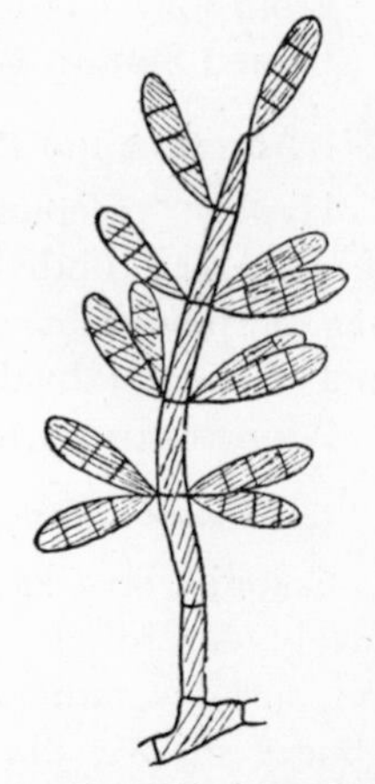

Fig. 114. Spondylocladium. conidiophore and conidia (after Lindau).

KEY TO THE SPECIES OF THE GENUS SPONDYLOCLADIUM

a. Conidia large, 25–38 × 12–15μ ... 1. *S. australe*
aa. Conidia smaller, 15–25 × 8.0–13.5μ ... 2. *S. xylogenum*

1. *Spondylocladium australe* Gilman and Abbott

Colonies on Czapek's agar spreading, floccose, aerial hyphae abundant; surface dark grayish to olivaceous-green, with an olive-gray floccose overgrowth in old cultures; reverse greenish-black to black. Conidiophores arise from aerial mycelium, erect, multiseptate, geniculate, dark colored, bearing conidia terminally and laterally, either single or in groups of two to six. Apex of conidiophore often slightly swollen. Conidiophores 80–250μ long. Conidia borne in a terminal whorl and laterally on the geniculations, 25–38 $\times$ 12–15μ, often slightly curved, smooth, three-septate. The two central cells are about twice as large as the end cells, and are dark colored, while the end cells are nearly hyalin.

From soil: China (81)

United States: Louisiana (50)

2. *Spondylocladium xylogenum* A. L. Smith

Colonies of Czapek's agar spreading, velvety, surface dark gray-green to black-green or black. Reverse black. Conidiophores arise from aerial mycelium, erect, dark colored, septate, geniculate, 75–150μ long. Conidia borne terminally and laterally on the conidiophores, very thickly on the stalks, 15–25 $\times$ 8.0–13.5μ, three-septate, curved slightly.

From soil: China (81)

United States: Louisiana (50)

(27) **Acrothecium** Preuss

Hyphae creeping, slightly raised. Conidiophores erect, undivided, dark colored. Conidia long or spindle-shaped, three- or more-celled, colored or almost hyalin forming a terminal head.

A single species treated.

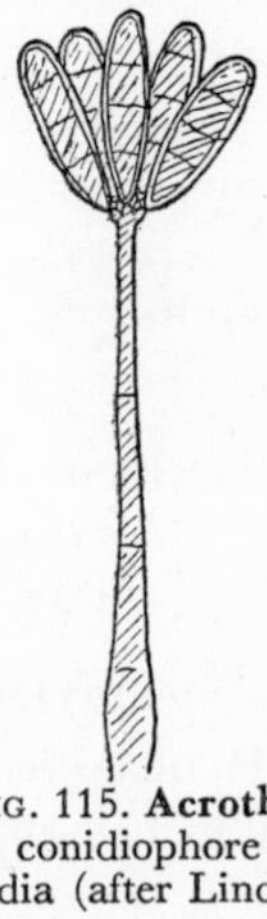

Fig. 115. **Acrothecium.** conidiophore and conidia (after Lindau).

1. *Acrothecium robustum* Gilman and Abbott

Colonies on Czapek's agar broadly spreading, velvety, consisting mostly of submerged mycelium and aerial conidiophores, with little aerial mycelium; surface black, reverse black. Conidiophores arise from submerged or aerial hyphae, multiseptate, dark colored, thick-walled, smooth, 50–150μ long, averaging about 100μ. Conidia borne typically in terminal heads, but are occasionally produced laterally on the conidiophores; apex of the conidiophores very slightly inflated. Conidia elongate, barrel-shaped, four- or five-septate, thick-walled, dark colored, smooth, 37–50 $\times$ 10–14μ.

From soil: China (81)

United States: Colorado (74) (75) (76), Louisiana (50), Texas (92), Utah (50)

(28) **Tetracoccosporium** Szabo

Conidiophores septate, branched, hyalin smoky. Conidia globose at the tips of short branches, black-brown, divided by two partitions at right angles to each other into four cells.

A single species treated.

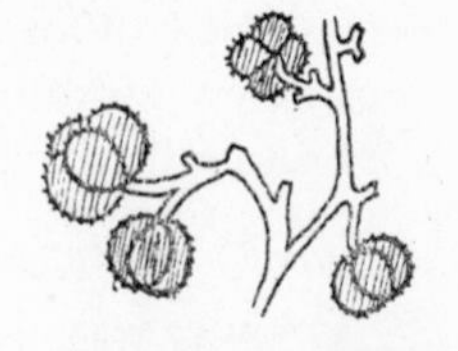

FIG. 116. **Tetracoccosporium.** conidiophore and conidia (after Lindau).

1. *Tetracoccosporium paxianum* Szabo
   Syn. *Stemphylium paxianum*

Colonies on bean agar broadly spreading, margins of the colony finger-like; surface velvety, gray to greenish-black or black; reverse uncolored. Aerial mycelium multiseptate, olivaceous, 4–6μ thick. Conidiophores arise from submerged or aerial mycelium, sometimes once branched. multiseptate, olivaceous, smooth, up to 75 $\times$ 4–5μ. Conidia borne terminally and laterally, singly or in heads of 3, 4, or 5; four-celled, cruciately septate, black, markedly verrucose. In old cultures conidial walls are so thick the septa are seen with difficulty. Conidia pear-shaped, 17–25 $\times$ 12–17μ.

From soil: United States: Idaho (106), Utah (50)

(29) **Stemphylium** Wallroth

Sterile hyphae, creeping, spreading, mostly dark colored, septate, floccose. Conidiophores arise as side branches, more or less erect, often very short, mostly unbranched and often nonseptate. Conidia borne singly and terminally, ovate, or almost club-shaped, often a little pointed, muriform, more or less dark colored to opaque.

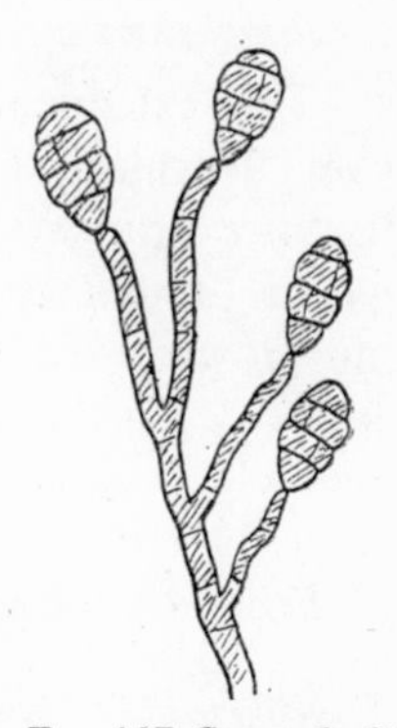

FIG. 117. **Stemphylium.** conidiophore and conidia (after Lindau).

KEY TO THE SPECIES OF THE GENUS STEMPHYLIUM

a. Colonies black, conidia smooth
  b. Pear-shaped or oval, constricted at apex — 1. *S. piriforme*
  bb. Elliptical to almost globose, not constricted at apex
    c. Spores up to 40μ long — 2. *S. botryosum*
    cc. Spores 14–25μ long — 3. *S. macrosporoideum*
aa. Colonies dark olive, conidia verrucose — 4. *S. verruculosum*

1. *Stemphylium piriforme* Bonorden

Colonies somewhat spreading, black. Hyphae freely branched, creeping, septate, smoky. Conidia terminal on the conidiophores, inverted pear-shaped or oval, muriform with three to four cross-walls, slightly constricted at the septa, black-gray, 25–30 × 12–15μ, smooth.

From soil: United States: Idaho (106)

2. *Stemphylium botryosum* Wallroth

Colonies very dark, orbicular; vegetative hyphae creeping, spreading, thin, irregularly branched, at first hyalin, then becoming light brown and finally dark brown, septate, flexuous, more or less moniliform. Conidiophores arise as branches, short, flexuous, simple or branched, hyalin or colored, more or less rough, simple or forked at the summit. Conidia terminating the primary as well as the secondary branches, with short pedicel, sometimes nearly globose, sometimes elliptical or oblong, divided horizontally into two to six compartments, of which one or several present a vertical or oblique septum, isabel-colored to brownish-black, 25–40 × 16–20μ; surface of conidia with age, finely dotted.

From soil: England (37), U. S. S. R. (108)

3. *Stemphylium macrosporoideum* (Berkeley and Broome) Saccardo

Turf extended, dark colored. Conidiophores thread-like, very delicate; irregularly branched, hyalin, interwoven, septate with short stem-like side-branches, which carry conidia at their tips. Conidia almost globose, mulberry-like, at first one-celled, hyalin, then four-celled, chest-nut-brown, 14–25μ in diameter; the division results either by two right-angled walls in four cells lying crosswise beside one another or so that one cell, becoming somewhat enlarged serves as a stalk-cell while the other three lie on it.

From soil: Canada (16), Egypt (115), U. S. S. R. (108)

4. *Stemphylium verruculosum* Zimmermann

Colonies spreading, dark olive-green. Hyphae curved, hyalin, branched, about 22μ thick, with short or long branches. Conidia inverted egg-shaped or elliptical, with two or three septa, muriform, verrucose, brown, non-transparent when mature, 17.5–22 × 11–13.5μ.

From soil: Japan (132)

(30) **Macrosporium** Fries

Sterile hyphae dark brown to almost black. Conidiophores seldom single, mostly arising in groups, erect, flexuous or almost straight, septate, often with the upper cells somewhat swollen, usually unbranched, brown to black, usually transparent, conidia formed acrogenously, then being displaced laterally. Conidia terminal and single, ovate or elongate, usually more or less club-shaped, sometimes drawn out to a light (hyalin) point, muriform, brown to black, often finely echinulate.

KEY TO THE SPECIES OF THE GENUS MACROSPORIUM

a. Conidia club-shaped
  b. Colonies greenish-brown, conidiophores 150–200μ long — 1. *M. cladosporioides*
  bb. Colonies black-brown, conidiophores up to 90μ long — 2. *M. commune*
aa. Conidia packet-like — 3. *M. sarcinaeforme*

1. *Macrosporium cladosporioides* Desmazieres

Colonies small, round, velvety, dark greenish-brown. Conidiophores arise in groups, erect, unbranched, gnarled, septate, almost hyalin, 150–220μ long, 5μ thick. Conidia ovate, elongate, or club-shaped, almost transparent, (sometimes torulose), muriform, with two to ten cross-walls, 15–75μ long and 8–14μ thick; ovate conidia finely granular, club-shaped ones smooth.

From soil: England (36) (37)

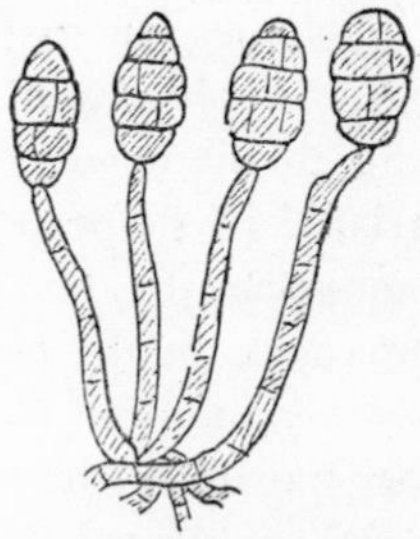

Fig. 118. **Macrosporium.** conidiophores and conidia (after Lindau).

2. *Macrosporium commune* Rabenhorst

Colonies dense, brown to black-brown. Conidiophores arise in groups, ascending, usually unbranched, septate, not constricted at the septa, 80–90μ long × 4–7μ thick. Conidia very variable, inverted egg-shaped, elongate, or club-shaped, narrowed at the base, with three to five cross-walls and several oblique transverse walls, olive-green or olive-brown, usually with finely granular surface, 18–35 × 8–14μ.

From soil: United States: Idaho (106)

3. *Macrosporium sarcinaeforme* Cavara

Hyphae branched, hyalin, septate, with short erect rigid brownish-olivaceous conidiophores sparingly septate and noded. Conidia sarciniform, with average constrictions at the transverse and longitudinal septa, concolorous with the conidiophores, 24–28 × 12–18μ.

From soil: Egypt (115)

(31) **Alternaria** Nees

Sterile hyphae creeping, septate. Conidiophores single or in groups, erect, septate, mostly unbranched, short. Conidia inverted club-shaped, mostly elongate at the tip, muriform in the lower portion, dark colored, lighter at the points, borne in more or less long, usually simple chains.

KEY TO THE SPECIES OF THE GENUS ALTERNARIA

a. Colonies black-green, conidia 50 × 16μ, rough	1. *A. humicola*
aa. Colonies brown	
b. Conidia 35–90 × 9–14μ, smooth	2. *A. fasciculata*
bb. Conidia 12–22μ long	3. *A. grisea*
aaa. Colonies brown-green, conidia 30–36 × 14–15μ, smooth	4. *A. tenuis*
aaaa. Colonies gray, conidia, 24 × 10μ	5. *A. geophila*

1. *Alternaria humicola* Oudemans

Colonies at maturity orbicular, black-green. Fertile hyphae well developed, hyalin, articulate, 3–5μ in diameter, racemosely branched. Conidia variable in shape, cylindrical, obclavate, oblong, lageniform, at first hyalin, later honey-colored, thin, dark, finally black-green and smoky, variable in size, maximum 16–50μ, three to seven times septate, muriform, in advanced age dense and very finely roughened, slightly or non-constricted at the septa.

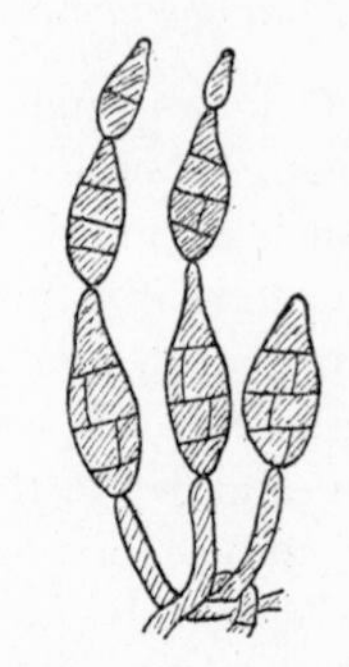

FIG. 119. **Alternaria.** conidiophores and conidia.

From soil: China (81), Czechoslovakia (91), England (36), Holland (100), India (21), Sumatra (130)

United States: Illinois (129), Iowa (3), Louisiana (2), New Jersey (146) (147), Utah (50)

2. *Alternaria fasciculata* Cooke and Ellis

Conidiophores brown, erect or ascending, irregularly curved, solitary or caespitose, septate, diameter uniform, 40–130 × 3μ; conidia dark brown, oblong ovate, minutely apiculate, 35–90 × 9–14μ, endochrome transversely two to seven times septate with usually several longitudinal septa, the apical cell short or elongated into a straight hyalin beak.

From soil: Egypt (115)

United States: New York (67)

3. *Alternaria grisea* Szilvinyi

Hyphae smooth, septate, brown 3–4.5μ in diameter. Conidia in chains or bush-like and then in chains. Conidia globose to pear-shaped, muriformly divided, bright brown to black-brown, often with a hyalin stipe, smooth, 12–22μ long.

From soil: Sumatra (130)

4. *Alternaria tenuis* Nees

Conidiophores short, septate, unbranched or branched, brown-green. Conidia in chains, muriform with three to five cross-walls constricted at the outer walls, olive-green or brownish-black, very variable in size and shape, 30–36 $\times$ 14–15μ.

From soil: Austria (131), Canada (15), England (37), Japan (132)

5. *Alternaria geophila* Daszewska

Mycelium dark brown, turf gray, cinereous, about 1 cm. high, composed of sterile prostrate filaments, brown, formed of short parts, filled with oil droplets. The parts of the filament measured 8–12μ long $\times$ 6–10μ in diameter. Fertile filaments formed from raised branches, septate, 4–6μ in diameter. The least distance between the septa 27.2μ. Spores borne on simple lateral branches, in chains of two to four spores, elongate, fusiform, brown and at maturity black, muriform. The size of the spores and the number of septa present, varies; 16–38μ long $\times$ 6–12μ wide, 24 $\times$ 10μ being the mode. Each cell of the spore contained a fat globule.

From soil: Egypt (115), Switzerland (38)

c. *STILBACEAE*

Hyphae prostrate, pale or dark colored. Conidiophores (and also sterile hyphae) bound together in parallel courses into erect simple or branched coremia. Coremia pale or dark, usually terminating in a head, either formed from a few parallel hyphae or thicker, fleshy and stroma-like. Spores terminal, single or in chains, hyalin or dark colored.

KEY TO THE GENERA OF THE STILBACEAE

a. Hyphae and conidia hyalin or bright-colored (Hyalo-stilbeae)
  b. Conidia single, not in chains
    c. Conidia in a head
      d. Stipe not branched (1) **Stilbella**
      dd. Stipe branched (2) **Tilachlidium**
    cc. Conidia not in a head, coremia sheaf-like (3) **Cilicipodium**
  bb. Conidia in chains (4) **Coremium**
aa. Hyphae and conidia dark-colored (Phaeostilbeae)
  b. Conidia one-celled
    c. Stipe firm, dry

d. Conidia not inclosed in mucus
e. Conidia not accompanied by sterile hairs (5) **Stysanus**
ee. Conidia accompanied by sterile hairs (6) **Trichurus**
dd. Conidia inclosed in mucus (7) **Metarrhizium**
cc. Stipe fleshy (8) **Stemmaria**

(1) **Stilbella** Lindau

Coremia with definite stipes and more or less definite heads; hyalin or bright colored. Stipes composed of parallel, branching hyphae, which diverge apically to form the heads. Final branches of the hyphae serve as conidiophores, not branching regularly, but usually unbranched, producing a single conidium terminally or many, one after another. Conidia single, not in chains, but held together in a mucus, ovate, elongate or globose, hyalin, guttulate, very small.

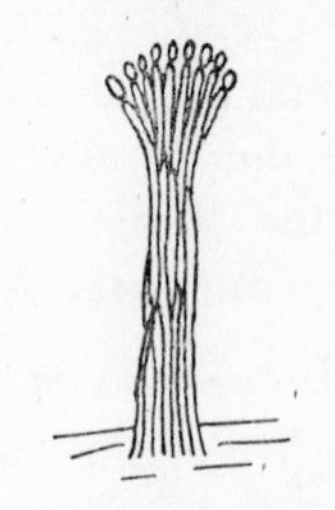

FIG. 120. **Stilbella.** coremium and conidia (after Lindau).

A single species treated.

1. *Stilbella bulbicola* P. Hennings

Colonies with radiating, submerged, hyalin mycelium with zonately arranged coremia. Coremia scattered singly or somewhat caespitose. Stipe, formed from hyphae 2–3μ in diameter, 0.3–0.5 mm. long, twisted, delicate, hyalin or yellow, 30–40μ in diameter, arising from a base, 40–80μ in diameter. Heads globose or subglobose, waxy, at first white, becoming yellow, 150–180μ in diameter. Conidiophores 1–1.5μ in diameter. Conidia ellipsoid to ovate, hyalin with one or two guttulae, 5–6 $\times$ 2.5–3.5μ.

From peat soil: United States: Iowa

(2) **Tilachlidium** Preuss

Coremia formed of fasciculated thread-like hyphae, branched, the secondary branches being sterigma-like, awl-shaped, somewhat club-shaped at the tip, furnished with little heads, usually consisting of a single hypha carrying on its end a conidial head. Heads slimy, later dry. Conidia one-celled, ovate, hyalin, occurring at the end of the conidiophores.

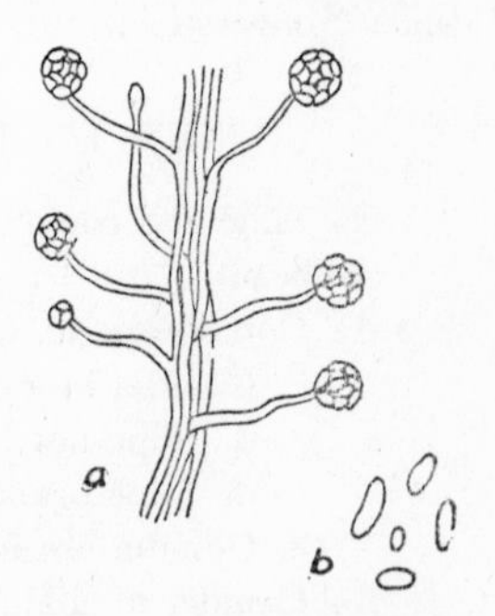

FIG. 121. **Tilachlidium.** *a* coremium; *b* conidia (after Lindau).

A single species treated.

1. *Tilachlidium humicola* Oudemans

Coremia circular, snow-white; turf woolly. Main stalk upright, cylindric, 35–40μ thick, made up of very delicate, articulate, closely interwoven hyphae, from all sides of which spring single hyphae as secondary branches, basidia-like, 40–80μ long, erect unbranched, nonseptate, curved, with an almost club-like head. Conidia bound together by mucus into a spherical, terminal, finally dry head, 15–18μ in diameter, elongate or ovate, very bright green, 6–7μ long, 3–5μ thick.

From soil: Holland (100)

Species of uncertain position:

Tilachlidium roseum Szilvinyi
From soil: Austria (131)

(3) **Ciliciopodium** Corda

Coremia stipe-like, without a definite head, rather large, bright colored. Stipe formed of parallel, unbranched or branched hyphae, which sheaf-like, separate from each other at the tip; externally rough or hairy. Conidial terminal at the hyphal tips, single, globose or ellipsoid, hyalin, without mucus.

A single species treated.

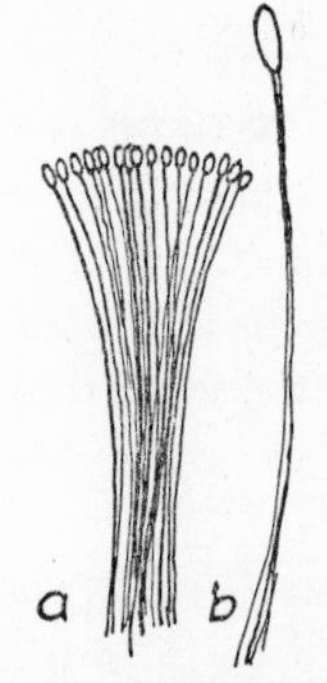

Fig. 122. **Cilicipodium.** *a* coremium; *b* conidiophore and conidium (after Lindau).

1. *Ciliciopodium hyalinum* Daszewska

Coremia only formed in slightly acid media. Mycelium whitish-yellow, forming a short-powdery white turf. Coremia yellow, with hyphae separating at the tip. Conidiophores with opposite branching, ending in two to four slender phialides, each bearing an elliptic conidium, hyalin, 4 × 2–3μ. Conidia are formed to the tip of the coremium.

From soil: Switzerland (38)

(4) **Coremium** Link

Coremia cylindric, carrying a conidial-bearing head at its apex. Conidia very small, continually without slime, in terminal chains.

A single species treated.

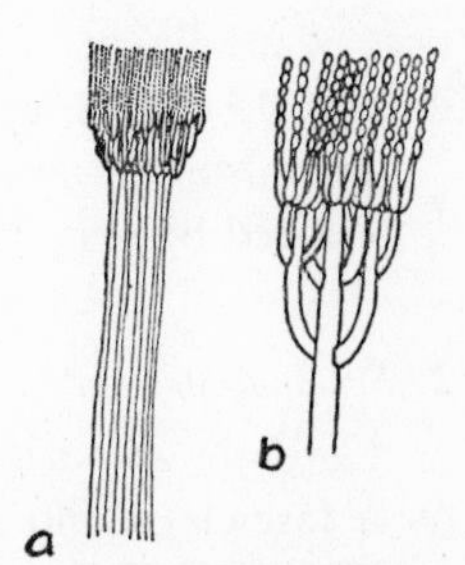

Fig. 123. **Coremium.** *a* coremium; *b* conidiophore and conidia (after Lindau.)

1. *Coremium arbuscula* H. Fischer

Turf matted, white, later dull red, 2–3 cm. broad to about 1 cm. high. Conidiophores either a single short cell which carries one or more phialides on its tip, or a longer cell carrying one or two whorls of branches (usually two or three) terminating in a bush of two to twelve phialides. Phialides flask-shaped, pointed, of various shapes, but always more or less generally tapering to a point. Conidia in chains, oval, hyalin, red to cinnamon brown, 4–6μ long, 2.5–3μ in diameter. Coremia simple, cone- or club-shaped, antler-, or tree-shaped, single or in groups, up to 2 cm. high, covered on their whole surface with conidiophores and red to reddish-brown from the conidia.

From soil: Austria (66)

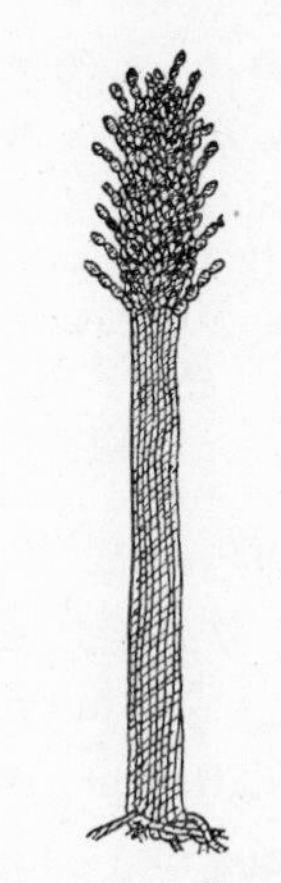

Fig. 124. **Stysanus.** coremium and conidia (after Lindau).

(5) **Stysanus** Corda

Coremia erect, clubbed-cylindric, dark colored, rigid. Conidia occurring in a loose, long or almost globose panicle, ovate or lemon-shaped, almost hyalin, formed in chains.

KEY TO THE SPECIES OF THE GENUS STYSANUS

a.	Conidia 5–6 × 3–3.5μ	1. *S. medius*
aa.	Conidia 6–8 × 4–5μ	2. *S. stemonites*

1. *Stysanus medius* Saccardo

Coremia gregarious to scattered, black, bristle-like. Stalk of close threads, 3μ in diameter, septate, smoke-colored, 300–400μ long, 30μ thick. Conidiophores looser above, paler and bearing spores at their tips. Conidia occurring in a narrow head, ovate, united into rapidly separating chains, green-black, 5–6μ long, 3–3.5μ thick.

From soil: England (11)
United States: Illinois (129)

2. *Stysanus stemonites* (Persoon) Corda

Coremia gregarious. Stalk thin, unbranched, brownish-black, formed by a fascicle of elongated, septate, green-brown hyphae, ending at the tip in a cylindric head. Conidia ovate to lemon-shaped, bluish-green, transparent, formed in chains, 6–8 × 4–5μ.

From soil: China (81), Denmark (68), India (49), U. S. S. R. (108)
United States: Maine (147), Michigan (52), Texas (147)

(6) **Trichurus** Clements and Shear

Similar to the genus Stysanus, but having the head furnished with long setae.

A single species treated.

1. *Trichurus terrophilus* Swift and Povah

Colonies irregular in outline, at first pale olive-gray with radial folds, becoming dark olive-gray and finally olivaceous-black, always with pale margin, at maturity forming a dense and powdery growth up to 1.5 mm. in height, with small droplets on the surface. Reverse greenish-black. Mycelium dark brown, septate, 2–3.5μ in diameter in early stage forming branched catenulate conidia on single hyphae. At maturity the mycelium adheres in rope-like strands from which arise vertically dark clavate fruit bodies 375–1,300μ tall, on stalks 95–800 $\times$ 20–70μ, the fertile portion 135–500 $\times$ 35–150μ, giving rise to simple or panicled chains of spores interspersed with bristle-like dark brown setae, 15–70μ long, 3μ wide at base, tapering gradually to the apex which is terminated in a sharp point. Setae nonseptate, or with one or two septa near the base, simple or forked, the branches commonly unequal and at an obtuse angle. Spores oval to elliptical, 3–6 $\times$ 2–3.5μ, pale green, greenish-black in mass.

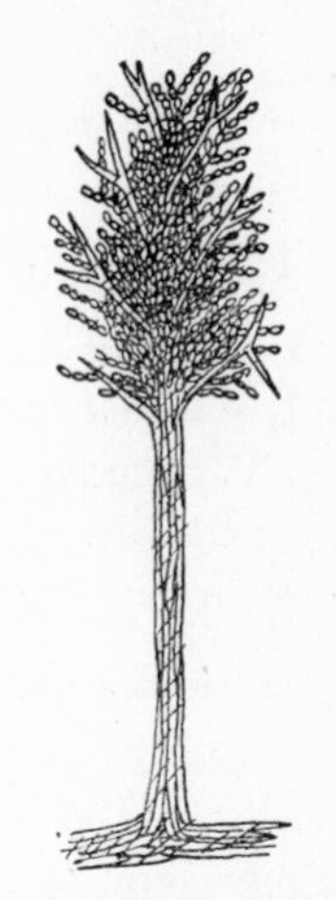

FIG. 125. **Trichurus.** coremium and conidia (after Swift).

From soil: United States: Illinois (129)

(7) **Metarrhizium** Sorokin

Mycelium floccose, white. Conidiophores clustered into broad coremia, bearing phialides in pairs or verticils at their tips. Conidia in chains forming glutinous masses, ovate to elongate.

A single species treated.

1. *Metarrhizium glutinosum* Pope*

Mycelial mat, grown on filter paper, pure white with dusky olive-green to olivaceous-black conidial masses in tufts of the mycelium. Conidiophores up to 75μ long, smooth, erect, septate, penicillately branched above, forming a palisade layer in tufts on which olivaceous black glutinous masses of conidia are produced. Conidia in chains, borne directly on finger-like phialides. Conidial chains distinguishable in masses only in

* POPE, SETH
1944. A new species of Metarrhizium active in decomposing cellulose. Mycologia 36: 343–350.

early stages of formation. Conidia elongate-ovate, smooth, dusky olive-green to olivaceous-black, 6–9.5 × 15–3.9μ.

From soil: United States: Maryland

(8) **Stemmaria** Preuss

Stalk erect, fasciculate, fleshy, branching broom-like above, nonseptate, forming a head. Conidia in chains, occurring at the tips of the branches.

A single species treated.

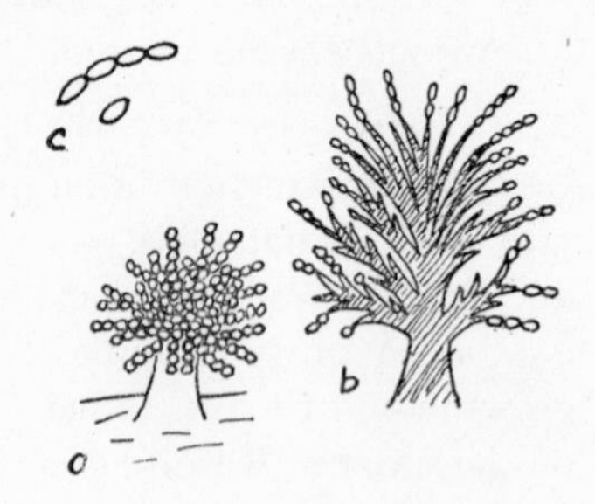

FIG. 126. **Stemmaria.** *a* habit; *b* section through coremium; *c* conidia (after Lindau).

1. *Stemmaria terrestris* Chaudhuri

Vegetative mycelium submerged; aerial mycelium in erect fascicles of hyphae, 2–2.5 mm. long. Conidia borne in chains in the brush-like heads of the coremia, elliptic, 3 × 2μ.

From soil: India (21)

d. *TUBERCULARIACEAE*

Hyphae wide-spread in or on the substrate. Conidiophores and sterile hyphae intermingle to form a fruiting layer (sporodochium) which is usually formed from thickly interwoven often radially arranged threads. Sometimes this layer rests upon a plectenchymatic stroma. The consistency of the layer is waxy or gelatinous, sometimes horny or cottony. Frequently the hyphae and conidia become embedded in mucus. The external form of the sporodochium is usually definite, occasionally extended as an unlimited crust. Conidiophores usually thickly crowded, and often forming a closed hymenium, branched or simple, rod-like. Conidia various, usually terminal, single, but also in chains and lateral.

KEY TO THE GENERA OF THE TUBERCULARIACEAE

a. Conidia and hyphae, hyalin or bright colored
  b. Conidia one- or two-celled, globose to elongate
    c. Sporodochia without hairs or bristles
      d. Sporodochia disk-like — (1) **Hymenula**
      dd. Sporodochia hemispherical
        e. Conidiophores with whorled branches — (2) **Dendrodochium**
        ee. Conidiophore branches not in whorls — (3) **Tubercularia**
    cc. Sporodochia covered with hairs or bristles — (4) **Volutella**
  bb. Conidia more than two-celled
    c. Conidia, fusiform, curved — (5) **Fusarium**

cc. Conidia, cylindric, with round ends (6) **Cylindrocarpon**
aa. Hyphae and conidia dark colored
b. Sporodochia surrounded by hyalin hairs or bristles (7) **Myrothecium**
bb. Sporodochia, punctate, gelatinous (8) **Epicoccum**

(1) **Hymenula** Fries

Conidial layer shield-shaped, regular, smooth, bright colored. Conidiophores simple, seldom branched. Conidia egg-shaped, terminal, single.

A single species treated.

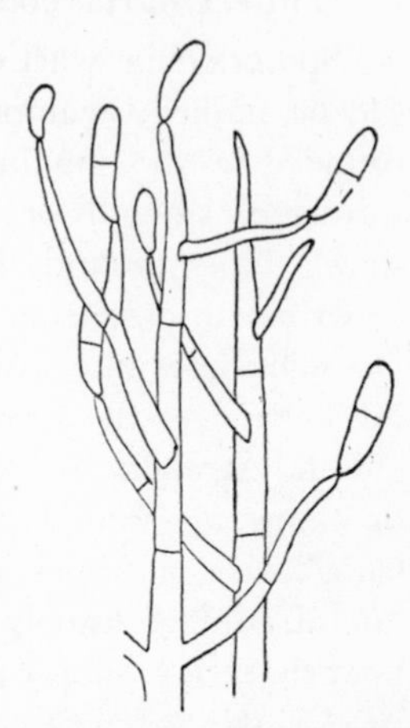

FIG. 127. **Hymenula.** conidiophore and conidia (after Sherbakoff).

1. *Hymenula affinis* (Fautrey and Lambotte) Wollenweber

Syn. *Fusarium affine* Fautrey and Lambotte

Conidia straight, somewhat dorsiventral near apex, apedicillate, typically one-septate, 10.2 × 2.8μ (9–11.4 × 2.6–3μ) usually in a continuous smooth or slightly roughened, slimy-layer, from hyalin to pale salmon-colored on a glucose agar. Conidiophores from simple to sparingly branched, septate. Mycelium hyalin. No chlamydospores.

From soil: Canada (15)
United States: Idaho (106)

(2) **Dendrodochium** Bonorden

Sporodochium cushion- or wart-like, of various sizes and with different margins, white or bright colored, in appearance somewhat resembling Tubercularia; smooth. Conidiophores in a hymenium, which covers the sporodochium, standing close beside one another, usually branched in whorls or trichotomous, but not irregularly as in Tubercularia, hyalin. Conidia acrogenous, ovate or elongate, hyalin.

A single species treated.

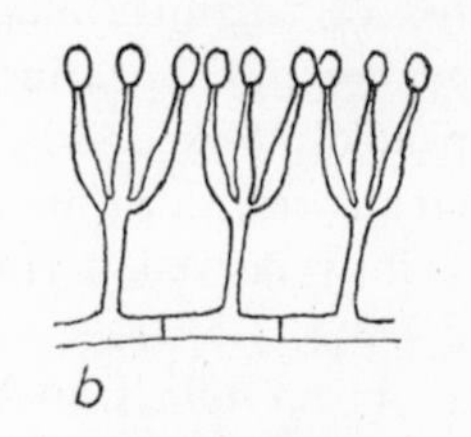

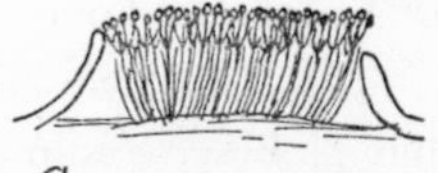

FIG. 128. **Dendrodochium.** *a* sporodochium; *b* conidiophores and conidia. (after Lindau)

1. *Dendrodochium gracile* Daszewska

Mycelium white, hyphae prostrate, branched, septate, 4–6μ in diameter. Sporodochia (on slightly acid media) globose, smooth, entangled in the much branched hyphae. Branching opposite or in verticils. Phialides short, swollen, bearing a single green conidium at their tips. Conidia 4–2 × 3μ.

From soil: Switzerland (38)

(3) **Tubercularia** Tode

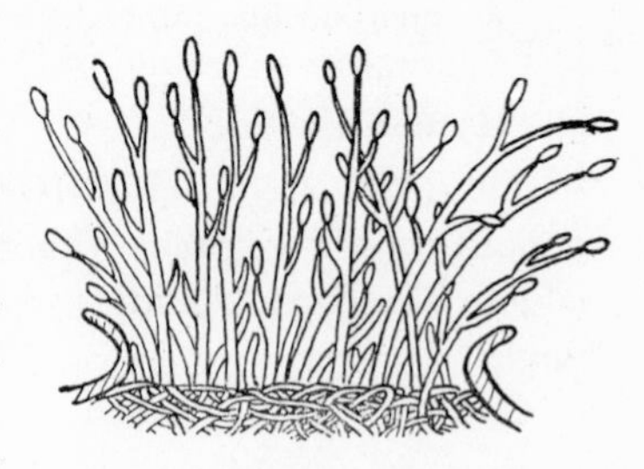

Fig. 129. **Tubercularia.** sporodochium and conidia.

Sporodochia, wart or cushion-like, sessile or stalked, erumpent and often surrounded by the remainder of the covering substrate, smooth or very seldom with simple hairs around the margin, smooth or wrinkled, possessing a firm inner tissue. Conidiophores interwoven to form a fruiting layer over the surface of the sporodochium, straight or curved, sometimes breaking up into single cells, branches elongate or at times very short. Conidia terminal, seldom appearing lateral, single, usually ovate or elongate cylindric or globose, seldom boat-shaped, hyalin, covering the conidial layer in a thick sheath.

A single species treated.

1. *Tubercularia vulgaris* Tode

Sporodochium rather large, erumpent, half-round, flat above or seldom somewhat conical, the margin ragged, typically shining cinnabar-red or brighter, often darker to black, smooth, usually single, seldom confluent, appearing somewhat stipitate. Conidiophores bushy, repeatedly forked, 50–250μ long, 1.5–3μ thick, primary branches straight, erect, with short alternating laterals. Conidia ellipsoid-elongate at times slightly curved, terminal on the lateral branches, hyalin, in mass red, 5.3–8 × 1.5–3μ.

From soil: Canada (16)

(4) **Volutella** Tode

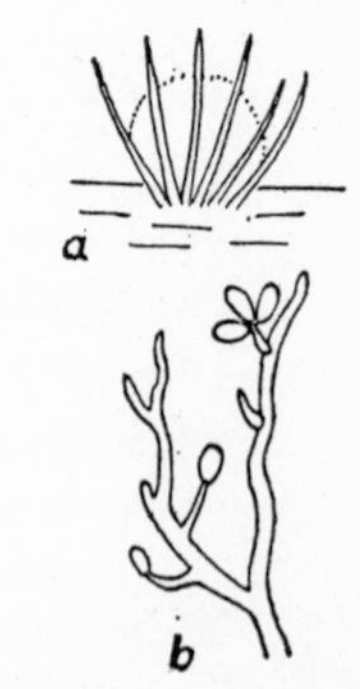

Fig. 130. **Volutella.** *a* habit; *b* conidiophores (after Lindau).

Fruit layer superficial, disc-shaped or somewhat globose, sessile or on a short stalk, regularly formed with long bristles or spines at the margin and sometimes in the middle of the disc. Conidiophores thickly gregarious, covering the entire disc, at the base united with the spines as branches, usually several times branched, the last branches forming a thick hymenium of fine, sterigma-like stalks. Conidia terminal, formed in masses, small, ovate or elliptical, hyalin.

KEY TO THE SPECIES OF THE GENUS VOLUTELLA

a. Sporodochia stipitate; setae hyalin	1. *V. roseola*
aa. Sporodochia sessile; setae dark	2. *V. piriformis*

1. *Volutella roseola* Cooke

Sporodochia subglobose, bearing conidia on surface, rosy; with a definite stipe, surrounded by long flexuous septate setae, tapering to their tips, hyalin. Conidia cylindric, thick.

From soil: Canada (16)

2. *Volutella piriformis* Gilman and Abbott

Colonies on Czapek's agar broadly spreading, brown or grayish-brown with superficial whitish aerial mycelium in old cultures. Reverse brownish-black. Mycelium largely submerged, dark brown. Sporodochium sessile erumpent from the subicle, dark brown to black, pyriform, with long, dark sheathed spines arising from the base and up the sides, 75–150 $\times$ 60–100μ in size; spines up to 175μ or 200μ long. Conidia brown, elliptical, smooth, 9.5–11.5 $\times$ 5.5–7μ.

From soil: United States: Louisiana (50)

(5) **Fusarium** Link

Conidial layer cushion-shaped or somewhat extended without a definite limit. Conidiophores branched. Conidia terminal, simple, spindle- or sickle-shaped, many-celled with indistinct cross-walls.

KEY TO THE SECTIONS OF THE GENUS FUSARIUM

a. Microconidia normally present, usually one-celled spindle-, egg-, pear-, or kidney-shaped	
b. Microconidia more or less pear-shaped	I. Sporotrichiella
bb. Microconidia not pear-shaped	
c. Chlamydospores lacking	
d. Macroconidia thin-walled	II. Liseola
dd. Large conidia comparatively thick-walled	III. Spicarioides
cc. Chlamydospores terminal and intercalary	
d. Conidia thin-walled, septate, tapering or constricted at the tip, with basal foot-cell; in mass brownish-white, rosy, salmon, at times quite bright	IV. Elegans

dd. Conidia relatively heavy-walled and strongly septate; allantoid, truncate or round at the tip with a short snout; with greater or less foot-cell; in mass brownish-white, cream, or golden-yellow, often infiltrated with the color of the stroma (verdigris, black-blue) — V. Martiella

aa. Microconidia usually lacking or one- to many-celled, kidney- or comma-shaped to spindle-shaped

b. Macroconidia without a foot-cell

c. Pionnotes typically present — VI. Eupionnotes

cc. Pionnotes scarce or lacking — VII. Arachnites

bb. Macroconidia more or less pedicellate

c. Terminal chlamydospores lacking

d. Intercalary chlamydospores lacking

e. Macroconidia almost cylindric in the middle, curvature of both sides similar, tip with a truncate curved snout — VIII. Lateritium

ee. Macroconidia almost cylindric in the middle, curvature of sides unlike, both ends tapering — IX. Roseum

dd. Intercalary chlamydospores present

e. Sporodochia usually lacking — X. Arthrosporiella

ee. Sporodochia present — XI. Gibbosum

cc. Terminal chlamydospores present

d. Intercalary chlamydospores present — XII. Discolor

dd. Intercalary chlamydospores lacking — XIII. Ventricosum

SECTION I. SPOROTRICHIELLA

KEY TO THE SPECIES

a. Sporodochia and pionnotes lacking

b. Microconidia typically globose to lemon-shaped, 5–6μ in diameter — 1. *F. poae*

bb. Microconidia typically spindle-ellipsoid, 3–4μ in diameter — 2. *F. chlamydosporum*

aa. Sporodochia and pionnotes abundant — 3. *F. sporotrichioides*

1. *Fusarium poae* (Peck) Wollenweber

Fruiting layer cobwebby, or felty-woolly, white to rose with various paired or whorled branches, of richly branched hyphae and conidiophores. Stroma carmine-purple, ochre-yellow and violet. Conidia usually one-celled (at most 4 per cent, one- to four-septate) in large numbers on the conidiophores, sometimes bound in chains, discharged and then appearing as isabellin or whitish powder scattered over the mycelium or giving a sandy or mealy appearance to the substrate. In addition to the typical globose, lemon-shaped or pear-shaped one- or two-celled microconidia; long spindle-shaped, ellipsoid to sickle-shaped conidia occur. Oval to lemon-shaped conidia; nonseptate 6.6–9 × 4.7–6μ; one-septate 12–26 × 2.7–5μ; sickle-shaped conidia three-septate 18–35 × 3.5–5μ. Chlamydospores mostly intercalary, in chains and knots, ochre-brown.

From soil: Canada (15)

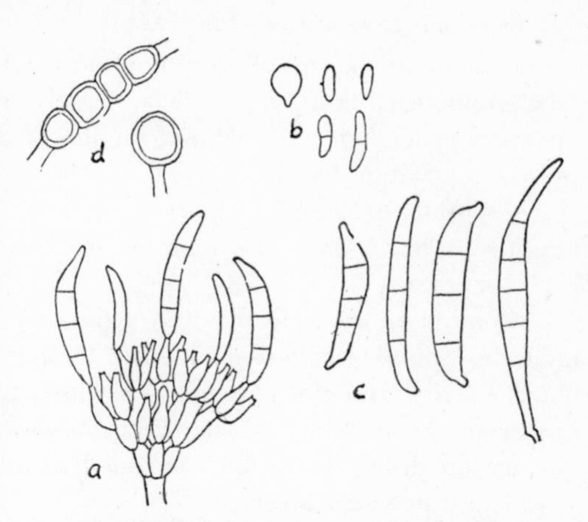

FIG. 131. **Fusarium.** *a* conidial head; *b* microconidia; *c* macroconidia; *d* chlamydospores.

2. *Fusarium chlamydosporum* Wollenweber and Reinking

Fruiting layer, floccose, hyalin or pink, plectenchymatic below, at times forming somewhat warty, sclerotial knots, variously colored, hyalin, carmine-purple-red, sulphur-yellow, ochre or dark brown. The dark color is caused by the occurrence in the mycelium of numerous, globose to pear-shaped, smooth to rough or spiny, intercalary or terminal, single, paired, catenulate or muriform, chlamydospores 10–16μ in diameter. Conidia small, spindle-ellipsoid, not globose, lemon-shaped, usually one-celled, seldom septate, three-septate spores sickle-shaped, weakly pedicellate.

0-septate, 4–11 × 2.5–4μ
1-septate, 11–16 × 3–4μ
3-septate, 27–32 × 3.5–4μ
In soil: Central America (164)

3. *Fusarium sporotrichioides* Sherbakoff

Conidia at the tips of conidiophores which branch irregularly or dichotomously, non-, one- (two- three-) septate, globose, ellipsoid or pear-shaped, often with a basal papilla of attachment, scattered dustily in the mycelium, hyalin:

0-septate, 6–12 × 3.5–7.1 μ
1-septate, 11–15 × 5.2–6.5 μ
3-septate, 17–33 × 4–8 μ

Conidia in sporodochia and pionnotes more or less three- to five-septate, spindle-sickle-shaped, with larger and smaller spores mixed, both narrow and thick, sometimes with parabolic curvature, both ends tapering, foot-cell real or slight, in mass ochre to salmon or orange-red, in age on drying becoming darkened in a resin-like layer or becoming paler in a powdery layer.

0-septate, 8–12 × 2.2–3.2 μ
1-septate, 15–22 × 3–3.8 μ
3-septate, 25–33 × 3.4–4 μ
5-septate, 42–56 × 3.7–4.5 μ

Stroma loam-colored, ochre-yellow, coffee to chestnut-brown or dull carmine. (The yellow tones are changed to violet by the addition of ammonia, while the red colors become yellow in the presence of hydrochloric acid.) Cinnamon to red-brown oval sclerotia (0.25–0.4 × 0.2–0.33 mm.) as well as intercalary chlamydospores, singly (7–14 μ thick), in chains or in knots of ochre to dark brown colors, occur frequently.

From soil: Canada (15)

SECTION II. LISEOLA

KEY TO THE SPECIES

a. Microconidia in more or less persistent chains
  b. Macroconidia scattered, neither sporodochial nor pionnotal: three-septate, 24 × 3.6 μ; five-septate, 31 × 4 μ — 5a. *F. moniliforme* v. *minus*
  bb. Macroconidia both in sporodochia and pionnotes
    c. Sickle spores rather small; three-septate 26 × 2.9 μ — 4. *F. lactis*
    cc. Sickle spores larger, three- to five-septate; three-septate, 36 × 3 μ — 5. *F. moniliforme*
aa. Microconidia not remaining in chains — 6. *F. neoceras*

4. *Fusarium lactis* Pirotta and Riboni

Microconidia in more or less long chains or in false heads, oval, cylindrical, straight, seldom somewhat curved, at times pear-shaped, one- or two-celled, later scattered as a bright powder in the hyalin pink to red mycelium. Later there occur on a plectenchymatic, violet to cherry-red (becoming blue with ammonia) stroma, the bushy much-branched conidiophores bearing the larger spindle or almost cylindrical, sickle spores, tapering at both ends, strongly bent, almost hooked at the tip with tenpin to pedicellate base, scattered or in false heads, seldom, in macroscopic bright-orange colored slime of sporodochial or pinnotal layers, one- to three-, exceptionally four- to five-septate:

0-septate, 3–16 × 1.5–6μ
1-septate, 9–32 × 2–6μ
2-septate, 15–35 × 2–4μ
3-septate, 16–40 × 2–4μ

The conidia become swollen, for example in germination to a diameter of 6μ.

5. *Fusarium moniliforme* Sheldon

Microconidia produced in chains and remaining connected, or held in false heads, later becoming scattered over the bright yellow to rosy-white aerial mycelium as a transparent shining powder, one- or two-celled, spindle-, egg-shaped. Macroconidia delicate, awl-shaped, slightly crescent-shaped or almost straight, tapering at both ends, often constricted at the tip, and sometimes hooked, at the base with a real or slight foot-cell, scattered or gathered into sporodochia or pionnotes, bright in mass, isabellin or salmon colored, drying to brick-red to cinnamon-brown or pale; three- to five-, seldom six- to seven-septate:

0-septate, 4–18 × 1.5–4μ
1-septate, 9–30 × 2–5μ
3-septate, 20–60 × 2–4.5μ
5-septate, 37–70 × 2–4.5μ
7-septate, 58–90 × 2.5–4.5μ

Chlamydospores lacking. Dark blue, spherical sclerotia 0.08 × 0.1 mm. may be present. Stroma more or less plectenchymatic, yellow, brown, violet, etc.

From soil: Canada (15)

5a. *Fusarium moniliforme* Sheldon var. *minus* Wollenweber

Differs from the type by absence of pionnotal and sporodochial slime. Microconidia in long chains or in false heads, then scattered in whitish-rosy aerial mycelium, oval-spindle form, one- to two-celled. Macroconidia few, sparingly scattered, (to 1 per cent) three- to five-septate, lanceform, seldom spindle-sickle-shaped, slightly curved, generally pointed at the tip, truncate at the base, tenpin-shaped, seldom with attachment papilla, not pedicellate:

0-septate, 3.5–15 × 1.5–4μ
1-septate, 8–20 × 2–4.4μ
3-septate, 19–30 × 2.5–4.6μ
5-septate, 24–42 × 3–6μ
6-septate, 29–43 × 3.5–4.5μ
Dark blue globose sclerotia sometimes present.

6. *Fusarium neoceras* Wollenweber and Reinking

Microconidia single or in false heads, not in chains, one-celled, oval-spindle-shaped, seldom two-celled, exceptionally three-celled, later scattered as dust in mycelium. Macroconidia in sporodochia and pionnotes, brownish white-cream to incarnate, at times becoming flecked with violet or blue tones, and varying in concentric zones of the stroma, and laid on it in rings, straight or weakly curved, tapering at both ends, slightly constricted at tip, with tenpin to slightly pedicellate base, three- (three- to five-) very seldom six- to nine-septate:

0-septate, 5–18 × 2.75–4.5μ
1-septate, 14–34 × 3.25–5.5μ
3-septate, 32–59 × 3.5–5μ
5-septate, 55–67 × 4.5–5.5μ
6- to 9-septate, 17–120 × 4–5μ
Chlamydospores and sclerotia lacking.

SECTION III. SPICARIOIDES

A single species treated.

7. *Fusarium decemcellulare* Brick

Syn. *Fusarium rigidiuscula*

Small, delicate oval to ellipsoid, one- or two-celled conidia occur in chains or false heads, on more or less branched conidiophores in the aerial mycelium, where they lie scattered as powder and easily become dispersed. In addition there occur larger, cylindrical, pedicellate, greatly curved, and beaked conidia, mostly five- to nine-septate, more infrequently three- to four- or ten- to twelve-septate.

0-septate, 5–11 × 2–4.5μ
1-septate, 10–28 × 2.5μ
3-septate, 20–67 × 3.5–6μ

5-septate, 42–72 × 4.5–8μ
7-septate, 60–95 × 4.5–8μ
9-septate, 68–114 × 4.5–8.5μ
11-septate, 75–131 × 5–9μ

Conidia in sporodochia in arched or also peg-like layers or flowing together in pionnotes-resembling slime, in mass white, cream, brownish-white or ochre colored. Stroma golden-yellow and carmine, covered by hyalin or slightly colored aerial mycelium, extended, forming thick plectenchymatic rough or toothed (stilboid) knots which break out in a star-shaped manner and on which are produced the ascus-stage. (*Calonectria rigidiuscula* (Berkeley and Broome) Saccardo).

SECTION IV. ELEGANS

KEY TO THE SPECIES

a. Typically without sporodochia (Subgroup *Orthoceras*)
  b. Pionnotes typically lacking
    c. Conidiophores with spiral branching — 8. *F. bostrycoides*
    cc. Conidiophores simple or branched in whorls
      d. Stroma pale, brownish-white to incarnate — 9. *F. conglutinans*
      dd. Stroma carmine-purple (blue with alkali) — 10. *F. orthoceras*
  bb. Pionnotes sparse or lacking
    c. Conidia rather long, three-septate, 46 × 3.5μ; five-septate, 60 × 4.2μ — 11. *F. angustum*
    cc. Conidia shorter, three-septate, 35 × 4μ; five-septate, 39 × 4μ — 12. *F. lini*
aa. Typically with sporodochia
  b. Macroconidia relatively slender, 3–3.7μ in diameter (Subgroup *Constrictum*)
    c. Conidia usually three-septate — 13a. *F. bulbigenum* v. *blasticola*
    cc. Conidia three- to five-septate
      d. Stroma more or less sclerotial, pale not blue — 13c. *F. bulbigenum* v. *lycopersici*
      dd. Stroma if sclerotial erumpent, blue to pale
        e. Conidia three-septate, 38 × 3.4μ; five-septate, 50 × 3.5μ — 13. *F. bulbigenum*
        ee. Conidia three-septate, 34 × 3.6μ; five-septate, 47 × 3.6μ — 13b. *F. bulbigenum* v. *niveum*
  bb. Macroconidia relatively thicker, 3.7–5μ in diameter (Subgroup *Oxysporum*)
    c. Sclerotia lacking

d. Sclerotial plectenchyma not erumpent
 e. Upper third of conidia not thicker than middle, but tip strongly hooked. Stroma red-violet — 14a. *F. oxysporum* v. *aurantiacum* f.1
 ee. Upper third of conidia thicker than the middle, tip frequently hooked, aromatic — 15. *F. redolens*
dd. Erumpent sclerotial plectenchyma present — 16a. *F. vasinfectum* v. *zonatum*

cc. Sclerotia present
 d. Sclerotia rather small, 0.2–3 mm. or less
  e. Conidia three-septate, 33 × 3.7μ; five-septate, 40 × 3.8μ; with lilac odor — 16. *F. vasinfectum*
  ee. Conidia three-septate, 34 × 3.8μ; five-septate, 42 × 4μ; with fungoid odor — 16b. *F. vasinfectum* v. *lutulatum*
 dd. Sclerotia large, 0.5–2, up to 12 mm.
  e. Five-septate conidia 42 × 4.2μ
   f. Stroma usually white to incarnate; not aromatic — 14. *F. oxysporum* f.6
   ff. Stroma rosy, violet to red (blue with alkali)
    g. Causing wilt diseases
     h. Pathogenic to peas (*Pisum sativum*) — 14. *F. oxysporum* f.8
     hh. Pathogenic to potato (*Solanum tuberosum*) — 14. *F. oxysporum* f.1
    gg. Not causing wilt diseases — 14. *F. oxysporum*
  ee. Five-septate conidia 45 × 4.2μ
   f. Stroma red-violet
    g. Sclerotia numerous; fungus not aromatic — 14b. *F. oxysporum* v. *nicotiana*
    gg. Sclerotia rather limited; seldom gregarious — 14c. *F. oxysporum* v. *cubense*
   ff. Stroma dark purple to red-violet
    g. Three-septate conidia 35 × 4.3μ — 14a. *F. oxysporum* v. *aurantiacum*
    gg. Three-septate conidia 34 × 4μ — 17. *F. dianthi*

8. *Fusarium bostrycoides* Wollenweber and Reinking

Stroma plectenchymatic, brownish-white, then green or violet, with yellowish-white aerial mycelium, sometimes rosy. Microconidia numerous, scattered or gathered into false heads in the mycelium, on simple to whorled or spirally branched conidiophores, one-celled, oval, 6–11 × 2.5–3.24μ, seldom one-septate, 15–22 × 2.5–3.75μ and macroconidia three-septate, 24–29 × 2.5–4μ, almost cylindrical to spindle-sickle-shaped, straight or slightly curved. Sporodochia and pionnotes lacking. Chlamydospores numerous, terminal and intercalary, globose, single or in chains, rough, 6–8μ in diameter.

In soil: Honduras, Central America (164)

9. *Fusarium conglutinans* Wollenweber

Stroma and mycelium pale white, then brownish to rosy-white. Conidia on simple or sparingly branched conidiophores, usually one-celled, seldom one-septate, exceptionally two- to three-septate: three-septate, 20–46 × 2.5–4.5μ. Macroconidia cylindric to spindle-sickle form, straight or slightly curved, tapering at the ends, the basal cell with an attachment papilla, sometimes approaching a foot-cell. Chlamydospores numerous, terminal or intercalary, globose to oval, one-celled 8–12 × 7–10μ, two-celled 14–18 × 6–10μ, usually smooth. Sclerotia lacking.

10. *Fusarium orthoceras* Appel and Wollenweber

Stroma turf-like, with rich aerial mycelium, below plectenchymatic, seldom sclerotial, pale, incarnate, green-flecked, purple-violet (becoming blue with alkali). Mycelium floccose, rosy-white, slightly submerged, and then tough-gelatinous, usually developing numerous one- or sparsely-septate microconidia, more infrequently macroconidia. Abstricts from free conidiophores single conidia, beside one another, soon falling, or united into false heads, exceptionally as evanescent thin pionnotes of incarnate red color, or sparingly sporodochial, and then with richer branching of the conidiophores. Conidia oval-cylindric, straight or curved, one-celled, 4–17 × 2–4μ, one-septate, 10–41 × 2.5–4.5μ, larger conidia almost straight to spindle-sickle-shaped; slender, delicate, falsely septate, with an attachment papilla at the base or inclined toward pedicellate: three-septate, 15–61 × 2.4–4.8μ, five-septate, 25–69 × 3–4.8μ. Chlamydospores terminal and intercalary, globose to pear-shaped, smooth or warted, one-celled, 6–14 × 5–13μ, seldom two-celled, 10–21 × 6–13μ.

From soil: Denmark (68), Egypt (115), England (11), Switzerland (38), U. S. S. R. (108)

United States: Iowa (50), Louisiana (50)

11. *Fusarium angustum* Sherbakoff

Conidia long, almost cylindrical, in the center almost straight or slightly curved, with the ends uniformly tapered, at times twisted, the base usually pedicellate, smaller conidia scattered, larger occurring on many substrates in thin pionnotes:

0-septate, 11 × 2.6μ
1-septate, 21 × 3μ
3-septate, 45.6 × 3.5μ
5-septate, 60 × 4.3μ
6- to 8-septate, 70–102 × 4–4.7μ
Chlamydospores 6–13μ in diameter, two-celled, 13–18 × 6–10μ.
From soil: China (81)
United States: New Jersey (146)

12. *Fusarium lini* Bolley

In nature conidia occur at times in sporodochia, with three septa, 21–41×2.5–4μ, cream to incarnate rosy in mass. In culture on favorable media microconidia predominate. Single isolations with sporodochial development show three- (three- to five-) septate macroconidia: three-septate, 21–41 × 2.5–4.5μ, five-septate, 33–50 × 3.5–4.5μ. Microconidia one-celled, 6–12 × 2–3μ, one-septate, 9–23 × 2–3μ. In form the macroconidia vary from those like *F. orthoceras* to those of *F. oxysporum*. The stroma is variously colored, hyalin, brownish-white, incarnate, greenish, rose to red (in alkaline conditions violet to blue), sclerotial erumpent stromata sometimes present, green to dark blue. Chlamydospores globose to pear-shaped, smooth or wrinkled 5–13μ, usually one-celled, terminal and intercalary, very numerous.

From soil: China (81)

13. *Fusarium bulbigenum* Cooke and Massee

Stroma sometimes flat, plectenchymatic, pale or rosy to red-violet, covered with bright, pink or lilac colored aerial mycelium sometimes rough, developing erumpent sclerotial bodies (0.5–5 mm. in diameter) bright brownish or green to dark blue. Conidia in sporodochia on flat or rounded stromata as well as scattered over the substrate, with easily dissolving ochre to lake colored slimy conidial membranes and later developing numerous submerged, or on the aerial mycelium, terminal and intercalary chlamydospores, 5–12μ in diameter, one- to two-celled or in chains. Smaller conidia one-celled or sparingly septate, larger in sporodochia and pionnotes, three- to five-septate, long, awl-shaped, straight, or slightly sickle-shaped, tapering at both ends, somewhat constricted and hooked at the tip or tapering evenly; the basal cell more or less pedicellate:

0-septate, 5–12 × 2–3.5μ
1-septate, 11–33 × 2–3.7μ
3-septate, 20–54 × 2.3–4μ
5-septate, 34–66 × 3–4.5μ

13a. *Fusarium bulbigenum* Cooke and Massee var. *blasticola* (Rostrup) Wollenweber

Syn. *F. blasticola* Rostrup

Differs from the type by somewhat thicker and less septate, three- (very seldom four- to six-) septate, conidia as well as by the infrequent occurrence of sporodochia and pionnotes:

0-septate, 4–9 × 2–4μ
1-septate, 8–33 × 2–4.3μ
3-septate, 20–48 × 2.7–4.5μ
5-septate, 30–60 × 3–4.8μ

Sclerotia sparingly present, relatively large, or lacking. Chlamydospores (5–12μ in diameter) typical.

From soil: United States: Louisiana (50)

13b. *Fusarium bulbigenum* Cooke and Massee var. *niveum* (E. F. Smith) Wollenweber

Syn. *F. niveum* E. F. Smith

Differs from the type by the somewhat thicker conidia and the saturated purple-red color (becoming blue in alkali) of the wide spread stroma. Mycelium white, carmine, rosy, or purple. Stroma sometimes sclerotially erumpent, dark blue. Sclerotial bodies relatively large, up to 3–6 mm in diameter occurring infrequently, vanishing in older cultures or becoming colorless. Microconidia one-celled, ellipsoidal, or little septate, straight or curved, numerous in the aerial mycelium. Macroconidia laid down in sporodochia or bright orange-red pionnotes, three- to five-septate, long, almost cylindrical to spindle-sickle-shaped, tapering at the ends, tip somewhat constricted, hooked or cone-shaped, truncate at the base, cone-shaped or pedicellate:

0-septate, 5–12 × 2–4μ
1-septate, 10–24 × 2.5–5μ
3-septate, 24–50 × 3–4.7μ
5-septate, 40–66 × 3–5μ

Chlamydospores typical, terminal or intercalary, globose or ovate, smooth, smaller (5–10μ) in conidia, larger (7–21 × 6–17μ) in mycelium. one- or two-celled.

13c. *Fusarium bulbigenum* Cooke and Massee var. *lycopersici* (Brushi) Wollenweber and Reinking

Differs from the type in that the sclerotial, erumpent stroma lacks pigmentation, never becoming blue. Aerial mycelium, floccose white to pink. Plectenchymatic stroma red-violet or pale, sclerotial stroma colorless, smooth, flat, later disappearing. Sporodochia point-like, rounded, later uniting more or less as pionnotes of red or bright orange color. Small conidia one-celled or sparingly septate, richly scattered in the aerial mycelium; larger conidia three- to five-septate:

0-septate, 8 × 2.5μ
1-septate, 18 × 2.8μ
3-septate, 25–66 × 2.3–4.3μ
5-septate, 32–68 × 2.8–4.5μ
Chlamydospores terminal and intercalary, typical.

14. *Fusarium oxysporum* Schlechtendahl

Stroma brownish-white to violet, plectenchymatic, smooth, extended or colored green to blue-black by erumpent, sclerotial hard bodies, and 0.5–3 mm. or 3–6 mm. in thickness, more or less wrinkled, under moist conditions, usually covered by fascicled, medium-high aerial mycelium, later forming sporodochia, more seldom pionnotes with three-septate spindle-sickle-shaped conidia, curved or almost straight, really or weakly pedicellate. Smaller conidia, one- or two-celled, oval to reniform, are numerous in the aerial mycelium but are lacking in the typical fruiting layers of the macroconidia:

0-septate, 5–15 × 2–4μ
1-septate, 10–26 × 2–4.5μ
3-septate, 19–45 × 2.5–5μ
5-septate, 30–60 × 3.5–5μ

Chlamydospores terminal and intercalary, globose, smooth or wrinkled, one-celled seldom two-celled, in hyphae and conidia, 5–15μ, in mycelium sometimes thicker, 10–15μ.

From soil: Canada (15), Egypt (115), India (21), Switzerland (38)
United States: Illinois (129), New Jersey (146), Texas (152)

form 1 Wollenweber
Conidia:
0-septate, 5–15 × 2–4μ
1-septate, 10–26 × 2–4.5μ
3-septate, 19–46 × 2.5–5μ
5-septate, 30–60 × 3–6μ

Macroconidia in sporodochia and pionnotes usually three- seldom four- to five-septate. Microconidia numerous in aerial mycelium. Stroma extended, smooth or sclerotial, erumpent, pale or green to dark blue (almost black). On rice-media aromatic. Chlamydospores typical.

form 6 Wollenweber

This fungus is characterized by pale, whitish-rosy (never violet-purple) color of the stroma, is not aromatic and has conidia in sporodochia and pionnotes, in mass brownish-white, isabellin, incarnate to pale orange-red, three-septate, seldom four- to five-septate: three-septate, 23–45 × 2.8–6μ, five-septate, 34–60 × 3.5–6μ. Microconidia numerous, typical. Sclerotia pale or with bluish cast, sometimes blue-black, tolerably small, 0.1–2 mm. Chlamydospores typical.

form 8 Snyder

Conidia in sporodochia and pionnotes three-, seldom four- to five-septate, exceptionally six- to seven-septate, smaller conidia one- to two-septate, and numerous microconidia scattered on the mycelium are present. Conidia three-septate, 25–59 × 2.8–5μ, five-septate, 35–71 × 3.4–5μ. Sclerotia or sclerotial bodies 0.5–2–5 mm. in diameter, and blue, green, or pale colors occur. Chlamydospores 4–14μ in diameter.

14a. *Fusarium oxysporum* Schlechtendahl var. *aurantiacum* (Link and sp.) Wollenweber

Differs from the primary species by the somewhat larger three- to five-septate conidia, by the increase in number of four- to five-septate conidia, the presence of extended sclerotial bodies which reach either 1–3 mm., or 4–6–16 mm. in size, and the deep purple-violet, often almost chestnut-brown color of the stroma on rice-media. The fungus produces no aromatic odor:

0-septate, 5.5–9.5 × 2.2–2.7μ
1-septate, 12–17 × 2.5–3.8μ
3-septate, 23–48 × 3–5.5μ
5-septate, 33–70 × 3–5.5μ
7-septate, 36–75 × 3.3–4.5μ

Chlamydospores more or less frequent, globose to oval, 5–12μ, two-celled, 11–14 × 7–9μ.

From soil: Canada (16)

form 1 Wollenweber

Differs from *F. oxysporum* var. *aurantiacum* by its nonerumpent stromata, lack of blue-black sclerotia, preponderance of slimy conidial masses (pionnotes), slightly longer conidia and scarcity of four- to five-septate conidia. The micro- and macroconidia measure:

0-septate, $7 \times 2.6\mu$
1-septate, $15 \times 3.2\mu$
3-septate, $24–50 \times 3–5.3\mu$
5-septate, $38–53 \times 3.5–5.3\mu$

Chlamydospores $6–12\mu$, globose, one-celled, smooth or rough, stroma plectenchymatic, expanded, pale or dark purple-red-violet, usually not sclerotial, erumpent. Sclerotia rare, bright golden-yellow. Not aromatic.

14b. *Fusarium oxysporum* Schlectendall var. *nicotianae* Johnson

Differs from forms 1 and 2 by its somewhat larger conidia, from *F. oxysporum* var. *cubense* by its relatively more numerous sclerotia and particularly by its parasitism on tobacco plants. Microconidia numerous, one-celled or sparingly septate, macroconidia in sporodochia and sometimes in pionnotes, three-, seldom four- to five-septate:

3-septate, $35 \times 4.2\mu$
5-septate, $44.3 \times 4\mu$

Chlamydospores $6–10.2\mu$. Sclerotia blue-black, tolerably numerous. Not aromatic.

14c. *Fusarium oxysporum* Schlectendahl var. *cubense* (E. F. Smith) Wollenweber and Reinking

Differs from the primary species by the somewhat longer conidia (especially in pionnotes) and extreme specialization as the cause of wilt disease of *Musa* sp. Conidia in sporodochia and pionnotes, three-, seldom four-, exceptionally five-septate:

3-septate, $17–51 \times 3–4.5\mu$
5-septate, $36–57 \times 3.5–4.7\mu$

Microconidia numerous in aerial mycelium, scattered, typical. Chlamydospores terminal and intercalary, globose or oval, one-celled or two-celled, developed in conidia, $4.5–10 \times 4–8\mu$, one-septate, $9–18 \times 4.5–7.25\mu$, in mycelium, $5.5–9\mu$. Sclerotia or sclerotial bodies blue-black, in a limited number, either 0.5–1 mm. or up to 4 mm. thick.

15. *Fusarium redolens* Wollenweber

Microconidia one-celled, $9 \times 3\mu$ or sparingly septate, one-septate, $16 \times 4.5\mu$. Macroconidia three-, seldom four-, exceptionally five-septate, spindle-sickle-shaped, curved, sometimes resembling *F. solani*, but in the upper third thicker than the middle, generally becoming narrowed at the base to a pedicel or an attachment papilla, occurring in sporodochia or pionnotes, in mass brownish-white, cream or bright incarnate, at first gelatinous, then powdery, fading: three-septate, 17–51 × 3–6.5μ, five-septate, 31–61 × 3.5–6.5μ. Chlamydospores terminal and intercalary, one-celled, 8μ or 3–12μ, two-celled, 11–24 × 5–14μ, smooth or wrinkled, in conidia or mycelium. Blue sclerotia lacking. Plectenchymatic stromata extended, pale, pink to lilac. The fungus is aromatic with an odor of lilac.

16. *Fusarium vasinfectum* Atkinson

This fungus differs from *F. oxysporum* by the somewhat smaller conidia, rich development of pionnotes, commensurate mass of small, 0.1–2 mm. in diameter, verdigris-blue sclerotial plectenchyma and purple-red (becoming blue with alkali) plectenchymatic stromata. Microconidia one-celled or rarely septate, scattered, macroconidia produced in sporodochia and pionnotes of isabellin to pale salmon color, spindle-crescent-shaped, slightly curved at both ends, tapering or beaked, at the base with a foot-cell or attachment papilla:

0-septate, 4–12 × 2–3μ
1-septate, 8–25 × 2–4μ
3-septate, 23–48 × 3–4.5μ
Seldom 4- to 5-septate; 5-septate, 30–50 × 3–5μ

Chlamydospores terminal and intercalary, one-celled, 7–13μ in diameter; two-celled, 12.6 × 7μ. On cooked rice, the fungus produces a strong aromatic lilac odor.

From soil: Canada (15)

16a. *Fusarium vasinfectum* Atkinson var. *zonatum* (Sherbakoff) Wollenweber

Differs from the foregoing fungus by the concentric zonation of its mycelial growth, by the lack of dark blue sclerotial stromata, by paler, somewhat honey-yellow color of the sporodochial and pionnotal conidial slime and by the slightly longer and thicker conidia:

3-septate, 37 × 3.9μ
5-septate, 42 × 4.1μ
An aromatic (lilac) odor is produced.
From soil: Canada (15)

16b. *Fusarium vasinfectum* Atkinson var. *lutulatum* (Sherbakoff) Wollenweber

Differs from the primary species by the somewhat longer conidia and by the occasional occurrence of numerous small (up to 0.5 mm. in diameter) blue-black sclerotial bodies. The conidia are three-, seldom four- and five-septate:

3-septate, 28–42 × 3.2–4.5μ
5-septate, 37–47 × 3.5–4.5μ
6- to 7-septate conidia, exceptionally observed, 50–66 × 3.5–5μ

Numerous microconidia, one- or two-celled occur in the aerial mycelium. Chlamydospores terminal and intercalary, one-celled (6–8 × 5–7μ), two-celled (8–12 × 4–7μ). The fungus produces an aromatic odor.

From soil: Canada (15)

17. *Fusarium dianthi* Prillieux and Delacroix

The conidia occur in bright orange colored sporodochial and pionnotal slimy masses and are spindle-sickle-shaped, pedicellate, often hooked at the tapered tip and somewhat thicker in the upper third than in the central portion, as well as strongly dorsiventral, usually three- (three- to five-) septate, exceptionally one- to two- and six- to eight-septate. Microconidia are numerous in white to rosy floccose aerial mycelium, one-celled or rarely septate:

0-septate, 5–15 × 1.5–4.5μ
1-septate, 10–30 × 2–4μ
3-septate, 16–63 × 2.5–5.5μ
5-septate, 30–80 × 3–5.5μ

The chlamydospores are globose, smooth or roughened, 6–12μ, two-celled long-oval, 13–16 × 5–13μ. The fungus is odorless on rice.

SECTION V. MARTIELLA

KEY TO THE SPECIES

a. Conidia almost cylindrical, curved only at tip; conidia in mass isabellin-ochre, or hyalin, at times blue from the stroma or verdigris — 18. *F. coeruleum*

aa. Conidia curved at both ends, sessile to slightly pedicellate; conidia in mass brownish-white, cream or butter-yellow, at times blue from the stroma or verdigris

  b. Average diameter of three- to five-septate conidia 4–5μ

    c. Conidia in masses usually three-septate

      d. Conidia three-septate, 35 × 4.2μ; five-septate 36 × 4.8μ — 19a. *F. javanicum* v. *radicicola*

      dd. Conidia three-septate, 30 × 4.5μ; five-septate 36 × 4.8μ — 20a. *F. solani* v. *minus* ddd. Conidia three-septate, $35 \times 4.6\mu$; five-septate, $50 \times 4.8\mu$ — 20b. *F. solani* v. *striatum*

cc. Conidia in masses usually three- to five-septate — 19. *F. javanicum*

ccc. Conidia in masses usually five-septate — 19b. *F. javanicum* v. *ensiforme*

bb. Average diameter three- to five-septate conidia $5–6\mu$

c. Conidia in masses usually three-septate

d. Conidia three-septate, $36 \times 5.5\mu$; five-septate, $48 \times 5.7\mu$ — 20. *F. solani*

dd. Conidia three-septate, $39 \times 5\mu$; five-septate, $49 \times 5.3\mu$ — 20c. *F. solani* v. *martii* f.1

ddd. Conidia three septate, $44 \times 5.2\mu$; five-septate, $55 \times 5.5\mu$ — 20c. *F. solani* v. *martii*

dddd. Conidia three-septate, $44 \times 5.2\mu$; five-septate, $56 \times 5\mu$ — 20c. *F. solani* v. *martii* f.3

ddddd. Conidia three-septate, $45 \times 4.9\mu$; five-septate, 54×5.1 — 20c. *F. solani* v. *martii* f.2

cc. Conidia in masses usually five-septate — 20d. *F. solani* v. *eumartii*

18. *Fusarium coeruleum* (Libert) Saccardo

Conidia in sporodochia, in slimy diffuse layers or scattered in the aerial mycelium, almost straight or slightly crescent-shaped, with obliquely cone-shaped, ellipsoidal or rounded tip and blunt-oval to teat-shaped base. Conidia in mass isabellin-ochre to brownish-white, frequently blue-violet to blue-black or verdigris of the stroma. Chlamydospores terminal or intercalary, one-celled, spherical (9μ) to pear-shaped ($8 \times 9\mu$) or two-celled ($14 \times 9\mu$). Stroma diffuse or warty, sclerotial, hyalin or violet to blue-black.

Conidia in masses, three-septate, $21–47 \times 3.5–6\mu$; seldom four- to five-septate, exceptionally non- to two-septate, six- to seven-septate.

From soil: Canada (15), Egypt (115)

19. *Fusarium javanicum* Koorders

Syn. *F. javanicum* var. *theobromae* (Appel and Strk.) Wollenweber

Conidia in mass brownish-white to pale brown, in age coffee-brown or tinted by the frequent olive-green to olive-brown leathery-gelatinous, seldom sclerotial stroma. Small one-celled or septate conidia, scattered in the aerial mycelium usually abundant and on drying are visible as a bright powdery layer on the substrata. Larger sickle spores, long, slightly curved, more strongly bent at the tip, constricted at both ends, more or less pedicellate at the base, three- to five-, exceptionally six- to eight-septate, in sporodochia and pionnotes:

0-septate, 8 × 3μ
1-septate, 18 × 3.6μ
3-septate, 22–54 × 3.5–6μ
5-septate, 35–60 × 4–6μ
7-septate, 60 × 5μ
Chlamydospores one- to two-celled, 5–8μ in diameter.

19a. *Fusarium javanicum* Koordevs var. *radicicola* Wollenweber
Syn. *F. radicicola* Wollenweber

The fungus almost always has a large number of one-celled or septate small conidia scattered in the mycelium or gathered into false heads. In addition it develops, here and there, brownish-white masses of larger sickle-spores in sporodochia or more seldom, in slimy (pionnotes-like) layers, three-, seldom four-, exceptionally five-septate, long, slightly curved, more strongly bent and constricted at the tip, and weakly pedicellate at the base:

0-septate, 4–15 × 1.7–4.5μ
1-septate, 8–27 × 3–5μ
3–septate, 20–50 × 3.5–5.3μ
5-septate, 38–59 × 4–5.3μ

In age these masses are darker or changed by the olive-green to coffee-brown stroma colors. Chlamydospores common, terminal or intercalary, one- or two-celled, in chains or knots: one-celled, 9–10 × 8.5–9μ; two-celled, 16–22 × 5–12μ, smooth or rough.

From soil: United States: Idaho (106), Texas (152)

19b. *Fusarium javanicum* Koorders var. *ensiforme* (Wollenweber and Reinking) Wollenweber

Stroma erumpent, sclerotia wrinkled, often blue-black. Conidia in mass (sporodochia, pionnotes) whitish to golden-yellow, long, slightly curved sickle-shape, somewhat constricted at the tip, truly pedicellate at the base five- (three- to six-) septate, exceptionally seven-septate, smaller one-celled forms scattered in the mycelium, seldom septate:

0-septate, 5–12 × 2–5μ
1-septate, 20–25 × 3.3–4.5μ

3-septate, 37–50 × 3.75–5μ
5-septate, 55–72 × 4.5–5μ
6-septate, 69–81 × 4.7–5μ

Chlamydospores one- to two-celled, round or oval, smooth or rough, 6–9μ in diameter.

20. *Fusarium solani* (Martius) Appel and Wollenweber
Syn. *F. alluviale* Wollenweber and Reinking

Conidia scattered in false heads, sporodochia or pionnotes, in mass brownish-white to loam-yellow or from green to dark brown, leathery or sclerotial stromata blue or green-flecked. Larger conidia strongly twisted spindle form, slightly curved, both ends rounded to tenpin-like base with a scarcely perceptible wart directed obliquely to the long axis, rarely slightly pedicellate, three- to five-septate:

0-septate, 11 × 3.8μ
1-septate, 20 × 4.3μ
3-septate, 19–50 × 3.5–7μ
5-septate, 32–68 × 4–7μ

Chlamydospores terminal and intercalary, brown, single, globose to pear-shaped, one-celled, 8 × 8μ, two-celled, 9–16 × 6–10μ; rarely in chains or knots, with smooth, sometimes finely warted when dry, walls.

From soil: Austria (66), Denmark (68), Egypt (115), England (36), Switzerland (38), U. S. S. R. (108)

United States: New Jersey (146)

20a. *Fusarium solani* (Martius) Appel and Wollenweber var. *minus* Wollenweber

Differs from *F. solani* by smaller, three-septate, seldom four-septate, exceptionally five-septate conidia:

3-septate, 20–41 × 3.5–6μ
4-septate, 35 × 4.7μ
5-septate, 30–50 × 3.7–6μ

Chlamydospores rough to warty or smooth, one-celled, 5–11μ, two-celled, 12 × 7.3μ, also catenulate and in knots.

20b. *Fusarium solani* (Martius) Appel and Wollenweber var. *striatum* (Sherbakoff) Wollenweber

The fungus belongs between *F. solani* and *F. javanicum*. It develops comparatively small sporodochia, later pionnotes, hyalin or colored by the stroma blue-green, olive, or sepia-brown. Conidia long, slightly curved, both ends constricted, more or less pedicellate at the base, usually three- (non- to five-) septate:

0-septate, 8.5–12.5 × 2.5–3.6μ
1-septate, 14–30 × 3–4.7μ
3-septate, 22–50 × 3.5–5.7μ
4-septate, 30–61 × 4–5.3μ
Chlamydospores, 7–11 × 6–9μ.

20c. *Fusarium solani* (Martius) var. *martii* (Appel and Wollenweber) Wollenweber

Conidia longer and more slender than *Fusarium solani*, in the center less, at the tip more strongly curved, with an attachment wart at the base, sometimes also a footcell, usually three- to four- (five-) septate, exceptionally six- to seven-septate:

0-septate, 8 × 3.5μ
1-septate, 18 × 4.2μ
3-septate, 31–60 × 4.6μ
5-septate, 36–70 × 4.5–6μ
7-septate, 62–90 × 4.5–6μ

Conidia in mass brownish-white, ivory to light brown, on greenish-blue or leather-brown stroma which may modify the conidia colors resulting in tone-mixtures of verdigris, gray, coffee-brown to black. Chlamydospores as in *F. solani*, 8–11 × 7.5–9.3μ.

From soil: Canada (15)

form 1 Wollenweber

Conidia three-septate, seldom four-septate, exceptionally five-septate, small non- to two-septate:

0-septate, 8 × 3.8μ
1-septate, 17 × 4.5μ
3-septate, 22–52 × 3.5–6μ
4-septate, 46 × 5.2μ
5-septate, 39–62 × 4.2–6μ

Chlamydospores in conidia, 8.5 × 6μ; in mycelium, 10 × 8μ (one-celled), 8–22 × 6–12μ (two-celled). Otherwise like *F. solani* var. *martii*.

form 2 Snyder

Conidia:
3-septate, 20–58 × 3.5–5.5μ
4-septate, 38–61 × 4.4–5.5μ, seldom
5-septate, 45–60 × 4.7–5.5μ, single
0-septate, 8 × 3.5μ, scattered
1-septate, 17 × 4.3μ

form 3 Snyder

Conidia:

3-septate, 33–52 × 4.5–5.5 μ

4-septate, 39–62 × 4.7–5.5 μ, seldom

5-septate, 52–62 × 5–5.5 μ, exceptionally

0-septate, 9 × 3.6 μ, scattered

1-septate, 17 × 4.3 μ

20d. *Fusarium solani* (Martius) Appel and Wollenweber var. *eumartii* (Carpenter) Wollenweber

Smaller conidia, one-celled or septate, rather scarce; larger conidia in sporodochia and pionnotes five- (three- to seven-) septate, exceptionally eight-septate:

0-septate, 5–14 × 2.5–5 μ

1-septate, 14–28 × 3–5.5 μ

3-septate, 22–63 × 4–6.6 μ

4-septate, 49 × 5.7 μ

5-septate, 36–77 × 5–8 μ

7-septate, 60–89 × 5–8 μ

Conidia in mass, brownish-white, sometimes colored by the stroma, (cf. *F. solani*) greenish or brown. Chlamydospores 7–8 μ in diameter as in *F. solani*, differing from that fungus chiefly by the larger number of septations and hence larger conidia.

SECTION VI. EUPIONNOTES

KEY TO THE SPECIES

a. Sickle-spores cylindric or slightly dorsiventral, allantoid, in mass isabellin-red — 21. *F. merismoides*

aa. Sickle-spores rather small, dorsiventral, pointed, little septate

- b. Red-brown sclerotia lacking — 22. *F. dimerum*
- bb. Red-brown sclerotia present — 22a. *F. dimerum* var. *nectrioides*

21. *Fusarium merismoides* Corda

Syn. *F. udum* (Berkeley) Wollenweber

Mycelium sparse, arachnoid, hyalin to pink, easily wilting, later forming a stroma, turf-like, gelatinous, smooth or wrinkled, above raggedly coremium-like, fascicled, club-like or radiating. Conidiophores sometimes branched in whorls. Conidia scattered, dusty, pale then rosy, or layered in evanescent slime, reddish to orange-red, later paler, cylindric-spindle-shaped, both ends tenpin-like to ellipsoidally narrowed, or rounded, not pedicellate, but sometimes somewhat constricted at the base, at the tip asymmetrical, with swollen back-line, slightly curved three- (non- to seven-) septate:

3-septate, 23–60 × 2.2–5μ

5-septate, 30–61 × 3–5μ

1-septate, 13–30 × 2.5–4μ

Chlamydospores globose 5–8μ, one- or two-celled (11–7μ) in old conidia or intercalary, single, paired, seldom in chains.

From soil: U. S. S. R. (108)

United States: Louisiana (50)

22. *Fusarium dimerum* Penzig

Stroma and mycelium white to incarnate-rosy. Conidia spindle-sickle-shaped, pointed at both ends, small, one-septate, 8–30 × 2–4μ, seldom non- to three-septate, in bright orange-colored pionnotes as well as in sporodochia or scattered as a dusty layer in the mycelium. Conidiophores various, often branched in whorls. Chlamydospores intercalary, one-celled, 7 × 7μ, globose, or two-celled, 12 × 6.5μ, or united in chains.

From soil: United States: Idaho (106)

22a. *Fusarium dimerum* Penzig var. *nectrioides* Wollenweber

Differs from the primary species by thinner one- to two-septate conidia and by the presence of numerous red-brown, globose sclerotia up to 0.1 mm. thick.

Conidia one-septate, 8–24 × 2–3μ, two-septate, 15–30 × 2–3.2μ.

In soil: Honduras (164)

SECTION VII. ARACHNITES

A single species treated.

23. *Fusarium nivale* (Fries) Cesati

Conidia scattered on cobwebby, hyalin to rosy, fasciculate, loose or bushy aerial mycelium, sometimes also in balls, or wide-spread, slimy, rosy, lake-orange masses which become darker, (cinnamon-brown) resinous on drying, paler (isabellin) in moist conditions, normally spindle-sickle-shaped, curved, tapered at the ends, and tenpin-shaped to rounded, not pedicellate, seldom somewhat constricted at the base, one- to three-septate, occasionally four- to seven-septate, mixed with scattered one-celled forms:

0-septate, 5–18 × 2–4μ

1-septate, 9–23 × 2.2–4.5μ

3-septate, 13–36 × 2.3–4.5μ
4- to 7-septate, 19–30 × 2.5–4μ

Chlamydospores and sclerotia lacking. Stroma delicate, thin, evanescent, or plectenchymatic, wrinkled, hyalin, isabellin, rosy, orange- to brick-red, later leather brown.

SECTION VIII. LATERITIUM

KEY TO THE SPECIES

a. Sporodochia small or not observed — 24b. *F. lateritium* var. *minus*

aa. Sporodochia typical and always present
- b. Conidia in sporodochia and pionnotes ± three- to five-septate — 24. *F. lateritium*
- bb. Conidia in sporodochia and pionnotes ± five-septate — 24a. *F. lateritium* var. *longum*

24. *Fusarium lateritium* Nees

Stroma fleshy, erumpent, smooth, arched, or rough, sclerotial, plectenchymatic, coarsely roughened to antler-like, forked or jagged, hyalin, rosy, yellow, orange, chestnut-brown to dark blue (becoming purple with acid), sometimes with globose bluish-black or pale sclerotia. Aerial mycelium hyalin, rose or yellowish, with the colors of the plectenchymatic part of the stroma. Conidia of various sizes scattered in the aerial mycelium, later in erumpent tubercle-like sporodochia which occur singly or in groups, or in a more or less evanescent pionnotes on flattened extended stromata, in mass, brick-red, golden-yellow, rosy to salmon, cut off from irregular or whorled conidiophores which grow in loose association or in thickly branching bush-like groups, three- to five-septate, seldom more or less septate, spindle-sickle-shaped, long, in their middle cylindric or slightly dorsiventral, at the tip often with a hooked back-line, with the tip cell constricted sometimes snout-shaped, at the base with a typical footcell, walls and septa slightly refractive. Conidia:

0-septate, 7–11 × 2.5–3.5μ, scarce
1-septate, 11–35 × 2–5μ
3-septate, 13–52 × 2–5μ
5-septate, 24–84 × 2.5–5μ
7-septate, 32–84 × 3–5μ

Chlamydospores rare, intercalary, in conidia and in mycelium. Sclerotial plectenchyma up to 5 mm. in diameter, brown or colorless.

24a. *Fusarium lateritium* Nees var. *longum* Wollenweber

The conidia are long, cylindrical, tapering at both ends, constricted at the tip, basal cell pedicellate, in sporodochial and pionnotal slimy masses, orange-red, usually five-septate, seldom three- to four- or six- to seven-septate, exceptionally nine-septate; mixed with small non- to one-septate spores scattered sparingly in the aerial mycelium:

0-septate, 8–16 × 2–3.3μ
1-septate, 9–20 × 2.5–3.5μ
3-septate, 19–54 × 3–4.2μ
5-septate, 45–80 × 3.5–5.5μ
7-septate, 56–90 × 4–6μ
9-septate, 80–94 × 4.5–6μ

In the stroma, small, globose, dark blue or pale sclerotia occur more or less frequently causing pale or blue flecks.

24b. *Fusarium lateritium* Nees var. *minus* Wollenweber

Differs from the primary species by narrower septation of the conidia, which lie in a flat layer of orange-red color on a flat stroma or as a loose powder scattered in the aerial mycelium. Conidia non- to one-septate, seldom up to 19 per cent of the total three-septate:

0-septate, 4–14 × 2–3.8μ
1-septate, 15–23 × 3–3.75μ
3-septate, 22–25 × 3–4μ
From soil: Egypt (115)

SECTION IX. ROSEUM

KEY TO THE SPECIES

a. Conidia prevailingly three-septate, 33 × 3.2μ — 25. *F. graminum*
aa. Conidia usually three- to five-septate; three-septate, 35 × 3.3μ
 b. Carmine-red pigment present in stroma — 26. *F. avenaceum*
 bb. Carmine-red pigment lacking in stroma — 26a. *F. avenaceum* f.1

25. *Fusarium graminum* Corda

Conidia on sporodochia in arched membrane on flattened stromata of various rough or smooth types, in gelatinous masses orange colored, in dry condition powdery and hyalin, rosy, in resinous drying of darker color, according to the place of occurrence and mycelial development. Stroma hyalin, rosy, citron-yellow, ochre, carmine- to brown-red. Conidia spindle-sickle-shaped, slender, thin, delicate walled, in the center almost cylindrical or somewhat dorsiventral, slightly curved, at times strongly bent at the tip, both ends gradually tapered, extended almost to a thread at the tip, in addition to shorter stouter conidia with thicker diameters; base typically pedicellate, conidia usually three-septate, seldom more or less septate:

3-septate, 20–52 × 2–5μ
5-septate, 32–62 × 2.5–5.3μ
7-septate, 40–80 × 3–5.3μ
0-septate, 11 × 3.0μ
1-septate, 11–24 × 2–5μ

26. *Fusarium avenaceum* (Fries) Saccardo
Syn. *F. subulatum* Appel and Wollenweber

Conidia seldom scattered, in false heads or balls, as layers in sporodochia and pionnotes, in gelatinous masses, orange, vermillion, scarlet, in resin-like drops becoming darker, in dry powdery condition lighter, becoming pink, changed by the coloring of the stroma below them. Stroma yellow, ochre, carmine to brown-red, aerial mycelium brighter, white or influenced by the stromatic color tones. Erumpent sclerotial, knotty, roughened stromata and true globose, single or gregarious sclerotia (60–80μ in diameter) of a dark blue or paler color are very seldom found. Conidia long, awl- or thread-like, proportionally circularly or ellipsoidally curved or both ends especially at the tip somewhat more strongly bent than the middle; section through the ends generally observed; basal cell more or less markedly pedicellate. Conidia three- to five-, seldom more or less septate:

3-septate, 22–61 × 2.3–6μ
5-septate, 35–80 × 2.5–6μ
7-septate, 61–74 × 3.4–5μ
0-septate, 6–17 × 2.5–4μ
1-septate, 10–25 × 2.4μ

Conidiophores simple or loosely to bushily branched with scattered or with two- to four-, seldom five-membered whorls of side branches.

From soil: Canada (15)

form 1 Wollenweber and Reinking

The fungus differs from the primary species by the absence of carmine-red color in the stroma. Yellow colors occur. They remain yellow upon the application of ammonia and do not become blue as does the primary species. Conidia in sporodochia and pionnotes orange, three- to five-septate, seldom more or less:

3-septate, 22–48 × 2–5.3μ
5-septate, 24–88 × 2.8–5.5μ
7-septate, 46–91 × 3–5.5μ
9-septate, 65–102 × 3-5μ
0-septate, 8–14 × 1.5–4.7μ
1-septate, 12–22 × 2–6μ

SECTION X. ARTHROSPORIELLA

KEY TO THE SPECIES

a. Sporodochia lacking
  b. Pionnotes lacking
    c. Macroconidia straight, seldom curved
      d. Conidia three-(four to five) septate, three-septate 26 × 4.2μ; five-septate 34 × 4.5μ ... 27. *F. semitectum*
      dd. Conidia five- (three to seven) septate, three-septate 24 × 3.6μ, five-septate 39 × 4.3μ ... 27a. *F. semitectum* var. *majus*
    cc. Macroconidia strongly curved ... 28. *F. camptoceras*
  bb. Pionnotes-like conidial slime present ... 29. *F. anguioides*
aa. Sporodochia and pionnotes present at times, conidia ± pedicellate
  b. Stroma and aerial mycelium of similar color, hyalin ... 30. *F. concolor*
  bb. Stroma of various colors ... 31. *F. diversisporum*

27. *Fusarium semitectum* Berkeley and Ravenel

Aerial mycelium white-incarnate or isabellin. The lower part of the stroma plectenchymatic, bright brown (in some isolations, violet-carmine). Chlamydospores intercalary. Sporodochia lacking. Conidia scattered on aerial mycelium, spindle-, lance- or sickle-shaped, straight or slightly curved, not pedicellate but often with an attachment-wart on the base. Smaller conidia non- to two-septate, larger three-septate, seldom four- to five-septate, occasionally six- to seven-septate:

0-septate, 4–16 × 2–4μ
1-septate, 8–20 × 2–4.5μ
3-septate, 23–50 × 3.2–6.25μ
7-septate, 36–77 × 4–6μ

27a. *Fusarium semitectum* Berkeley and Ravenel var. *majus* Wollenweber
Syn. *F. incarnatum* (Robin) Saccardo

Differs from the primary species by the larger, more septate conidia. Conidia, powdery in white to rosy or brown, floccose aerial mycelium or united into balls in false heads, in mass lake-rose, five-septate, seldom three- to four-septate, occasionally six- to ten-septate with smaller non- to two-septate forms intermixed. Larger conidia spindle-lance form, straight or slightly curved, usually tenpin-like, at times with attachment wart at the base, which exceptionally forms a footcell:

0-septate, 5–15 × 2–4μ
1-septate, 9–24 × 2.5–4μ
3-septate, 13–40 × 2.5–4.8μ
5-septate, 29–52 × 2.5–6μ
7-septate, 45–70 × 3.7–6.2μ
9-septate, 50–70 × 4–6μ

28. *Fusarium camptoceras* Wollenweber and Reinking

Aerial mycelium carmine-pink or isabellin, the lower part of the stroma leathery brown or rosy. Chlamydospores intercalary. Sporodochia lacking. Conidia scattered in the aerial mycelium, sickle-shaped, strongly curved, tapering at both ends, tip more or less constricted, basal cell rounded, or conical-pointed, not pedicellate, but at times with an attachment papilla, smaller conidia non- to two-septate, larger three- to five- (six- to seven-) septate:

0-septate, 7–12 × 2.5–3.5μ
1-septate, 11–18 × 3–4μ
3-septate, 17–28 × 3.5–5.5μ
5-septate, 32–37 × 4.5–6μ
7-septate, 36–46 × 4–5μ
In soil: Honduras, Central America (164)

29. *Fusarium anguioides* Sherbakoff

Conidia of various form, short, spindle-, lance-, and wedge-shaped, rounded or truncate at both ends, non- to three- or more septate, and typically slender, resembling *F. avenaceum*, slightly curved or almost straight to twisted, one- to fifteen-septate conidia with more or less pedicellate base:

3-septate, 20–38 × 3.9–5.3μ
5-septate, 47–68 × 3.9–4.6μ, usually
6- to 7-septate, 65–86 × 4.2–5.2μ, seldom

8- to 9-septate, 80–102 × 4.3–5.8μ, exceptionally

The color of the slimy conidial mass is rosy, orange-yellow to cinnamon-brown, that of the stroma leathery-brown, ochre and brick-red (the ochre to brown color developed on rice does not turn violet upon the application of ammonia but remains yellow or becomes yellow- to olive-green.) Chlamydospores intercalary in and on the conidia as well as intercalary in the mycelium, single (8μ in diameter) and in chains or knots, leathery brown.

From soil: England (11)

30. *Fusarium concolor* Reinking

Aerial mycelium floccose, white to incarnate, the under part of the stroma more or less plectenchymatic, leather-yellow to bright orange, seldom rosy-red. Conidia scattered or in slimy red masses, rarely found in sporodochia. Smaller conidia numerous, long-ellipsoid-oval, one-celled or septate, large conidia, long spindle-sickle-shaped, strongly curved, both ends usually tapered, tip cell slightly constricted almost snout-like, basal cell conical truncate, rounded or almost pedicellate, three- to five-seldom six- to seven-septate:

0-septate, 10–16 × 2.5–3.3μ
1-septate, 13–30 × 2.5–4.3μ
3-septate, 24–70 × 3–5.5μ
5-septate, 46–77 × 3.7–5.3μ
7-septate, 68–80 × 4–5μ

Chlamydospores terminal or intercalary, globose, smooth, becoming wrinkled, one-celled (7–12 × 7–11μ), two-celled (13–15 × 9–10μ) sometimes in long chains. Sclerotia lacking.

31. *Fusarium diversisporum* Sherbakoff

Conidia of various shapes, short, spindle-form, septate (three-septate, 28 × 4.3μ) to sickle-shaped in sporodochia and extended slimy masses, usually five-septate, 41–61 × 2.9–4.4μ, slightly curved, or somewhat twisted, pointed at the tip and pedicellate at the base, rosy-cinnamon-brown in mass. Chlamydospores intercalary in the mycelium and in the conidia. Sporodochia few but well formed. Aerial mycelium rich, of medium fineness, white. Yellow-rose, gray or amber-brown colors may occur in the stroma. By the pedicellate, small, awl-shaped conidia the fungus resembles *Roseum*, by the thick lanciform conidia, *F. semitectum*. Small one- or two-celled conidia occurring quite frequently in the younger parts of the colony measure: non-septate, 10 × 3.25μ and one septate, 12 × 3μ.

From soil: America and Asia (164)

SECTION XI. GIBBOSUM

KEY TO THE SPECIES

a. Curvature of sides of macroconidia more or less parabolic
- b. Conidia three-septate, 33 × 4μ; five-septate, 46 × 4.6μ — 32. *F. equiseti*
- bb. Conidia three-septate, 33 × 3.75μ; five-septate, 42 × 4.3μ — 32a. *F. equiseti* var. *bullatum*

aa. Curvature of sides of macroconidia more or less hyperbolic
- b. Stroma not carmine-red to golden-yellow
  - c. Conidia five-septate, 43 × 4.4μ; seven-septate, 53 × 4.7μ — 33. *F. scirpi*
  - cc. Conidia five-septate, 47 × 4μ; seven-septate, 55 × 4.3μ — 33a. *F. scirpi* var. *caudatum*
- bb. Stroma carmine-red to golden-yellow; conidia five-septate, 45 × 4.1μ; seven-septate, 55 × 4.3μ — 33b. *F. scirpi* var. *acuminatum*

32. *Fusarium equiseti* (Corda) Saccardo
Syn. *F. falcatum* Appel and Wollenweber
F. viticola Thuemen

Conidia at first sparingly scattered in whitish to yellow to pink aerial mycelium, one-celled or septate, oval or long to spindle-sickle-shaped, sometimes comma-like, disappearing with the occurrence of typical large sickle-spores. Macroconidia in tubercular sporodochia on plectenchymatic, pale or brown, never carmine-red stroma of various extent, or in slimy easily dissolving masses, also in balls, seldom scattered as dust, at first paler, almost mealy-white, then ochre to lake-rose, on drying becoming honey to cinnamon-brown when resinous, of brighter colors as a dry powder; typically of a twisted spindle-shape with parabolic curvature in the thicker central part, and gradually tapered toward the ends, extended at the tip into a thin, straight or strongly curved point, and with a foot-cell at the base; back-line more or less bent outwards, belly-line flattened inwards, equally divided by septa, usually five-septate, seldom three- to four-septate, exceptionally up to twelve-septate:

0-septate, 7–18 × 2.5–6μ
1-septate, 12–24 × 2–4μ
3-septate, 12–44 × 2.3–5.5μ

5-septate, 26–74 × 2.8–5.7μ
7-septate, 42–80 × 4.6μ

Conidiophores simple or branched. The lateral branches extended tree-like, or compressed bush-like; conidiophores in two or three, exceptionally more, many times multiplied whorls, and have at their ends small sterigma-like papillae in ones, twos or threes, and also longer spore-bearing members. Sterile and fertile hyphae are irregularly septate and 3–6μ thick. Chlamydospores (6–14μ in diameter) round; smooth or rough, on mycelium, on old conidiophores as well as in conidia, more often intercalary then terminal, more frequently many-celled in chains or knots than one-celled, in mass brown.

From soil: Canada (15) (16), Denmark (68), U. S. S. R. (108)

32a. *Fusarium equiseti* (Corda) Saccardo var. *bullatum* (Sherbakoff) Wollenweber

Syn. *F. bullatum* Sherbakoff

The conidia of this form are usually somewhat less curved than those of the primary species and other typical members of the group, the pedicel of the conidial base is not always so sharply expressed, the septation is less rather than greater and in the mycelium often there occur lanciform conidial forms resembling Arthrosporiella—or subnormal forms resembling Discolor-fusaria. Typical sporodochial and pionnotal conidia measure: five-septate, 31–47 × 4.1–4.9μ and three-septate, 30–36 × 3.7–3.8μ. They are cream to lake color in mass. Chlamydospores as in the primary species are usually intercalary in chains and in small or large heads. Aerial mycelium is usually abundant and of medium height and thickness. The color is almost pure white, while the stroma is leather colored and this color is also present at times in the substrata.

From soil: United States: New Jersey (146)

33. *Fusarium scirpi* Lambotte and Fautrey

Sporodochia occurring as pale yellowish-red or ochre to lake, when dry rust-brown, pinhead-like points occur, either fine powdery sand-like, dissolving in moisture and slimy, when dry incrusted bright brown or paler, at times cinnamon-brown. Aerial mycelium loose, downy to felty, later disappearing, at first, hyalin or brownish, lower parts of stroma leather-brown to dark brown, rarely with blue-black globose sclerotia. Conidia as in *F. equiseti*, but the tips longer extended and sharper, the back-line more strongly hyperbolic, the crosswalls of greater number and closer together in the central portion than at the ends. Smaller scattered conidia at first rather frequent, non- to three-septate, oval, spindle-shaped, kidney- or comma-shaped, also club- to lance-shaped, larger sporodochial and pionnotal spores typical sickle-shaped, five- (five- to seven-) septate, seldom three- to four-septate or eight- to eleven-septate:

0-septate, 5–12 × 2.4μ
1-septate, 8–20 × 2–4μ
3-septate, 10–55 × 2.5–7.3μ
5-septate, 20–73 × 3–6μ
7-septate, 30–75 × 3.8–6μ
9-septate, 51–83 × 4.5–6μ

Chlamydospores intercalary, seldom terminal, usually in chains or knots of brown color, seldom single, 7–14μ in diameter, globose. Sclerotia if present, globose, brown to dark blue, 60–80μ in diameter.

From soil: Canada (15)

33a. *Fusarium scirpi* Lambotte and Fautrey var. *caudatum* Wollenweber
Syn. *F. caudatum* Wollenweber

Conidia comparatively small, typically with a somewhat thread-like elongated, or variously strongly curved tip-cell and pedicellate basal cell, five-septate, seldom more or less septate:

1-septate, 14 × 3μ
3-septate, 13–33 × 2.3–5μ
5-septate, 22–80 × 3–5μ
7-septate, 54–80 × 3.5–4.5μ

Chlamydospores numerous. Stroma hyalin or brown, never carmine or blue.

From soil: United States: New Jersey (146)

33b. *Fusarium scirpi* Lambotte and Fautrey var. *acuminatum* (Ellis and Everhart) Wollenweber
Syn. *F. acuminatum* Ellis and Everhart
F. sanguineum Sherbakoff (non Fries)
F. lanceolatum Pratt

Stroma plectenchymatic, variously colored, blood-red, purple, yellow, at times erumpent sclerotial, dark-blue, brown or pale. Aerial mycelium white or pink. Conidia in sporodochia and pionnotes orange, sickle-shaped, tapered at both ends, the tip more or less elongated into a thread, the base pedicellate or only with an appendage, often rounded to truncate, five-septate, seldom three- to four-septate, exceptionally non- to two- or six- to seven-septate:

0-septate, 4–12 × 2–5μ
1-septate, 11–12 × 2.2–4μ
3-septate, 16–44 × 2.5–4.5μ
5-septate, 28–61 × 3–5.3μ
7-septate, 45–84 × 3.2–5.2μ

Chlamydospores intercalary, usually in chains or knots, rarely terminal, in conidia often one- or two-celled, globose, 7–20μ, one-septate, 20–30 × 10–18μ, brown in mass.

From soil: China (81)
United States: Idaho (106), New Jersey (146)

SECTION XII. DISCOLOR

KEY TO THE SPECIES

a. Conidia with foot-cells not predominating; fruiting layer floccose (Subgroup *Trichothecioides*) — 34. *F. trichothecioides*
aa. Conidia with foot-cells predominating in pionnotes and sporodochia (Subgroup *Saubineti*)
  b. Tri-septate conidia 3–4μ in diameter — 35. *F. reticulatum*
  bb. Tri-septate conidia 4-5μ in diameter
    c. Macroconidia up to 5μ in diameter
      d. Stroma carmine-purple, chestnut-brown, yellow, rosy
        e. Conidia relatively stout, in sporodochia and pionnotes
          f. Conidia usually three-, seldom four- to five-septate — 36. *F. sambucinum* f.2
          ff. Conidia ± three- to five-septate
            g. Blue sclerotia and plectenchyma lacking — 36. *F. sambucinum*
            gg. Blue sclerotia and plectenchyma present — 36. *F. sambucinum* f.1
        ee. Conidia relatively long and slender
          f. Conidia five(three to five)-septate — 41a. *F. sublunatum* v. *elongatum*
          ff. Conidia three- to five-septate
            g. Conidia three-septate, 41 × 4.3μ; five-septate, 51 × 4.9μ; seven-septate, 73 × 5.4μ — 37. *F. graminearum*
            gg. Conidia three-septate, 30 × 4.3μ; five-septate, 36 × 4.5μ; seven-septate, 41 × 4.7μ — 38. *F. flocciferum*
      dd. Stroma not becoming carmine-purple

e. Stroma various shades of yellow; conidia three-septate, 28 × 4.5μ; five-septate, 38 × 5.1μ — 36. *F. sambucinum* f.6

ee. Stroma ochre-yellow, brown, blue; blue color in the pionnotes; conidia three-septate, 25 × 5.2μ; five-septate, 29 × 5.5μ — 36. *F. sambucinum* f.5

cc. Macroconidia 5–9μ in diameter

d. Macroconidia three-septate, seldom four- to five-septate — 39a. *F. tumidum* var. *humi*

dd. Macroconidia three- to five-septate

e. Stroma pale, yellow to brownish-white — 39. *F. tumidum*

ee. Stroma rosy but not carmine — 34. *F. trichothecioides*

ddd. Macroconidia five(three- to five- to seven-) septate

e. Blue-black, globose sclerotia lacking

f. Conidia three-septate, 29 × 6.1μ; five-septate, 38 × 6.6μ — 40. *F. culmorum*

ff. Conidia three-septate, 31 × 5.1μ; five-septate, 44 × 5.9μ — 40.a *F. culmorum* var. *cereale*

ee. Blue-black sclerotia present — 41. *F. sublunatum*

34. *Fusarium trichothecioides* Wollenweber

Conidia rarely in sporodochia, free or arranged in false heads, scattered in pink aerial mycelium as rosy powder or in bright orange gelatinous masses, smaller one-celled or one- to three-septate, ellipsoid, cylindric-spindle-shaped, rounded at both ends or truncate, straight or slightly curved, resembling larger conidia of *F. sambucinum*, spindle-sickle-shaped, more strongly curved at the constricted snout-like, almost hook-like tip than in the middle, with pedicellate base, three- to five-, seldom fewer septate:

0-septate, 7–17 × 3.1–5.5μ
1-septate, 12–26 × 3.5–7μ
3-septate, 19–42 × 4–7μ
5-septate, 30–52 × 4–7μ
6-septate, 39–47 × 5–7μ, exceptionally
Chlamydospores scarce, intercalary.
From soil: United States: Idaho (106)

35. *Fusarium reticulatum* Montagne

Fruiting-layer pink with rich aerial mycelium, plectenchymatic base, golden-yellow to purple-red, and sometimes with erumpent sclerotial bodies of dark blue. Conidia in sporodochia and pionnotes as gelatinous bright orange colored masses, on richly branched conidiophores, sickle-shaped, both ends alike, abruptly narrowed, pedicellate, three- (four- to five-) septate exceptionally more or less septate:

3-septate, 17–40 × 2.5–4μ
5-septate, 28–40 × 3–4μ
From soil: Canada (15)

36. *Fusarium sambucinum* Fuckel

Syn. *F. aridum* Pratt
F. discolor Appel and Wollenweber

Conidia spindle-sickle-shaped, curved, both ends bent inwards in a hook, constricted at the tip or tenpin-like, at the base pedicellate, heavy-walled, sometimes intermixed with smaller non- to two-septate subnormal spores of various shapes, scattered in white then yellow or rosy aerial mycelium. Sickle forms in sporodochia and pionnotes rosy to lake-orange, in slimy masses, sometimes changed by turning into stromatic carmine, chestnut-brown or ochre color of the plectenchymatic or sclerotial stroma, three- to five-, very rarely six- to seven-septate:

3-septate, 16–40 × 3–6μ
5-septate, 22–55 × 3.5–6μ
7-septate, 35–56 × 3.7–6μ

The sclerotial stroma often becomes cauliflower-like, up to 1 cm. in height, knotty, meshed or vesicular, stilboid bodies usually dark brown. Chlamydospores intercalary, globose, single in chains or knots, relatively few.

From soil: Canada (15)
United States: Idaho (106)

form 1 Wollenweber

Syn. *F. elegantum* Pratt

Differs from the primary species by the dark blue color of the globose 0.1–0.15 mm. in diameter sclerotia and irregular form of the erumpent sclerotial stromata as well as the somewhat narrower conidia, three-septate, 20–35 × 3–5.5μ and five-septate, 24–46 × 4–6μ, exceptionally six- to seven-septate conidia (seven-septate, 35–50 × 4–6μ) occur. The stroma is like the primary species, carmine, chestnut-brown or ochre, often paler, but by the intervention of the sclerotia coloring becomes greenish-black or olive.

From soil: United States: Idaho (106), Texas (92)

form 2 Wollenweber

Syn. *F. subpallidum* var. *roseum* Sherbakoff

This rather pale fungus grows with hyalin, white, yellowish or rosy mycelium mold-like on a stroma which forms little or no red color, and never becomes blue. The conidia occur in slimy sporodochial or pionnotal masses of rosy to bright orange-red or ochre color and are usually three-septate, 25 × 4.9μ, seldom four- to five-septate; five-septate, about 30 × 5.3μ.

form 5 Wollenweber

Syn. *F. subpallidum* Sherbakoff

In this fungus strongly developed blue or verdigris colored stroma is covered with a slimy bright orange colored conidial mass reminding one of the members of the section Martiella. Conidia three- to five-septate; three-septate, 25 × 4.6μ, and five-septate, 35 × 4.9μ.

From soil: Austria (131)

United States: Idaho (106)

form 6 Wollenweber

Syn. *F. sulphureum* Schlechtendahl

F. discolor Appel and Wollenweber var. *sulphureum* Appel and Wollenweber

F. genevense Daszewska

Differs by the sulphur-yellow color of the plectenchymatic part of the stroma as well as the aerial mycelium and the lack of carmine color from the primary species and other forms. Dark blue, globose sclerotia may be present or absent. The conidia occur in sporodochia and pionnotes and in mass are bright orange. Sclerotial plectenchyma leather- to sepia-brown. Chlamydospores intercalary, typical. Conidia three- to five-septate; three-septate, 28 × 4.5μ, five-septate, 38 × 5.1μ.

In acid soils: cosmopolitan.

From soil: Austria (66) Switzerland (38)

37. *Fusarium graminearum* Schwabe

Fruiting-layer variously colored, white-rosy, golden-yellow, ochre, (becoming blue-violet when treated with ammonia) or carmine-purple, sometimes plectenchymatic, extended, more or less covered with floccose aerial mycelium, sometimes limited, erumpent, sclerotial, on which lies a conidial mass resembling pionnotes or (seldom) sporodochia, ochre or bright orange-red. Conidia at times compact as in *F. culmorum*, at times longer than in that variety, spindle-sickle-shaped, strongly curved, tapering at both ends, with a rounded or constricted tip and pedicellate base, three- to five-septate, seldom one- to two-septate or six- to nine-septate:

3-septate, 25–66 × 3–6μ
5-septate, 28–72 × 3.2–6μ
7-septate, 50–88 × 4–7μ
9-septate, 55–106 × 4–8μ

Chlamydospores lacking or very rare, intercalary.

38. *Fusarium flocciferum* Corda
Syn. *F. idahoanum* Pratt
F. nigrum Pratt

Sporodochia somewhat arched, dusty or gelatinous. Mycelium white, yellow, rosy, plectenchymatic stroma golden-yellow, carmine-red, to chestnut-brown, drying dark brown, more or less erumpent, sclerotial. Conidia spindle-sickle-shaped, long somewhat more slender than in *F. sambucinum*, three- to five-septate, very seldom six- to seven-septate and non- to two-septate:

3-septate, 17–39 × 3–5.4μ
5-septate, 29–54 × 3.7–5.7μ
7-septate, 36–55 × 4–6μ

Sporodochia and pionnotes bright rosy, isabellin, ochre to pale orange-red. Chlamydospores numerous, globose, 6–8μ in diameter, at times twice as large, intercalary, single, in chains or knots, brown.

From soil: United States: Idaho (106)

39. *Fusarium tumidum* Sherbakoff

Fruiting-layer yellow- to leather-brown, sometimes orange, but never carmine, conidial masses sporodochial or pionnotal layers, isabellin, incarnate to bright orange, in age cinnamon-brown. Aerial mycelium white to pink or yellowish. Conidia spindle-sickle-shaped, curved, with flask neck, often hooked tips and pedicellate base, heavy walled, strongly septate, the septa at right angles or often also oblique, typically three- to five-septate, seldom six- to eight-septate, exceptionally nine- to ten-septate, one-celled oval and one- to two-septate conidia very sparingly present:

0-septate, 12–21 × 4.5–8μ
1-septate, 12–35 × 5–8.5μ
3-septate, 12–50 × 4–11.5μ
5-septate, 25–88 × 5.5–11μ
7-septate, 35–100 × 6–14μ

9-septate, 70–112 × 6–14μ

Sclerotia lacking. Chlamydospores rare, intercalary.

39a. *Fusarium tumidum* Sherbakoff var. *humi* Reinking

Differs from the primary species by somewhat longer and usually three-septate, exceptionally four- to five-septate conidia colored as above, in mass cream, pink to bright orange, in sporodochia and pionnotes or in small droplets, then powdery, scattered, spindle-sickle-shaped:

1-septate, 30 × 8.5μ

2-septate, 28–37 × 10.5μ, few

3-septate, 43–56 × 7.3–12μ

5-septate, 40–67 × 7.5–12.5μ

Chlamydospores intercalary, single or paired to knotty, smooth or wrinkled, about 15μ in diameter.

In soil: Honduras, Central America (164)

40. *Fusarium culmorum* (W. G. Smith) Saccardo

Syn. *F. rubiginosum* Appel

Conidia at first scattered in the aerial mycelium, free or cut off in false heads, later as slimy masses covering the stroma or covering tubercularia-like sporodochia, of various colors, yellow, red, then ochre to coffee-brown, brighter in dry powdery layers, colored more or less by the purple-red, or golden-yellow to ochre-brown (becoming blue with ammonia) tones of the stroma. Conidia spindle-sickle-shaped, gradually or sharply tapered at both ends, tip cell sometimes constricted into a snout or flask-like neck, basal cell pedicellate, walls thick, often browned, with five-septa, seldom three- to four-septa, or six- to eight- septa, exceptionally less than three-septa:

3-septate, 19–40 × 4–7.6μ

5-septate, 23–74 × 4–8.8μ

7-septate, 36–75 × 4–9μ

Chlamydospores more frequently intercalary than terminal, globose or oval in the mycelium as well as in the conidia, one- to two-celled, in chains or knots, brown in mass, one-celled, 9–14μ in diameter, two-celled, 13–27 × 7–19μ in diameter.

From soil: Austria (66), Denmark (68)

United States: Idaho (106), Iowa (50), Louisiana (50), Utah (50)

40a. *Fusarium culmorum* (W. G. Smith) Saccardo var. *cereale* (Cooke) Wollenweber

Differs from *F. culmorum* by the slimmer, longer five-septate, seldom three- to four- or six-septate, exceptionally non- to two- or seven- to nine-septate conidia:

3-septate, 18–44 × 3.7–8.5μ
5-septate, 26–60 × 4–9μ
7-septate, 40–63 × 4–8μ
9-septate, 50–74 × 5–6μ
From soil: Canada (15)

41. *Fusarium sublunatum* Reinking

Stroma plectenchymatic, hyalin, rosy-cinnamon-brown with wine-red streaks, covered by sparse white aerial mycelium, with dark blue to olive sclerotial bodies (stromata erumpentia) or globose, radiating sclerotia, very soon slimy with incarnate-rose to lake-orange conidial masses whose bright colors fade in age, or are changed by the stroma. Conidia spindle-sickle-shaped, curved, both ends narrowed, at the tip constricted into a flask-neck, at the base pedicellate, thicker or thinner according to the moisture present, five- (three- to five-) septate, seldom non- to two- or six- to eight-septate:

0-septate, 10–16 × 3.5–4μ
1-septate, 20–29 × 3.5–4.5μ
3-septate, 12–50 × 4–6.5μ
5-septate, 41–81 × 4.8–7.5μ
7-septate, 66–90 × 5.8–6.8μ
8-septate, 81 × 6.8μ

Chlamydospores numerous, terminal and intercalary, globose, smooth, one-celled, 7–9μ, or two-celled, 9–5 × 7–9μ. Sclerotia globose, single or gregarious, wrinkled, smoky dark olive, 0.25–0.5 mm.

From soil: Cosmopolitan (121)

41a. *Fusarium sublunatum* Reinking var. *elongatum* Reinking

This fungus differs from the primary species chiefly by narrower conidia:

Stroma brownish-white, thin membranous, covered by delicate white to pink aerial mycelium, then dark blue to green, (with ammonia peach colored) with globose sclerotia, and slimy rosy to lake-orange conidial masses, which later become green to black-brown. Conidia are long, cylindric to spindle-sickle-shaped, five- (three- to five-) septate, exceptionally six- to eleven-septate, smaller non- to two-septate, long, ellipsoid conidia very sparsely scattered:

0-septate, 14–19 × 3–3.8μ
1-septate, 19–24 × 3.5–3.8μ

3-septate, 33–47 × 3.8–5.8μ
5-septate, 43–78 × 3.8–6μ
7-septate, 80–89 × 4.3–5.8μ
9-septate, 89 × 5μ
11-septate, 116 × 5μ

Chlamydospores terminal or intercalary, one-celled, 7–12 × 7–10μ, one-septate, 9–18 × 6–10μ. Sclerotia present or lacking, globose, at first small, 0.03–0.05 mm. then some increase in size or all become more or less blue-black heads, 0.25–1.5 mm.

From soil: United States: Texas (92)

SECTION XIII. VENTRICOSUM

A single species treated.

42. *Fusarium argillaceum* (Fries) Saccardo

Stroma turf-like, gelatinous below, above wadded or moss-like, floccose, loose, buttressed and ragged, white to cream-colored, tinted or gray by the decomposition of the substrata. Conidia at the ends of long twisted or regularly branched conidiophores which usually twine around one another, wedge to spindle-sickle-shaped, bellied at the center, with tenpin, truncate, or rounded ends, generally tapered, at times with a plump pedicellate base somewhat constricted or with a peg, thick-walled, strongly septate, usually three-septate, seldom one- (non-, two-, four- to six-) septate:

0-septate, 10 × 4.5μ
1-septate, 14–43 × 5–7.5μ
3-septate, 24–67 × 4–11μ
5- to 6-septate, 44–107 × 5–9μ

They gather in droplets or at the base of the mycelial turf as a cream-colored underlayer. At the tip of old conidiophores or on lateral buds of thicker hyphae occur round to oval chlamydospores, 5–12μ thick, with smooth, at maturity, wrinkled or toothed walls.

In soil: China (81), Europe and North America (164)
United States: Louisiana (50)

(6) **Cylindrocarpon** Wollenweber

Microconidia usually continuous, ovate, fusiform or pyriform. Macroconidia, non- to many-septate, cylindric, cylindric-fusiform, claviform, straight or curved, sometimes obtuse, obtuse-conic, more or less convex, base at times apiculate, at other times pedicellate, scattered, in false-heads, tuberculate sporodochia, pionnotes, erect columns and erumpent; white, cream, yellow, pink, rose, usually formed laterally at the apices of the conidiophores or on short phialides. Conidiophores simple or irregularly, alternately, or verticillately branched, free, floccose, caespitose or sporodochial. Stromata more or less plectenchymatic, effuse, thalloid or tuberculate, globose, smooth or rough, radially erumpent, hyalin, yellow, golden, red, violet, ochraceous, brown, rarely green, or black. Mycelium white, yellow, citrine, rose carmine-violet. Chlamydospores in some species lacking, in others globose, simple, in chains or nodal clusters, usually intercalary, rarely terminal, brown.

FIG. 132. **Cylindrocarpon.** conidiophores and conidia (after Wollenweber).

KEY TO THE SPECIES OF THE GENUS CYLINDROCARPON

a. Chlamydospores lacking or rare in the mycelium	
b. Sporodochia typically lacking	1. *C. heteronemum*
bb. Sporodochia present	
c. Macroconidia curved, typically five-septate. Conidia moderately curved	2. *C. candidum*
cc. Macroconidia curved, typically five- to seven-septate	3. *C. candidum* var. *majus*
aa. Chlamydospores typically present in the mycelium	
b. Conidia typically one-septate	4. *C. didymum*
bb. Conidia typically one- to three- septate	5. *C. radicicola*

1. *Cyli-drocarpon heteronemum* (Berkeley and Broome) Wollenweber

Mycelium white, then pale yellow, floccose. Conidia scattered or in false heads, at first ellipsoid-ovate, one-celled, then cylindric, straight or slightly curved, two-celled:

0-septate, $8–12 \times 2.3–2.5\mu$

1-septate, $11–17 \times 2.3–3.5\mu$

No sporodochia.

From soil: Canada (16)

2. *Cylindrocarpon candidum* (Link) Wollenweber

Aerial mycelium white, rarely light yellow-greenish. Plectenchyma or stroma, ochraceous, brown, or dark olive. Microconidia ovate-cylindric, free or in false heads, rarely in sporodochia, $3.5–19 \times 1.3–3\mu$. Macroconidia usually in sporodochia and pionnotes, cream-white or yellow, cylindric-claviform, straight or slightly curved, obtuse or subglobose at the tip; five-septate rarely three-, four-, six-septate, very rarely seven-septate.

3-septate, 44 × 5.1μ

5-septate, 50–80 × 4.5–6.5μ

7-septate, 71 × 5.3μ

Conidiophores at first simple, then verticillately branched. No chlamydospores.

From soil: Canada (16)

3. *Cylindrocarpon candidum* var. *majus* Wollenweber

Aerial mycelium white, plectenchyma ochraceous. Microconidia at first ovate-cylindric, free or in false heads, rarely in sporodochia, non-septate, 4–12 × 2.5μ, one-septate, 9–22 × 3–4μ. Macroconidia in sporodochia and pionnotes, cream-white or yellow, cylindric-claviform, curved, inequilateral, ellipsoid or obtuse at the base, five- to seven-septate, rarely three- to four- or eight- to ten-septate:

3-septate, 20–54 × 3.5–5μ

5-septate, 51–75 × 5–6μ

7-septate, 75–100 × 5–6.5μ

9-septate, 82–110 × 5–7μ

Conidiophores at first simple, then verticillately branched. Conidio-chlamydospores sometimes present.

From soil: Canada (16)

4. *Cylindrocarpon didymum* (Hartung) Wollenweber

Syn. *Ramularia eudidyma* Wollenweber

Aerial mycelium white becoming yellow. Plectenchyma brown. Primary conidia scattered, then in sporodochia and pionnotes, at times in columns more or less erumpent, deposited consecutively, white or cream, cylindric, somewhat inequilateral-obtuse-conic, apiculate at the base, rarely two- to three-septate:

1-septate, 21–29 × 4–5.5μ

3-septate, 35–43 × 5.5–6.3μ

0-septate, 7–12 × 3–5μ

Chlamydospores brown, more rarely terminal than intercalary, smooth or spiny, one-celled or in chains, 8–11μ in diameter.

From soil: Canada (16)

5. *Cylindrocarpon radicicola* Wollenweber

Syn. *Ramularia macrospora* Wollenweber (non Fresenius)

Conidia free, in false heads, in pionnotes, rarely in sporodochia, in mass white to cream, at first small, continuous, ovate or ellipsoid-cylindric, 7–20 × 2–5μ, then typically one- to three-septate, large, cylindric, straight, rarely-slightly curved, subapiculate at the base, one-septate, 24–29 × 4.5–6.5μ, three-septate, 30–38 × 4.7–7.5μ. Chlamydospores intercalary, numerous, commonly in chains or in knots, brown, globose, 10–16μ in diameter. Aerial mycelium floccose, yellowish-white, reverse ochraceous-badium. Conidiophores at first slightly, later penicilliate, or verticillately branched.

From soil: Canada (15), Europe (160)

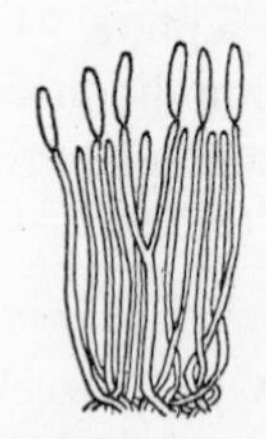

FIG. 133. **Myrothecium.** conidiophores and conidia (after Lindau).

(7) **Myrothecium** Tode

Conidial layer, shield or cushion shaped, black, surrounded at the edge by fine hyalin hairs. Conidiophores short rod-shaped. Conidia very small, ovate or cylindric.

KEY TO THE SPECIES OF THE GENUS MYROTHECIUM

a. Sporodochia shield-shaped, conidia 8–10 × 2μ — 1. *M. roridum*
aa. Sporodochia papillaeform, conidia 7 × 3μ — 2. *M. convexum*

1. *Myrothecium roridum* Tode

Sporodochia shield-shaped, then confluent and sessile, black, with a white rim, 2–6 mm. in diameter. Conidiophores unbranched or forked, bush-like, 30–40μ long, 2μ wide. Conidia cylindric, truncated at both ends, with two oil drops, smoky olive-green, 8–10μ, seldom 14μ long, by 2μ thick.

From soil: India (49)
United States: North Dakota (147)

2. *Myrothecium convexum* Berkeley and Curtis

Sporodochia papillaeform, formed by a white integument. Conidia short filiform, 7 × 3μ.

From soil: United States: Illinois (129)

(8) **Epicoccum** Link

Sporodochia spherical or arched, usually very small, usually with a stromatic underlayer of flat or hemispheric cells. Conidiophores rising from this stroma, usually very short, dark colored, abstricted from all sides. Conidia singly at the tips of the conidiophores, globose or ellipsoid, dark, smooth or spiny or knobby, often reticulate on the surface, one-celled or divided by cross-walls.

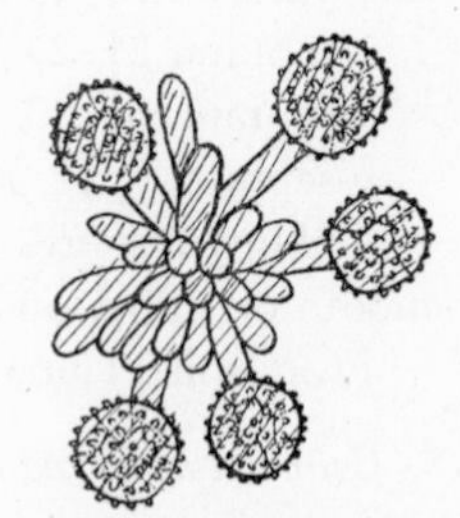

FIG. 134. **Epicoccum.** sporodochium and conidia(after Lindau).

1. *Epicoccum nigrum* Link

Sporodochia scattered, punctate, black on a hemispherical, somewhat depressed, black stroma. Conidiophores club-shaped, nonseptate, black, 12–14 × 5–7μ. Conidia globose, scarcely stipitate, dark black-brown, finely warted, reticulately wrinkled, 21–25μ in diameter.

From soil: Canada (16)

IV. MYCELIA STERILIA

KEY TO THE GENERA OF THE MYCELIA STERILIA

a. Sclerotia present
  b. Sclerotia of indefinite form — (1) **Rhizoctonia (Pellicularia)**
  bb. Sclerotia definite globose bodies — (2) **Sclerotium**

(1) **Rhizoctonia** de Candolle

Sclerotia without definite form, often grown together, horny-fleshy, with thinner undifferentiated edges, frequently imbedded in the mycelium and bound together by mycelial strands.

A single species treated.

1. *Rhizoctonia solani* Kühn

This species was connected with *Corticium vagum* B. and C. of the Basidiomycetes which was recently transferred to the genus Pellicularia (114), the description of which follows:

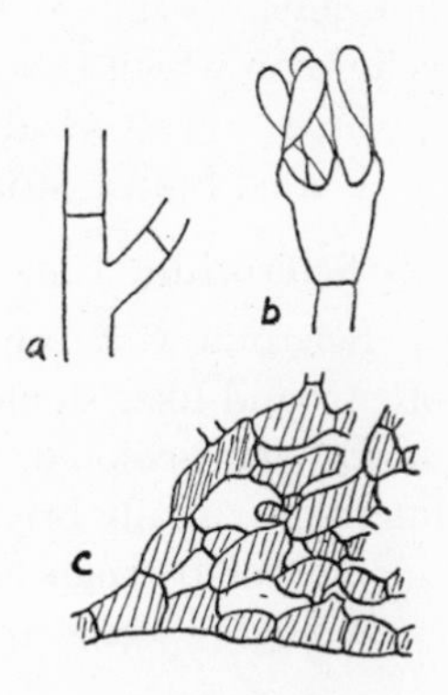

FIG. 135. **Rhizoctonia.** *a* hypha; *b* basidium and spores; *c* sclerotial hyphae.

Pellicularia Cooke

Fructification resupinate, mucedinoid or hypochnoid, reticulate-pelliculate, finely granulose, under the lens more or less tufted, or even and loose-membranous. Hyphae strongly stainable in aniline blue, thick, short-celled except for basal strands, branching at right angles and often with the formation of cruciform cells, the ascending hyphae usually several times cymosely divided, bearing the terminal basidia in more or less candelabrum-like clusters, or in parasitic species sometimes relatively short and little divided. Basidia subcylindric, not greatly exceeding in diameter the supporting cells, relatively short, bearing four or in several species six to eight sterigmata. Spores smooth-walled or rarely asperulate or spinulose, colorless or pale ochraceous. Cystidia wanting, or present and of various forms. Saprobic or facultative parasites.

Pellicularia filamentosa (Patouillard) Rogers

Syn. *Corticium vagum* var. *solani* Berkeley and Curtis

Fructification forming a delicate separable, flaky or thin-hypochnoid pellicle, when dry, white to buffy, or semitranslucent; hyphae branching at right angles and with some cruciform cells, without clamps, 4.5–14 (–17)μ in diameter, the basal long-celled, with often barrel-shaped segments, branching abundantly and bearing the basidia in small imperfectly symmetrical cymes. Basidia subcylindric and barrel-shaped (widest toward the middle) or obpyriform or clavate (widest at the summit), (10–) 12–18 (–23) $\times$8–11 (–12.5)μ, bearing four sterigmata which arise as blunt knobs and become later horn-shaped, (3–) 5.5–12 (–20) $\times$ 1.5–3.5 (–4.5)μ. Spores ellipsoid or oblong-ellipsoid, thin-walled, flattened on the inside, a little the broadest below the middle, truncate-apiculate, 7–12.5 $\times$ 4–7μ, occasionally germinating by a stout promycelium on which is borne a similar secondary spore.

From soil: Canada (15)

United States: Idaho (106)

(2) **Sclerotium** Tode

Sclerotia variously formed, globose, elongate, swollen or flattened, often band-like, single or confluent, sometimes covering wide surfaces, mostly dark-colored, commonly black, hard, particularly when dry, internally usually bright-colored. Rind-tissue sharply differentiated from the interior by color and cell structure.

A single species treated.

1. *Sclerotium rolfsii* Saccardo

Mycelium densely floccose, not ropy, and bearing numerous, pinkish-buff to olive-brown to clove-brown, globose sclerotia, 0.8–2.5 mm. in diameter.

From soil: Southern United States

EXCLUDED SPECIES

The following species were not included in the keys because of insufficient data concerning their position or occurrence.

Acrotheca lunzinense Szilvinyi (131)
Bispora pusilla Saccardo (67)
Catenularia fuliginea var. *lunzinense* Szilvinyi (131)
Coccospora agricola Goddard (52)
Dematium lunzinense Szilvinyi (131)
Discocolla pirina Prillieux and Delacroix (131)
Fumago vagans Persoon (130) (131)
Geomyces vulgaris Traaen (144)
Hadrotrichum lunzinense Szilvinyi (131)
Haplotrichum roseo-flavum Szilvinyi (131)
Myceliophthora sulphurea Goddard (52)
Oedocephalum lunzinense Szilvinyi (131)
Oidiodendron flavum Szilvinyi (131)
Rhinotrichum roseo-flavum Szilvinyi (131)
Sachsia albicans Bay (67)
Septogloeum propinquum (Bud. et Vleugl.) Wr. (108)
Sphaeronema fagi Oudemans (67)
Toruloidea tobaica Szilvinyi (130)

PERTINENT LITERATURE

1. ABBOTT, E. V.:
1923. The occurrence and action of fungi in soils. Soil Sci. 16:207–216.

2. ———
1926. A study of the microbiological activities in some Louisiana soils: A preliminary survey. La. Agr. Exp. Sta. Bul. 194, 25 pp.

3. ———
1926. Taxonomic studies on soil fungi. Iowa State Coll. Jour. Sci., 1:15–36.

4. ———
1927. Scolecobasidium, a new genus of soil fungi. Mycologia 19:29–31.

5. ADACHI, M. and S. HIRABAYASHI:
1932. Mikrobiologische Untersuchungen über die Böden in Taiwan (Formosa). 4. Studien über die mikrobiologische Eigenschaften der Humussäureböden bie Zôzan in der Nähe von Taihoku. II. Über die Schimmelpilzflora. Jour. Soc. Trop. Agr. 4(4):389–401.

6. ADAMETZ, L.
1886. Untersuchungen über die niederen Pilze der Ackerkrumme. Inaugural Dissertation, Leipzig. 78 pp.

7. ALCORN, G. D., and C. C. YEAGER
1938. A monograph of the genus Cunninghamella with additional descriptions of several common species. Mycologia 30:653–658.

8. APPEL, O. and H. W. WOLLENWEBER
1910. Grundlagen einer Monographie der Gattung Fusarium. Arb. K. biol. Anst. Land- und Forstw. 8:1–207.

9. BAINIER, G.
1907. Scopulariopsis. Bul. Soc. Myc. France 23:98.

10. ———
1919. Mycothéque de l'Ecole de Pharmacie. XXX. Monographie des Chaetomidium et des Chaetomium. Bull. Soc. Myc. France 25:191–237.

11. BAYLISS-ELLIOTT, J. S.
1930. The soil fungi of the Dover salt marshes. Ann. Appl. Biol. 17:284–305.

12. BECKWITH, T. D.
1911. Root and culm infections of wheat by soil fungi in North Dakota. Phytopath. 1:169–176.

13. BEYMA, F. H. VAN
1933. Beschreibung einiger neuer Pilzarten aus dem Centralbureau voor Schimmelcultures-Baarn (Holland.) Zentralbl. f. Bakt. Parasit. u. Inf. (Abt. II) 88:132–141.

14. BIOURGE, PH.
1923. Les moissisures du groupe Penicillium Link. La Cellule 33:331 pp.

15. BISBY, G. R., N. JAMES AND M. TIMONIN
1933. Fungi isolated from Manitoba soil by the plate method. Canadian Jour. Res. 8:253–275.

16. BISBY, G. R., M. TIMONIN AND N. JAMES
1935. Fungi isolated from soil profiles in Manitoba. Canadian Jour. Res. (Sec. C. Bot. Sci.) 13:47–65.

17. Boedijn, K. B.
1933. Ueber einige phragmosporen Dematiazeen. Bull. Jard. Bot. Buitenzorg (Ser. III) 13:120–134.

18. Bolley, H. L.
1901. Flax wilt and flax-sick soil. North Dakota Agr. Exp. Sta. Bull. 50: 27–58.

19. Burt, E. A.
1926. Thelephoraceae of North America XV. Ann. Mo. Bot. Gard. 13:295.

20. Butler, E. J.
1907. An account of the genus Pythium and some Chytridiaceae. Mem. Dept. Agr. India. (Bot. Ser.) 15:1–160.

21. Chaudhuri, H. and G. S. Sachar
1934. A study of the fungus flora of the Punjab soils. Ann. Mycol. 32:90–100.

22. Chivers, A. H.
1915. A monograph of the genera Chaetomium and Ascotricha. Mem. Torrey Bot. Club. 14:155–240.

23. Christenberry, G. A.
1940. A taxonomic study of the Mucorales in the Southeastern United States. Jour. Elisha Mitchell Sci. Soc. 56:333–366.

24. Ciferri, R. and P. Redaelli
1934. *Sporendonema epizoum* (Corda) Cif. et Red., an entity including *Hemispora stellata* and *Oospora d'Agatae*. Jour. Trop. Med. Hyg. 37:167–170.

25. Coker, W. C.
1923. The Saprolegniaceae. Univ. North Carolina Press, Chapel Hill, pp. 1–201.

26. ———
1927. Other water molds from the soil. Jour. Elisha Mitchell Sci. Soc. 42:207–226.

27. ——— and Braxton, H. H.
1926. New water molds of the soil. Jour. Elisha Mitchell Sci. Soc. 42:139–149.

28. ——— and J. N. Couch
1923. A new species of Thraustotheca. Jour. Elisha Mitchell Sci. Soc. 39:112–115.

29. ——— and J. N. Couch
1924. Revision of the genus Thraustotheca, with a description of a species. Jour. Elisha Mitchell Sci. Soc. 40:197–201.

30. ——— and V. D. Matthews
1937. Blastocladiales, Monoblepharidales, Saprolegniales. N. Amer. Flora 2(1):1–76.

31. Cookson, I.
1937. On *Saprolegnia terrestris* sp. nov. with some preliminary observations on Victorian soil Saprolegniales. Proc. Royal Soc. Victoria 49:235–243.

32. Couch, J. N.
1927. Some new water fungi from the soil, with observations on spore formation. Jour. Elisha Mitchell Sci. Soc. 42:227–242.

33. ———
1931. Observation on some species of water molds connecting Achlya and Dictyuchus. Jour. Elisha Mitchell Sci. Soc. 46:225–229.

34. ——— and A. J. Whiffen
1942. Observations on the genus Blastocladiella. Amer. Jour. Bot. 29:582–591.

35. Currie, J. N. and C. Thom
1915. An oxalic acid producing Penicillium. Jour. Biol. Chem. 22:287–293.

36. Dale, Elizabeth
1912. On the fungi of the soil. Ann. Mycol. 10:452–277.

37. ———
1914. On the fungi of the soil. Ann. Mycol. 12:32–62.

38. Daszewska, W.
1912. Étude sur la désagrégation de la cellulose dans la terre de Bruyére et la Tourbe. Bull. Soc. Bot. Geneva. II, 4:294.

39. Dierckx, F.
1901. Essai de revision du genre Penicillium Link. Soc. Scientif., Bruxelles.

40. Dissmann, E.
1931. Zur Kenntnis einer neuen Isoachlya-Art aus dem Erdboden. Botan. Centralbl. Beih. 48(2):103–111.

41. Dixon, Dorothy
1928. The micro-organisms of cultivated and bush soil in Victoria. Australian Jour. Exp. Biol. and Med. Sci. 5:

42. ———
1930. The micro-organisms of the tertiary red sands near Melbourne, Victoria. Australian Jour. Exp. Biol. and Med. Sci. 7:161–169.

43. Dixon-Stewart, D.
1932. Species of Mortierella isolated from soil. Trans. Brit. Mycol. Soc. 17: 208–220.

44. Emerson, R.
1938. A new life cycle involving cyst-formation in Allomyces. Mycologia 30:120–132.

45. ———
1941. An experimental study of the life cycles and taxonomy of Allomyces. Lloydia 4:77–144.

46. Emmons, C. W.
1930. *Coniothyrium terricola* proves to be a species of Thielavia. Bull. Torrey Bot. Club. 57:123–126.

47. Engler, A. and K. Prantl
1897–1900. Die näturlichen Pflanzenfamilien. Teil 1, Abt. 1 and **, Leipzig.

48. Fischer, Hugo
1910. Ueber *Coremium arbuscula* n. sp. Zentralbl. f. Bakt. Parasit. u. Inf. (Abt. II) 26:57–58.

49. Galloway, L. D.
1936. Indian soil fungi. Indian Jour. Agr. Sci. 6 (pt. 1):578–585.

50. Gilman, J. C. and E. V. Abbott
1927. A summary of the soil fungi. Iowa State College. Jour. Sci. 1:225–344.

51. Goddard, H. N.
1911. A preliminary report of fungi found in agricultural soil. 13th Rept. Mich. Acad. Sci., 208–213.

52. Goddard, H. N.
1913. Can fungi living in agricultural soil assimilate free nitrogen? Bot. Gaz. 56: 249–305

53. Griffiths, D. and F. J. Seaver
1910. Fimetariaceae, N. Amer. Flora 3(1): 65–88.

54. Grimes, M., M. O'Connor and H. A. Cummins
1932. A study of some Phoma species. Trans. British Mycol. Soc. 17: 97–111.

55. Hagem, O.
1907. Untersuchungen über Norwegische Mucorineen, I. Vidensk. Selsk., I Math. Naturw. Klasse, 7: 1–50.

56. ———
1910. Untersuchungen über Norwegische Mucorineen, II. Vidensk. Selsk., I. Math. Naturw. Klasse, 10: 1–152.

57. ———
1910. Neue Untersuchungen über Norwegische Mucorineen. Ann. Mycol. 8: 265–286.

58. Harvey, J. V.
1925. A study of the water molds and pythiums occurring in the soils of Chapel Hill. Jour. Elisha Mitchell Sci. Soc. 41: 151–164.

59. ———
1927. *Brevilegnia diclina* n. sp. Jour. Elisha Mitchell Sci. Soc. 42: 243–246.

60. ———
1927. A survey of water molds occurring in the soils of Wisconsin, as studied during the summer of 1926. Trans. Wisconsin Acad. Sci. 23: 551–565.

61. ———
1930. A taxonomic and morphological study of some members of the Saprolegniaceae. Jour. Elisha Mitchell Sci. Soc. 45: 319–332.

62. ———
1942. A study of western water molds. Jour. Elisha Mitchell Sci. Soc. 58: 16–42.

63. Hoehnk, W.
1932. A new parasitic Pythium. Mycologia 24: 489–507.

64. Ivimey-Cook, W. R. and Enid Morgan
1934. Some observations on the Saprolegniaceae of the soils of Wales. Jour. Bot. 72: 345–349.

65. Jamiesen, C. O. and Wollenweber, H. W.
1912. An external dry rot of potato tubers caused by *Fusarium trichothecioides* Wollenweber. Jour. Washington Acad. Sci. 2: 146–152.

66. Janke, A., and H. Holzer
1929. Ueber die Schimmelpilzflora des Erdbodens. Zentralbl. f. Bakt. Parasit. u. Inf. (Abt. II): 79: 50–74.

67. Jensen, C. W.
1912. Fungus flora of the soil. N. Y. (Cornell) Agr. Exp. Sta. Bull. 315: 414–501.

68. Jensen, H. L.
1931. The fungus flora of the soil. Soil Science 31: 123–158.

69. Jones, F. R., and C. Drechsler
1925. Root rot of peas in the United States caused by *Aphanomyces euteiches* (n. sp.). Jour. Agr. Res. 30: 293–325.

70. KNIEP, H.
1929. *Allomyces javanicus*, n. sp., ein anisogamer Phycomycet mit Planogameten. Ber. Deut. Bot. Ges. 47: 199–212.

71. KOMINAMI, K.
1915. *Zygorhynchus japonicus*, une nouvelle Mucorinee heterogame. Mykol. Zentralbl. 5: 1–4.

72. KUBIENA, W.
1931. Mikropedologische Studien. Wiss. Arch. Landwirtsch. Abt. A. Arch. Pflanzenbun. 5(4): 613–648.

73. LANGERON, M., AND R. V. TALICE
1932. Nouvelles methodes d'etude et essai de classification des champignons levuriformes. Ann. Parasitol. 10: 1–80.

74. LE CLERG, E. L.
1930. Cultural studies of some soil fungi. Mycologia 22: 186–210.

75. ———
1931. Distribution of certain fungi in Colorado soils. Phytopathology 21: 1073–1081.

76. ——— AND F. B. SMITH
1928. Fungi in some Colorado soils. Soil Science 25: 433–441.

77. LENDNER, A.
1908. Les Mucorinees de la Suisse. Beitr. Kryptogamenfl. Schweiz. 3(1): 1–180.

78. LING-YOUNG, M.
1930–31. Etude des phénoménes de la sexualité chez les Mucorinees. Rev. Gen. Bot. (Paris) 42: 144, 205, 283, 348, 409, 491, 535, 618, 681; 43: 30.

79. LINNEMANN, G.
1935. Beitrag zu einer Flora der Mucorineae Marburgs. Flora N. F. 30: 176–217.

80. LUND, A.
1934. Studies on Danish freshwater Phycomycetes. K. Danske Videnskab. Selsk. Skrifter. Nat. og Math. Ad., Ser. 9, 6(1): 1–98.

81. MA, R. M.
1933. A study of the soil fungi of the Peking district. Lingnan Sci. Jour. Supp. 12: 115–118.

82. ———
1933. Seasonal variations of fungi in soils in the vicinity of Peiping. Peking Nat. Hist. Bull. 1932–33. 7: 293–297.

83. MCLEAN, H. C., AND G. W. WILSON
1914. Ammonification studies with soil fungi. N. J. Agr. Exp. Sta. Bull. 270: 1–39.

84. MASON, E. W.
1927. On species of the genus Nigrospora Zimmermann recorded on monocotyledons. Trans. Brit. Mycol. Soc. 12: 152–165.

85. ———
1933. Annotated account of fungi received at the Imperial Bureau of Mycology. List II (Fascicle 2) pp. 1–67. Imperial Mycological Inst., Kew, Surrey.

86. MATTHEWS, V. D.
1931. Studies on the genus Pythium. Univ. North Carolina Press. Chapel Hill p. 1–136.

87. MEREDITH, C. H.
1938. Phycomycetes in Iowa soil. Phytopathology 28: 15–16 (Abstract).

88. ———
1940. A quick method of isolating certain phycomycetous fungi from the soil. Phytopathology 30: 1055–1056.

89. MES, M.
1929. On the identity of *Dematium scabridum* Gilman and Abbott with the conidial form of *Ceratostomella adiposum* (Butl.) Sartoris. Verhandel. K. Akad. Wetenschap. Amsterdam Afd. Natuurk. 26(4): 30–31.

90. MIDDLETON, J. T.
1943. The taxonomy, host range and geographic distribution of the genus Pythium. Mem. Torrey Bot. Club 20: 1–171.

91. MOLLIARD, M.
1902. *Basisporium gallarum*, n. gen., n. sp. Bull. Soc. Mycol. France 18: 167–170.

92. MORROW, M. B.
1931. Correlation between plant communities and the reaction and microflora of the soil in south central Texas. Ecology 12: 497–507.

93. ———
1932. The soil fungi of a pine forest. Mycologia 24: 398–402.

94. NAMYSLOWSKI, B.
1910. Studien über Mucorineen. Bull. Acad. Sci. Carcovie, Ser. B. Sci. Nat. 477–520.

95. ———
1910. *Zygorhynchus vuilleminii* ,une nouvelle Mucorinée isolée du sol et cultivée. Ann. Mycol. 8:152–155.

96. NIELSEN, N.
1927. Fungi isolated from soil and from excrements of arctic animals derived from Disko and North Greenland. Meddel. om Gronland 74: 1–8.

97. NIETHAMMER, ANNELIESE
1933. Studien über die Pilzflora bögmischer Böden. Archiv. Mikrobiol. 4: 72–98.

98. ———
1935. Die Mucorineen des Erdbodens. Zeitschr. Pflanzenkr. 45: 241–280.

99. ———
1937. Die mikroskopischen Boden-pilze. W. Junk. The Hague 193 pp.

100. OUDEMANS, C. A., AND J. A. AND C. J. KONING
1902. Prodrome d'une flore mycologique obtenu par la culture sur gelatine preparee de la terre humeuse du Spanderswoud pres Bussum. Arch. Neerl. Sci. Nat., ser. 2, 7: 286–298.

101. PAINE, F. S.
1927. Studies on the fungus flora of virgin soils. Mycologia 19: 248–266.

102. PALLISER, HELEN L.
1910. Chaetomiaceae, North Amer. Flora 31: 59–64.

103. PISPEK, P. A.
1929. Edafske mukorineje Jugoslavije. Acta Inst. Bot. Univ. Zagreb 4: 1–36.

104. PISTOR, R.
1929. Beitrage zur Kenntnis der biologischen Tatigkeit von Pilzen in Waldoboden. Zentralbl. f. Bakt. Parasit. u. Inf. (Abt. II)80: 169–200, 378–410.

105. Povah, A. H. W.
1917. A critical study of certain species of Mucor. Bull. Torrey Bot. Club 44: 241–259, 287–321.

106. Pratt, O. A.
1918. Soil fungi in relation to diseases of the Irish potato in southern Idaho. Jour. Agr. Res. 13: 73–100.

107. Rabenhorst, L.
1884–1910. Kryptogamen-Flora von Deutschland, Oesterreich und der Schweiz, Bd. 1. Abt. 1–9 (2nd Aufl.) Leipzig.

108. Raillo, A. I.
1928. Microflora of the soil. Jour. All. Russ. Congr. Bot., Leningrad, 1928. 181–182.

109. ———
1929. Beitrage zur Kenntnis der Boden-Pilze. Zentralbl. f. Bakt. Parasit. u. Inf. (Abt. II)78: 515–524.

110. Rathbun, Annie E.
1918. The fungous flora of pine seed beds. Phytopathology. 8: 469–483.

111. Reinking, O. A., and M. M. Manns
1934. Parasitic and other fusaria counted in Colombia soils. Zentralbl. f. Bakt. Parasit. u. Inf. (Abt. II)89: 502–509.

112. Richter, W.
1935. Vorarbeiten zu einer Saprolegniaceenflora von Marburg. Flora 131 (n. f. 31): 227–262.

113. Ridgway, R.
1912. Color standards and color nomenclature. Washington. Published by the author.

114. Rogers, D. P.
1943. The genus Pellicularia (Thelephoraceae). Farlowia 1: 95–118.

115. Sabet, Y. S.
1935. A preliminary study of the Egyptian soil fungi. Egyptian Univ. Bull. Faculty Sci. 5: 1–29.

116. Saccardo, P.
1882–1926. Sylloge Fungorum. Padua.

117. Sartoris, G. B.
1927. A cytological study of *Ceratostomella adiposum* (Butl.) comb. nov., the black-rot fungus of sugar cane. Jour. Agr. Res. 35: 577–585.

118. Scales, F. M.
1914. The enzymes of *Aspergillus terricola*. Jour. Biol. Chem. 19: 259–272.

119. Seaver, F. J.
1928. The North American cup-fungi (Operculates). New York pp. 1–284.

120. Shennikov, A. P.
1927. Data on surface soil fungi in different plant associations. Bull. Jard. Bot. Princ. U. S. S. R. 26: 205–208.

121. Sherbakoff, C. D.
1915. Fusaria of potatoes. N. Y. (Cornell) Mem. 6: 87–270.

122. Sideris, C. P., and G. E. Paxton
1929. A new species of Mortierella. Mycologia 21: 174–177.

123. Smith, Annie L., and C. Rea
1908. New and rare British fungi. Trans. Brit. Mycol. Soc. 3: 34–46.

124. Snyder, W. C., and H. N. Hansen
1940. The species concept of Fusarium. Amer. Jour. Bot. 27: 64–67.

125. Sopp, O. J. H.
1912. Monographie der Pilzgruppe Penicillium. Vidensk. Selsk., I Math. Naturw. Klasse, 1912, II 208 pp.

126. Sparrow, F. K., Jr.
1940. Phycomycetes recovered from soil sample collected by W. R. Taylor on the Allan Hancock 1939 expedition. Univ. Southern Calif. Publ. Allan Hancock Pacific Expedition 3: 101–112.

127. ———
1943. Aquatic Phycomycetes. Univ. Michigan Studies, Scient. Ser. 15: 1-785.

128. Stratton, Robert
1921. The Fimetariales of Ohio. Ohio Biol. Surv. Bull. 3: 75–144.

129. Swift, M. E.
1929. Contributions to a mycological flora of local soils. Mycologia 21: 204–221.

130. Szilvinyi, A. V.
1936. Zur mikrobiologischen Kenntnis der Tobaheide auf Sumatra. Archiv. f. Hydrobiol. Supp. Bd. 14: 512–552.

131. ———
1941. Mikrobiologische Bodenuntersuchungen in Lunzer Gebiet. Zentralbl. f. Bakt. Parasit. u. Inf. (Abt. II)103: 133–189.

132. Takahashi, R.
1919. On the fungous flora of the soil. Ann. Phytopath. Soc. Japan 1: 17–22.

133. Taubenhaus, J. J.
1920. Wilts of the watermelon and related crops. Texas Agr. Exp. Sta. Bull. 260: 1–50.

134. Thom, C.
1910. Cultural studies of species of Penicillium. United States Department of Agriculture Bur. Animal Ind. Bull. 118: 109 pp.

135. ———
1915. The Penicillium luteum-purpurogenum group. Mycologia 7: 134–142.

136. ———
1930. The Penicillia. Baltimore. 644 pp.

137. ——— and M. B. Church
1918. *Aspergillus fumigatus*, *A. nidulans*, *A. terreus* n. sp., and their allies. Amer. Jour. Bot. 5: 84–104.

138. ——— and ———
1921. *Aspergillus flavus*, *A. oryzae* and associated species. Amer. Jour. Bot. 8: 103–126.

139. ——— and ———
1926. The Aspergilli. Baltimore. pp. 272.

140. ——— and K. B. Raper
1939. The Aspergillus nidulans group. Mycologia 31: 653–669.

141. ——— and ———
1941. The Aspergillus glaucus group. United States Department of Agriculture Misc. Publ. 426: 1–46.

142. Tieghem, Ph. van, and G. Lemonier
1873. Recherches sur les Mucorinees. Ann. Sci. Nat.(5 ser.)17: 261–399

143. Todd, Romona
1932. Phycomycetes, Ascomycetes, and Fungi Imperfecti in Oklahoma soil. Science 76: 464.

144. Traaen, A. E.
1914. Untersuchungen über Bodenpilze aus Norwegen. Nyt Magaz. f. Naturvidensk., 52: 19-120.

145. Vuillemin, P.
1910. Materiaux pour une classification naturelle des Fungi Imperfecti. Comp. Rend. Acad. Sci. (Paris) 150: 882–884.

146. Waksman, S. A.
1916. Soil fungi and their activities. Soil Sci., 2: 103–156.

147. ———
1917. Is there any fungous flora of the soil? Soil Sci. 3: 565–589.

148. ——— and E. B. Fred
1922. A tentative outline for the plate method for determining the number of microorganisms in the soil. Soil Sci. 14: 27–28.

149. Ward, M. W.
1939. Observations on a new species of Thraustotheca. Jour. Elisha Mitchell. Sci. Soc. 55: 346–352.

150. ———
1939. Observations on *Rhizophlyctis rosea*. Jour. Elisha Mitchell Sci. Soc.. 55: 353–359.

151. Wehmer, C.
1893. Beitrag zur Kenntniss einheimischer Pilze I. Zwei neue Schimmelpilze als Erreger einer Citronensauregahrung. Hannover and Leipzig (Hahn) 92 pp.

152. Werkenthin, F. C.
1916. Fungous flora of Texas soils. Phythopathology. 6: 241–253.

153. Westling, R.
1912. Über die grünen Spezies der Gattung Penicillium. Ark. f. Bot. 11: 1–156.

154. Whiffen, A. J.
1943. New species of Nowakowskiella and Blastocladia. Jour. Elisha Mitchell Sci. Soc. 59: 37–44.

155. Wolf, F. T.
1939. A study of some aquatic Phycomycetes isolated from Mexican soils. Mycologia 31: 376–387.

156. ———
1941. A contribution to the life history and geographic distribution of the genus Allomyces. Mycologia 33: 158–173.

157. ———
1941. A new species of Achlya from Costa Rica. Mycologia 33: 274–278.

158. ——— and F. A. Wolf
1941. Aquatic Phycomycetes from the Everglades Region of Florida. Lloydia 4: 270–273.

159. Wollenweber, H. W.
1913. Studies on the Fusarium problem. Phytopathology. 3: 24–50.

160. Wollenweber, H. W.
1913. Ramularia, Mycosphaerella, Nectria, and Calonectria. Phytopathology. 3: 197–240.

161. ———
1914. Identification of species of Fusarium occurring on the sweet potato *Ipomoea batatas*. Jour. Agr. Res. 2: 251–285.

162. ———
1917. Fusaria autographice delineata. Ann. Mycol. 15: 1–56.

163. ———
1928. Ueber Fruchtformen der Krebserregenden Nectriaceen. Zeitschr. f. Parasitenk. 1: 138–173.

164. ——— and O. A. Reinking
1935. Die Verbreitung der Fusarien in der Natur. Berlin. 80 pp.

165. ——— and ———
1935. Die Fusarien. Berlin. pp. 1–355.

166. ———, C. D. Sherbakoff, O. A. Reinking, Helen Johann, and Alice A. Bailey
1925. Fundamentals for taxonomic studies of Fusarium. Jour. Agr. Res. 30: 833–843.

167. Younkin, S. G.
1938. *Pythium irregulare* and damping off of watermelons. Phytopathology 28: 596.

168. Zaleski, K.
1927. Ueber die in Polen gefundenen Arten der Gruppe Penicillium Link. I, II, III. Bull. Internat. Acad. Polonaise Sci. et Lett. Cl. Sci. Math. et Nat., Ser. B. Sci. Nat. 1927 (6B): 417–457; 459–513; 515–563.

169. Zycha, H.
1935. Mucorineae. Kryptogamenfl. Mark Brandenburg. 6a: 1–256.

GLOSSARY

absciss: cut off, detach.

abstriction: forming spores by cutting off of successive sections of sporophore through growth of septa.

acervulus (i): a determinate fruiting-body without a covering of fungous tissue; usually a discoid or flat mass of conidiophores producing conidia in a moist mass.

acicular: needle-shaped.

acrogenous: borne at the tip.

acuminate: gradually tapering to a point.

adnate: broadly attached.

agglutinate: firmly attached together.

allantoid: sausage-shaped.

alveolate: pitted like a honey-comb.

amoeboid: exhibiting a creeping movement.

amphigynous: with the oogonium growing through the antheridium so that the latter sits like a collar on the oogonial stalk.

anastomose: forming a net-work.

androgynous: with the antheridia arising on the same hyphae as the oogonia which they fertilize

angular: not rounded, showing angles in section.

anisogamous: with dissimilar gametes.

antheridium: a cell containing male gametes.

antherozoid: motile male cell.

aparaphysis (es): sterile filament resembling a paraphysis but of different origin.

apical: concerning end farthest from the base.

apiculate: terminating in a short, abruptly-pointed tip.

apiculus (i): a short, often sharp, papilla at the end of a spore.

apedicellate: without support of pedicel.

aplanogamete: nonmotile gamete.

apogamous: a condition in which sex cells develop vegetatively in the absence of copulation.

apophysis (es): a swelling, often just below the sporangium in the Mucorales.

apophysate: borne on a swollen portion.

apothecium (a): the cup-shaped, or saucer-shaped fruit-body containing asci.

appendage: a filamentous process.

appressed: closely flattened down, as leaves against a stem.

arachnoid: cobweb-like.

areolate: marked out in little areas.

articulate: jointed.

ascocarp: fruiting bodies of the Ascomycetes.

ascomata: fruiting bodies of the Ascomycetes.

ascus (i): a spore sac in which nuclear fusion and meiosis precedes spore formation in the Ascomycetes, usually containing eight spores.

ascigerous: bearing asci.

ascospore: spore resulting from meiosis in the Ascomycetes, borne in a sac, the ascus

ascogenous: producing asci.

asexual: without sex.

asperulate: slightly roughened.

asterigmate: without stalks (sterigmata).

attenuate: tapering, drawn-out.

avellaneous: variously interpreted from drab to hazel

azygospore: zygospore formed without conjugation.

bacillate: rod-shaped.

basidium (a): spore-bearing cell, the spores of which result from a meiosis, usually four.

basidiospore: spore resulting from meiosis, formed on a basidium.

basipetal: developing toward the base.

biciliate: with two cilia.

biflagellate: with two flagella.

bifurcate: forking by twos.

bilabiate: having two lips.

blastospore: spore which arises by budding.

botryose: clustered like grapes.

budding: arising by extrusion.

bursiform: bag or pouch-like.

caducous: falling off, deciduous.

caespitose: in dense groups forming tufts.

calyptrate: bearing a cap or lid.

calyptriform: shaped like a lid or candle-snuffer.

campanulate: bell-shaped.

canaliculate: grooved, channelled.

carbonaceous: black and brittle.

carotinoid: classes of yellow and red pigments found in various plants.

C. d. C.: Code des Coleurs, Klincksieck, Paris.

catenulate: in chains.

cellulin: animal cellulose.

cerebriform: with brain-like folds.

chlamydospores: thick-walled, nonsexual accessory spores.

cilium: vibratile, whip-like process; usually one of a number of relatively short organells.

cinereous: ash gray.

circinate: coiled.

cirrhus (i): a tendril-like curl of extruded spores.

citriform: lemon-shaped.

clavate: club-shaped.

cleistothecium (a): a closed fruit-body, without an ostiole, containing asci.

coenocyte: a multinucleate cell.

collarette: a little collar, usually left behind by dissolving sporangial wall.

columella (ae): sterile, inflated end of sporangiophore, extending into the sporangium.

concolorus: uniform in color.

concrescent: growing together.

conic: cone-shaped.

conidium (a): an asexual cell or cells, abscissed from the end of a hypha for dissemination.

conidiophore: hypha bearing conidia.

continuous: without septa, one-celled.

copulation: union of sex cells.

coremium: a fascicle of parallel conidiophores.

coriaceous: leathery.

cortical: pertaining to a cortex, bark, or rind.

corymb: a convex or flat-topped cluster of fruiting branches with branchlets arising from different points on the axis.

crenulate: scalloped.

cruciform: cross-shaped.

cruciately: in a cross-shaped manner.

cuneiform: wedge-shaped.

cuticularize: forming a firm cover or skin.

cutinize: cover with a skin.

cyme: a convex or flat-topped cluster of fruiting branches, the central spores or sporangia maturing first.

cyst: sac or cavity, thick-walled resting cell.

cystidium (a): a thin-walled, specialized, light colored, sterile cell.

deciduous: falling off at maturity, in season.

decussate: arranged in pairs, each at right angles to the next pair above or below.

dehiscence: splitting.

deliquescing: dissolving, melting away, liquifying (at maturity)

dendroid: tree-like.

denticulate: finely toothed, dentate.

depauperate: starved, underdeveloped.

desmids: members of family of Algae.

diatoms: members of family of Algae.

dichotomous: branching in twos (pairs).

diclinous: with antheridia on other hyphae than the oogonia which they fertilize.

dictyosporangium (a): sporangia in which the spores germinate without emerging (Saprolegniales).

diffluent: readily dissolving

dioecious: sometimes used of heterothallic fungi, especially when male and female structures are on distinct thalli.

diplanetic: with two morphologically different phases of the swarm period (of zoospores).

discoid: disc-shaped, rayless.

distal: away from the point of origin.

divaricate: extremely divergent

doliform: jar-shaped.

dorsiventral: having a front and back.

echinate: spiny.

echinulate: *ibid.*

eccentric: asymmetrical in growth.

ejaculate: throw out.

ellipsoid: more or less elliptical.

encyst: to form cysts or invest with a thick wall.

endobiotic: within living bodies.

endospore: inner wall of a spore.

epibiotic: on living bodies.

epispore: outer coat of a spore.

epithecium (a): the layer above the asci, usually formed by the tips of the paraphyses.

erumpent: breaking out.

eucarpic: that condition in which only a portion of the thallus is used in reproduction.

evanescent: vanishing early.

excipulum (a): tissue surrounding the hymenium laterally.

exine: outer wall of a spore.

exogenous: arising on the outside.

extramatrical: outside the substrate.

fasciculate: crowded in bundles.

ferruginous: rust-colored.

fertile: bearing spore or spores.

fetid: stinking, putrid.

filament: thread.

filamentous: thread-like, stringy.

filiform: thread-like.

fimbriate: with edges minutely fringed.

flaccid: soft and limber.

flagellum (a): whip-like process conferring motility; usually borne singly or in small numbers and relatively long.

flexuous: bent alternately in opposite directions.

floccose: having tufts of soft woolly hairs which are often deciduous.

fructification: the act of fruiting.

fugacious: disappearing early.

fuliginous: smoky, sooty.

fulvous: tawny.

funiculose: in ropes or bundles.

furfuraceous: covered with bran-like scales, scurfy.

fuscous: brown or grayish-black, dusky.

fusiform: spindle-shaped.

fusoid: *ibid.*

gametangium: cell containing gametes.

gamete: sex cell.

gametophyte: gamete-bearing plant.

gemma: reproductive cell, resembling a chlamydospore; bud.

geniculate: bent, knee-like.

globose: spherical.

glomerulus (a): a little mass.

gregarious: scattered closely over a small area, herd-like.

guttulate: containing one or more oily globules.

hathi-gray: elephant-gray.

haustorium (a): special branch of a hypha serving for absorption.

helicoid: coil-like.

hemispheric: half-sphere.

heterothallic: a condition in certain fungi in which two distinct thalli must be brought together for the production of sexual spores.

hirsute: hairy.

holdfast: special hypha for attachment.

holocarpic: the whole thallus becomes a sporangium.

homothallic: a condition in certain fungi in which sexual spores are produced without the intervention of a second thallus.

hülle: (from the German) cover; special thick-walled cells surrounding the cleistothecia in the Aspergillaceae.

hyalin: colorless, transparent.

hypochnoid: fruiting layer supported on a loose web of hyphae as in the genus Hypochnus

hymenium (a): fruiting-layer.

hypha (ae): branch of the mycelium.

hypogynous: with the antherids formed below the oogonium and on the same hypha.

hypothecium (a): layer of tissue immediately below the hymenium.

incrusted: covered with a layer of mineral matter.

indeterminate: indefinite, unlimited.

infundibuliform: funnel-shaped.

inoperculate: without a lid.

in situ: in its natural position.

interbiotic: between living bodies.

integument: an outer, natural covering.

intercalary: growth between apex and base.

internode: space between nodes or joints.

interstitial: between spaces.

intine: inner wall of a spore.

intramatrical: within the substrate.

isabellin: pinkish-cinnamon (color)

isogamous: with like gametes.

lanose: woolly.

lageniform: flask or gourd-shaped.

lanceolate: lance-shaped.

lenticular: shaped like a double convex lens.

lobulate: having lobes.

lumen: cavity of a cell bounded by cell walls.

lunate: crescent-shaped.

macroconidium (a)**:** large conidium.

macrospore: a large spore where there are spores of two sizes.

mammiform: breast-shaped.

meiosis: reduction division of chromosomes.

merosporangium: spore chains which arise by the simultaneous division of an elongated cell.

metula (ae)**:** branch from which phialides arise.

microconidium (a)**:** small conidium.

moniliform: chain-like, a string of beads.

monocentric: a thallus with a single center of growth and differentiation.

monoclinous: with antheridia on the same hyphae as the oogonium.

monecious: having both sexes on the same plant.

monoplanetic: having a single swarming period (of zoospores)

monopodial: growing at the apex in the direction of previous growth.

monoverticillate: with one whorl.

motile: having motion.

mucedinoid: musty, mouldy.

mucilaginous: slimy.

mucus: a gummy adhesive substance found in plants, soluble in water, insoluble in alcohol.

multisporus: many-spored.

muriform: with both longitudinal and tranverse septa.

mycelium (a)**:** vegetative body of a fungus, collectively.

node: joint, enlarged place on hypha, point of origin of branches of hypha.

nodulate: with intermittent thickenings (nodes)

nonmotile: lacking independent means of motion.

nucleate: containing microscopically visible, oil-like globules, or nuclei.

obclavate: club-shaped with the broad end at base.

obconic: reversely conical, broad end outward.

obovate: egg-shaped with broad end outward.

obovoid: *ibid.*

oidiospore: spores formed by breaking-up of the hypha.

oidium (a)**:** conidium formed by the breaking-up of the hypha.

oleaginous: oily.

olivaceous: (color) with an olive shade.

oogamous: with a large nonmotile egg and a small motile sperm.

oogonium (a)**:** cell containing one or more eggs.

oosphere: naked mass of protoplasm which becomes fertilized to form the oospore.

oospore: immediate product of fertilization of an egg or oosphere.

operculate: furnished with a lid.

operculum (a)**:** lid.

orbicular: circular in outline.

organell: small organ.

ostiole: mouth.

oval: egg-shaped with broad end at base.

ovate: *ibid.*

ovoid: *ibid.*

panduriform: fiddle-shaped.

panicle: a compound sporophore of indeterminate type with each spore or sporangium with a pedicel.

paniculate: branched.

papilla (ae)**:** small, nipple-shaped elevation.

pars sporif.: spore-bearing part.

paraphysis (es)**:** sterile filaments accompanying fructifications.

parthenogenetic: form of apogamy in which an egg or equivalent female structure develops without fertilization.

parum distincte: not at all clear.

pedicel: a slender stalk.

pellicle: a thin, outer layer of cortex in lichens and certain fungi.

pelliculate: with a differentiated thin layer of hyphae on the surface.

penicillus (i): a complex system of conidial-bearing branches forming a brush.

penultimate: last but one.

peridium: the outer wall of the fruit-body.

periphery: the outer boundary or surface.

periplasm: spongy, peripheral layer of cytoplasm in the oogonium.

persistent: remaining intact.

perithecium (a): a rounded, oval, or pyriform receptacle having an ostiole and containing asci.

phialide: spore-bearing cell with tubular tip from which spores are extruded.

pileus (i): the cap of a mushroom.

pilose: covered with long, soft, hairy filaments.

pionnotes: a slimy or gelatinous mass in which conidiophores and conidia are embedded.

pip-shaped: shaped like an apple seed.

planohaplont: motile gametophyte.

planont: a motile cell.

planozygote: motile zygote.

plectenchyma: a thick tissue-like structure, formed by intertwining and adhering of hyphae.

plicate: folded like a fan.

pluriform: with many different forms.

polycentric: with many centers of growth and differentiation.

polyhedric: many sided.

polymorphic: with several forms.

proliferate: producing other stipes on itself near the base.

promycelium (a): short and short-lived product of tube germination of a spore which abstricts acrogenously a small number of spores.

prosenchymatous: a plectenchyma in which the single hyphal elements are still recognizable as such.

prosporangium (a): the initial cell which forms the vesicle which functions as a sporangium.

prostrate: lying flat.

proximal: near the central portion of the body.

pruinose: covered with whitish dust or bloom; powdery.

pseudo: false.

pseudoparenchyma: a false parenchyma looking like a true parenchyma with isodiametric cells but formed from true hyphae.

punctate: with points.

pycnidium: a variously shaped receptacle for asexual spores.

pyriform: pear-shaped.

racemose: branching in racemes, elongated cluster of sporophores with each spore or sporangium on a pedicel.

recurve: bend back.

refractive: possessing quality of bending light rays.

refringent: *ibid.*

reniform: kidney-shaped.

resupinate: with the fruiting structure reclining on the substrate and facing outward.

reticulate: netted.

rhizoid: root-like.

rugose: coarsely wrinkled.

saprobic: living on dead organic matter.

sarciniform: packet-like.

scabrous: rough, with short rigid projections.

sclerotium (a): resting bodies, composed of a hardened mass of hyphae.

scorpioid: coiled like the tail of a scorpion.

scutellate: like a small shield.

septate: divided, partitioned.

septum (a): cross-wall.

sessile: without a stalk.

seta (ae): bristle.

sexual: having sex, resulting from the fusion of gametes.

siliquiform: shaped like a silique, spindle-shaped.

simple: unbranched.

sinuous: wavy, serpentine.

spinescent: with spines.

spinulose: with small spines.

sporangiferous: bearing sporangia.

sporangiole: small sporangium reduced in number of spores.

sporangiophore: hypha bearing sporangia.

sporangium (a): cell containing spores.

spore: cell or cells differentiated for dissemination or reproduction.

sporodochium (a): a fruiting-body consisting of a compact mass of interwoven conidiophores.

sporophyte: spore-bearing plant.

squamulose: scaly.

stellate: star-shaped.

sterigma (ata): the small spicule-like extension on a basidium bearing a basidiospore.

sterile: not producing spores.

stilboid: with a stalked head.

stipe: stalk or stem.

stipple: dot.

stolon: a runner; a hypha which sprouts where it touches the substrate to form rhizoids or sporophores, or both.

striate: with minute furrows or lines.

stroma (ata): a cushion-like body on or in which fruiting-layers are developed.

stylospore: spore borne on a filament or hypha.

subcentric: slightly eccentric.

subglobose: almost spherical.

subicle: a tufted or matted mycelium under the sporophores.

subiculum: a more or less dense felt of hyphae covering the substrate.

subjacent: just below.

sublanose: slightly woolly.

subpulverulent: slightly powdery, dusty.

suspensor: portion of hypha suspending a gamete or gametangium (Mucorales)

swarmspore: zoospore, motile spore.

sympodial: with an axis simulating a simple stem but made up of a number of axes which have arisen successively as branches one from another.

sympodium (a): an axis made up of the bases of a number of successive axes.

tenuous: delicate.

terricolous: living on soil.

thallus: vegetative body of undifferentiated cells.

thyrsiform: with middle branches longer than those below or above.

tomentum (a): a layer composed of long, soft, entangled fibrils.

torulose: cylindric, with swollen portions at intervals.

translucent: transmitting light.

trichogyne: receptive filament of the female organ.

trifurcate: three-forked.

truncate: cut-off, with face of ends at right angles to long axis.

tuberculate: wart-like.

tubercle: wart-like protuberance.

tuberous: round and swollen.

turbinate: top-shaped.

turgid: filled out, rigid.

umbel: a cluster of sporophores with pedicels all arising at the same point.

umbelliferous: bearing umbels.

umber: dark brown, reddish-brown.

umbo: a raised knob or mound on the center.

undulant: wavy.

undulate: *ibid.*

unicellular: one-celled.

uniciliate: with one cilium.

uniflagellate: with a single flagellum.

uniguttulate: with a single oil-drop.

uniseriate: in one line or series.

vacuolation: formation or development of vacuoles.

vacuole: a minute cavity containing air, watery fluid, chemical secretion, or protoplasm found in an organ, tissue, or cell.

velutinous: velvety.

ventricose: swollen or enlarged in the middle, bellied.

verdigris-green: a bluish-green color.

vermiform: worm-shaped.

verruciform: wart-shaped.

verrucose: warty.

verruculose: minutely verrucose.

verticil: a whorl of sporophores, or phialides.

vesicle: a bladder-like sac.

vesiculose: composed of vesicles.

vinaceous: wine-colored.

weft: felt-like mat of hyphae on certain fungi.

whorl: a cluster of radiating branchlets.

zonate: with concentric bands.

zoosporangium (a): sporangium containing zoospores.

zoospore: a motile spore.

zygote: cell resulting from fusion of gametes.

zygospore: spore resulting from fusion of equal gametes.

INDEX

Synonyms appear in Roman type. Species of uncertain position are designated by an asterisk.